The Personality of Henry Cavendish - A Great Scientist with Extraordinary Peculiarities

Archimedes

NEW STUDIES IN THE HISTORY AND PHILOSOPHY OF SCIENCE AND TECHNOLOGY

VOLUME 36

Archimedes has three fundamental goals; to further the integration of the histories of science and technology with one another: to investigate the technical, social and practical histories of specific developments in science and technology; and finally, where possible and desirable, to bring the histories of science and technology into closer contact with the philosophy of science. To these ends, each volume will have its own theme and title and will be planned by one or more members of the Advisory Board in consultation with the editor. Although the volumes have specific themes, the series itself will not be limited to one or even to a few particular areas. Its subjects include any of the sciences, ranging from biology through physics, all aspects of technology, broadly construed, as well as historically-engaged philosophy of science or technology. Taken as a whole, *Archimedes* will be of interest to historians, philosophers, and scientists, as well as to those in business and industry who seek to understand how science and industry have come to be so strongly linked.

For further volumes:
http://www.springer.com/series/5644

Russell McCormmach

The Personality of Henry Cavendish - A Great Scientist with Extraordinary Peculiarities

Springer

Russell McCormmach

ISSN 1385-0180 ISSN 2215-0064 (electronic)
ISBN 978-3-319-35064-6 ISBN 978-3-319-02438-7 (eBook)
DOI 10.1007/978-3-319-02438-7
Springer Cham Heidelberg New York Dordrecht London

Printed on acid-free paper

Springer is part of Springer Science+Business Media (www.springer.com)

For Lewis Pyenson

Preface

When I was five, my mother gave me a copy of Jonathan Norton Leonard's *Crusaders of Chemistry*. I still have it, signed and dated by her, 1938. Long after I had forgotten the other chemists, I remembered the one who measured. He is the subject of the chapter in Leonard's book entitled "Henry Cavendish, the Measuring Machine," which concludes: "So lived and died the coldest, most unhuman mortal who ever wrote his name large in the history of science .... His sole interest was to measure the objects in the material universe." Many years later, my boyhood fascination with the measuring machine returned, transformed into a scholarly interest in a scientific genius. I entered the field of history of science with a dissertation on his work, and for many years after, I saved every fact I could find about him. Popular histories of science like Leonard's take material from scholarly works. We do not have to look far to discover the source of Leonard's characterization of the man who was a measuring machine. It was the nineteenth-century chemist George Wilson's fascinating biography, *The Life of the Honble Henry Cavendish.*

Although I have reservations about the measuring machine, I greatly admire Wilson's portrait of Cavendish, and I am in its debt for my understanding of the man. I took up the study of the history of science in part to learn how the scientific view of the world came about. If such a view does actually exist, I thought it probably applied to me, an initial reason for my curiosity. From my study of scientists of the past, I came to think that Cavendish came as close as any to holding a scientific view of the world. Even in his case, it was not exclusive. He came into the world with an aristocratic take on it.

The origin of this book is a diagnosis of Cavendish by the eminent neuropsychologist Oliver Sacks, which appeared as a brief communication in the scientific journal *Neurology* in 2001. Skeptical of recent claims of Asperger's syndrome, a form of autism, for historical figures, Sacks considers Cavendish an exceptional case, finding the evidence for his autism "almost overwhelming."[1] In his memoir the same year, *Uncle Tungsten*, Sacks says that upon rereading Wilson's biography

[1] Oliver Sacks, "Henry Cavendish: An Early Case of Asperger's Syndrome?" *Neurology* 57 (2001): 1347.

of Cavendish, he has concluded that Cavendish was a "unique autistic genius."[2] The same nineteenth-century biography was the origin both of the man as measuring machine and of the man with a psychological disorder. Below I examine these characterizations and their connection.

In 1996, Christa Jungnickel and I published a biography of Cavendish. We pointed out that because of his strange behaviors he invites a psychological approach, but that it was not the approach we took, as we explained. At the end of the biography, we briefly mentioned possible psychological descriptions of his behavior such as social anxiety, shyness, and embarrassment. We pointed out that he also showed "autistic-like traits,"[3] which we listed, citing Sacks, who had introduced us to the world of autism with his moving account of the autistic scientist Temple Grandin.[4]

Three years later, we brought out an improved version of our biography, in which we again briefly brought up psychological descriptions, but this time we omitted any mention of autism. Autism is a disorder that begins in childhood, and almost nothing is known about Cavendish's childhood; also certain criteria for autism seemed a questionable fit, and we wanted the biography to be solid. Since then I have found in recent writings on the subject a growing acceptance of a more inclusive understanding of autism together with a trend in clinical thinking that favors an autistic continuum approach. In the present book, I consider Sacks's diagnosis of Cavendish's autism, which has been on my mind for the past dozen years.

Besides Cavendish, a number of eighteenth-century scholars – for example, Jeremy Bentham, John Howard, Carl Friedrich Gauss, and Emmanuel Kant – have been diagnosed with a form of autism. Without doubt, they all showed autistic-like traits. They pursued their interests obsessively, with little regard for what people might have thought. They also leave us wondering. Might their non-social behaviors have come about through choice? Might they have avoided society and ignored convention in pursuit of their goals because they could afford to? Did they show the same traits as autistic persons with the significant difference that the latter had no choice in the matter?

---

[2] Oliver Sacks, *Uncle Tungsten: Memories of a Chemical Boyhood* (2001; New York: Vintage Books, 2002), 121.

[3] Hugo Lidbetter writes that Jungnickel and McCormmach "got very close to suggesting" that Cavendish may have had Asperger's syndrome. We got closer than that: we say it. Since this book went to press, an article by Lidbetter was brought to my attention. The author's purpose is to make a "systematic exploration" of Sacks's claim that Cavendish had Asperger's Syndrome. His article consists of matching Cavendish's behaviors with the Gillberg diagnostic criteria for Asperger's syndrome. He thinks Cavendish had this disorder. "Henry Cavendish and Asperger's syndrome: A New Understanding of the Scientist," *Personality and Individual Differences* 46 (2009): 784–93, on 784. I thank Steve Silberman for this reference. Lidbetter brings Christa Jungnickel and me into his discussion. In this connection, he makes some mistakes. The reference to Asperger's syndrome is: Christa Jungnickel and Russell McCormmach, *Cavendish* (Philadelphia: American Philosophical Society, 1996), 368. This edition is given the short title *Cavendish* (1996). The later edition is *Cavendish, the Experimental Life* (Lewisburg: Bucknell University Press, 1999), 304–9. The short title is *Cavendish* (1999).

[4] Oliver Sacks, *An Anthropologist on Mars* (New York: Vintage, 1995). Temple Grandin, *Thinking in Pictures and Other Reports from My Life* (New York: Doubleday, 1995).

I re-examine Cavendish in light of Sacks's suggestion and the questions it raises. I look for evidence that supports it as well as evidence that does not. I look at possible alternative ways of understanding Cavendish. Shortly after his death, and probably before, the word "eccentricity" was used to describe his behavior. What did his eccentricity consist of? In the society of his time, and for a person of his rank, how exceptional was his eccentricity? Behaviors that are disquieting to us may have been well tolerated his setting, in which case we may ask, Is there any reason to look for a neurological abnormality? A proper study of these and related questions requires the length of a small book.

Sacks says that Cavendish had an "astounding" achievement and life, and that having reread his biography he is "if anything more mystified."[5] I too have unanswered questions. I understand some things about Cavendish, enough to have written a biography about him, but it is an incomplete biography. These many years later, I still look for a fuller understanding, which I equate with explanation. This book has an additional motivation. Without an understanding of Cavendish's behavior, he appears simply strange, an object of curiosity at best, of moral judgment at worst, drawing pity or scorn. To leave him that way unnecessarily is a shame. He was an outstanding scientist, and one of the most baffling personalities in the history of science. A fuller understanding of him benefits both his biography and the history of science.

At the end of a recent biography of the physicist Paul Dirac, the author Graham Farmelo suggests that his subject was autistic. In a review of the book, the distinguished physicist and science writer Freeman Dyson takes issue with its claim of autism, citing Dirac's friendships, which he finds hard to reconcile with autism. The physicist Jochen Heisenberg, son of Werner Heisenberg, agrees with Dyson's criticism and goes beyond it, associating the claim of autism with "a certain kind of facile, if interesting, deconstruction of character."[6] There are reasonable grounds to question Farmelo's, as most any, historical diagnosis, and there are strong feelings about diagnosing the great and the dead. I realize that any biographer who approaches his subject from a psychological perspective and is in a mental state other than fear and trembling is in a deep sleep.

In the days following Cavendish's death, his once close colleague Sir Charles Blagden told the president of the Royal Society that Cavendish was "a true

---

[5] Sacks, *Uncle Tungsten*, 120.

[6] Graham Farmelo, *The Strangest Man: The Hidden Life of Paul Dirac, Mystic of the Atom* (New York: Basic Books, 2009). Based on Dirac's thinking and behavior, Farmelo is persuaded that Dirac was autistic, and further that his autism was important for his work: "Yet again, it is possible that this correlation [visual thinking] between autistic characteristics and Dirac's behavior is a coincidence, but, in light of other such correlations, this seems unlikely. I believe it to be all but certain that Dirac's behavioral traits as a person with autism were crucial to his success as a theoretical physicist." Other correlations referred to here are Dirac's "systematic" ordering of information from physics and mathematics, "self-centredness," and "concentration and determination." These traits do not explain Dirac's talent, but "give some insight into his unique way of looking at the world." Ibid., 425. Freeman Dyson, "Silent Quantum Genius," *New York Review of Books*, 25 February 2010. Jochen Heisenberg, "Plain Talk about Dirac," ibid., 13 May 2010.

anchor: could always depend on knowing what was right for him."[7] Blagden was perceptive. The course of life Cavendish chose for himself was surely right for him. He made no major mistakes in his work or in his other activities. It took vigilance, for the world that was right for him was a small portion of a greater world, which could upset his life, and which he confronted with painful awkwardness. Some people may find it difficult to empathize with a person who is as complete within himself as Cavendish appeared to be, but they probably will not find it difficult to sympathize.

In the history of science, we may find no better fit between science and character and temperament than we do in Henry Cavendish. The fit was eminently efficient, there being practically nothing left over. It was Cavendish's singular desire to seek truth and knowledge of the physical universe, and science offered this truth-and-knowledge seeker a lifetime of opportunity. Endowed with a fine reasoning mind, keen sensory perception, and skilled hands, he exercised the full measure of his faculties in his pursuit of the laws of nature. Marked traits of his personality — caution, persistence, thoroughness, objectivity, accuracy, and precision — were engaged in the pursuit. Further marked traits of his caused him great distress in certain social situations. He entered fully into a company of scientific men, which held some of the social risks of any company, knowing it was right for him, as it was for anyone who was serious about science. It was fortunate for Cavendish that such an activity, science, existed, for it enabled him to make a nearly complete life for himself. It is hard to think of any alternative that would have served him half as well.

The book is divided into three Parts. For the heading of Part I, I have selected Blagden's quotation, slightly altered: He always knew what was right for him. The first chapter introduces Cavendish the aristocrat and Cavendish the natural philosopher. Chapter 2 discusses the psychology of scientists and the biographical problem of Cavendish. Chapter 3 describes his behaviors in everyday life. His peculiarities take center stage here, necessarily, for they have been the basis for accounts of his personality. The intent here is to organize the evidence, not to put him on display. Chapter 4 treats his behaviors at work. Chapter 5 treats aspects of his personality. Chapter 6 summarizes a number of earlier interpretations of Cavendish. The next three chapters analyze Cavendish's personality. Readers will encounter repetitions in the book. Descriptions of Cavendish's behaviors, which are introduced as evidence in Chaps. 3 and 4 are reintroduced for comparison in the chapters dealing with his personality. Cavendish's luminous thinking on scientific problems is treated fully in existing biographies and other studies and is not repeated here.

Part II is about journeys Cavendish and Blagden took in 1785–87 to observe the industry and geology of several regions of Britain. The journals they kept are probably the best single source of information about the kind of interest Cavendish

---

[7] 24 February 1810, Charles Blagden Diary, Royal Society 5: 426.

took in the wider world, and as such they illustrate facets of his personality. They also have an interest of their own as descriptions of technologies of the early Industrial Revolution.

Part III contains various sorts of materials that supplement Parts II and III, which are the substance of the book. The materials consist of a transcript of the journal of the 1785 journey, improved family trees for Cavendish, accounts of Cavendish's houses and his claim to own Holker Hall,[8] and discussions about the disorder autism.

For their generous help in reading and commenting on the manuscript of this book, I thank Geoffrey Cantor, Uta Frith, Ioan James, Lewis Pyenson, and Michael Sokal.

Eugene, OR, USA Russell McCormmach

[8] The two chapters about Cavendish's houses are largely taken from Jungnickel and McCormmach, *Cavendish, the Experimental Life*, rev. ed., in press.

Contents

List of Figures

Part I
He Always Knew What Was Right for Him

Chapter 1
The Person

1.1 The Aristocrat

Henry Cavendish was born in 1731 in Nice, where his parents had gone because of his mother's health, which continued to fail. She died 2 years later, after giving birth to a second son. Henry's father never remarried. When Henry was 11, he was enrolled in Hackney Academy, a progressive school outside London. From there, he proceeded to St. Peter's College (Peterhouse), Cambridge University in 1749, leaving after 3 years without a degree. For the next 30-odd years, he lived at his father's house on Great Marlborough St., London. Freed from the need to support himself, he followed his inclinations, which were to study and carry out researches in the physical sciences. Around the time his father died, in 1783, he acquired two houses of his own, one in and one outside London, both of which he adapted to his scientific habit. He was a prominent member and administrator of the Royal Society of London. His manner of living was modest, and over time he accumulated an immense fortune. He died in 1810, at age 78.

Henry Cavendish was descended from dukes, the highest rank of the peerage: these were the duke of Kent on his mother's side and the duke of Devonshire on his father's (Chap. 12). The branch that bore his name was the richer and more powerful of the two, and he owed his recognition in the wider society largely to it. William Cavendish, Henry's great grandfather, had been a moving force in the Glorious Revolution of 1688 – for his services, he was elevated from earl to duke of Devonshire – and his descendants remained active in politics in support of the Whig cause. Carried out by aristocrats like William Cavendish, the Revolution shifted the balance of power in the nation from the court to the aristocracy. The empowered aristocracy proved quite successful: under its leadership, liberties were enlarged, new industries arose, trade prospered, wars were won, the empire expanded, and the

R. McCormmach, *The Personality of Henry Cavendish - A Great Scientist with Extraordinary Peculiarities*, Archimedes 36, DOI 10.1007/978-3-319-02438-7_1,

arts and sciences achieved admirable goals. The aristocracy's influence in the nation increased through the century.[1]

As heads of one of the most powerful families of the nobility, the dukes of Devonshire had few equals. Indeed, there were few peers of all ranks, perhaps 200 in a given year, but these peers had a large number of relatives, who counted among the aristocracy. Charles Cavendish, Henry Cavendish's father, was a younger son of the second duke of Devonshire. By common law, children of peers were commoners, but by courtesy they were distinguished from commoners. Charles Cavendish had the right to be called "the right honourable," and he was usually called "lord," the same as nobility. Lord Charles was sent to Eaton, one of the great public schools, where boys read Latin writers from whom they learned the classical virtues of order, symmetry, balance, and restraint, and a code of values, chief among which were the "duty of service" and the right of the aristocracy to govern. Henry Cavendish's first cousin the politician Lord George Augustus Cavendish told Horace Walpole that he "liked an aristocracy and found it right that great families with great connections should govern."[2] William Cavendish, fourth duke of Devonshire, another first cousin of Henry Cavendish's, assumed that "Great Britain should be governed by an aristocracy, with himself a principal."[3] This is the society Henry Cavendish took his rightful place in, though he performed his duty in the Royal Society instead of in Parliament.

A talented politician, the fourth duke of Devonshire served as cabinet minister and briefly as prime minister. In personality, he was very like Henry Cavendish, detached, withdrawn, unimpassioned, cautious, supremely objective, and devoted to work and duty. He kept to himself, an observer not a joiner; he had no intimate friends in politics; everything he did he did well. According to the editors of his diary, he showed "complete self-assurance as to his place in the order of the world."[4] Henry conducted his affairs in the Royal Society and in the scientific world at large with the same self-assurance, and for the same reasons.

The head of the family for most of Henry Cavendish's adult life was William Cavendish, fifth duke of Devonshire. Like Henry, he was intelligent, and he had some of the same personality traits. He was introverted, reclusive, given to routine, and awkward, but there the parallel ends. He lay in bed until the middle of the afternoon, upon which time he went to his club to gamble through the night. He was dissolute, unfaithful, and passive. Henry Cavendish detested the indulgent fifth duke, who brought shame to the family name, of which he was very proud (Figs. 1.1, 1.2, 1.3, 1.4, 1.5, 1.6, 1.7, 1.8, 1.9, 1.10, 1.11, 1.12, 1.13, 1.14, and 1.15).[5]

---

[1] John Cannon, *Aristocratic Century: The Peerage of Eighteenth-Century England* (Cambridge: Cambridge University Press, 1984), 6, 175, 179.

[2] Ibid., 3, 10, 15, 34–35.

[3] William Cavendish, *The Devonshire Diary: William Cavendish, Fourth Duke of Devonshire, Memoranda on State of Affairs, 1759–1762*, eds. P. D. Brown and K. W. Schweizer, Camden Fourth Series, vol. 27 (London: Royal Historical Society, 1982), 19.

[4] Ibid., 19–21.

[5] John Pearson, *The Serpent and the Stag: The Saga of England's Powerful and Glamorous Cavendish Family from the Age of Henry the Eighth to the Present* (New York: Holt, Rinehart and Winston, 1983), 122–23. Lady Sarah Spencer quoted in Hugh Stokes, *The Devonshire House Circle* (London: Herbert Jenkins, 1917), 315.

Cavendish's Family

Fig. 1.1 The Honorable Henry Cavendish. Graphite and gray wash sketch by William Alexander (Courtesy of the Trustees of the British Museum). Cavendish refused to sit for a portrait. Alexander succeeded by attending a dinner as a guest of the Royal Society Club. There he surreptitiously sketched Cavendish and separately he sketched his coat and hat hanging on the wall. At home, he combined the two sketches into one. Persons who were shown the sketch recognized Cavendish (Courtesy of the Trustees of the British Museum)

Fig. 1.2 Henry de Grey, Duke of Kent. By Jacopo Amiconi? (Courtesy of the Bedfordshire Record Office)

Fig. 1.3 Jemima (Crewe), Duchess of Kent. First wife of the duke. By Riley (Courtesy of the Bedfordshire Record Office)

Fig. 1.4 Kent Family. *Conversation Piece at Wrest Park.* At the duke of Kent's country house at Silsoe in Bedfordshire. Probably painted by Charles Phillips, around the year Anne de Grey, Henry Cavendish's mother, was born. From *left* to *right*: Mary de Grey, William Bentinck, Barbara Godolphin, Lord Berkeley, Charles Bentinck, Earl of Clanbrassil, Countess of Portland, Duke of Kent, Jemima Campbell (later Marchioness de Grey), Sophia de Grey, Duchess of Kent, Elizabeth Bentinck, Countess of Clanbrassil, and Countess Middleton (Courtesy of the Bedfordshire Record Office)

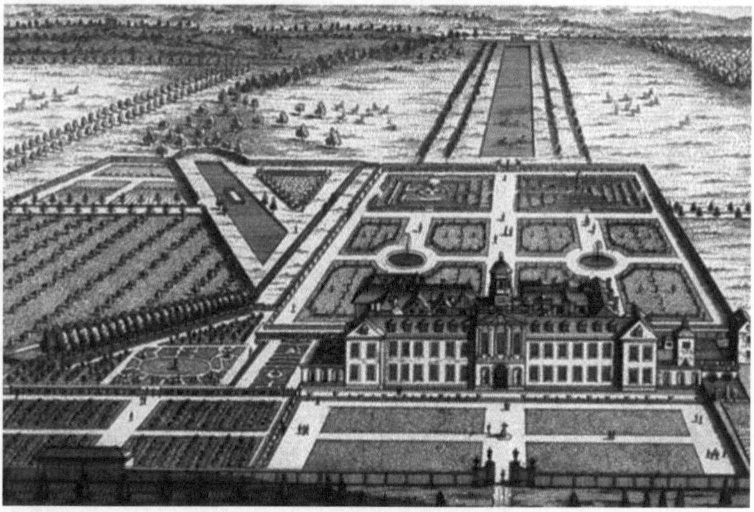

Fig. 1.5 Wrest House. By Pieter Van der Aa. This shows the house, garden, and park as they appeared around 1708. The present Wrest House was built in the nineteenth century (Wikimedia Commons)

Fig. 1.6 St. James Square, No. 4. The Duke of Kent's London house (Courtesy of the Greater London Record Office)

Fig. 1.7 William Cavendish, Second Duke of Devonshire. By Charles Jervas. Devonshire Collections, Chatsworth (Courtesy of the Chatsworth Settlement Trustees. Photograph Courtauld Institute of Art)

Fig. 1.8 Rachel (Russell), Duchess of Devonshire. By M. Dahl. Devonshire Collections, Chatsworth (Courtesy of the Chatsworth Settlement Trustees. Photograph Courtauld Institute of Art)

Fig. 1.9 Chatsworth House and Gardens. By Pieter Tillemans. The duke of Devonshire's country estate in Derbyshire. Turn of the eighteenth century (Wikimedia Commons)

Fig. 1.10 Devonshire House. Engraving by unknown artist around 1800 (Wikimedia Commons)

Fig. 1.11 William Cavendish, Fifth Duke of Devonshire. By Pompeo Girolamo Batoni. Henry Cavendish disliked the duke (Wikimedia Commons)

Fig. 1.12 Georgiana (Spencer), Duchess of Devonshire. By Thomas Gainsborough (Wikimedia Commons)

Fig. 1.13 "A Gambling Table at Devonshire House." By Thomas Rowlandson in 1791. Georgianna, duchess of Devonshire who is seated at the *left*, loved to gamble. She took an interest in Henry Cavendish's scientific work (Wikimedia Commons)

Fig. 1.14 Lord Charles Cavendish. By Enoch Seeman. Devonshire Collections. Father of Henry Cavendish (Courtesy of the Chatsworth Settlement Trustees. Photograph Courtauld Institute of Art)

Fig. 1.15 Lady Anne de Grey. By J. Davison. Mother of Henry Cavendish (Courtesy of the Bedfordshire Record Office)

The fifth duke forbade his wife Georgiana to visit Henry Cavendish's laboratory because "he is not a gentleman – *he works*."[6] A gentleman was not accused of working if he entered public life as a politician, and the family is said to have expected Henry to follow that path, as his father had done for a time. They were "much displeased with his steady refusal to quit the studies which he loved," according to a contemporary.[7] We do not know what distant members of the family thought about him, and it is possible that one or more of them were disappointed, but anyone who spent time with Cavendish would have seen that he was not cut out for public life. When a conversation turned to politics, he usually remained silent, and when he made an exception it became gossip. Richard Kirwan, a scientific colleague, wrote to the president of the Royal Society that he heard "Mr Cavendish talks Politics," surprising him because Cavendish had been "silent" during "Ld North's Rump Parliament, in wch his family were so much engaged," and which had "agitated the whole Nation."[8] Cavendish's home away from home was the Royal Society, which formally banished political, religious, and other divisive issues. He was intended for what he chose to do, pursue his studies.

Like the wider society, the Royal Society acknowledged rank, and its leadership regularly had an aristocratic element. The president of the Royal Society James Douglas, fourteenth earl of Morton, praised his fellow aristocrat Lord Charles Cavendish for his scientific work, referring to him as that "Noble Member of the Society."[9] George Parker, second earl of Macclesfield, a later president of the Royal Society, said that Lord Charles was equally distinguished for his "high Birth and eminent Station in life" as for his promotion of natural knowledge. In the age of aristocracy, Lord Charles who was also an accomplished researcher was an "Ornament" of the Royal Society.[10] Henry Cavendish could not be called "lord" or "noble," but he had the right to be addressed with the courtesy title "the honourable," as he was in the Royal Society, where he too was an ornament for the same pair of qualities.

1.2 The Natural Philosopher

When the duke of Devonshire said that Henry Cavendish "*works*," he spoke truthfully. He was a natural philosopher, and his work was natural philosophy. These terms have long gone out of use. Natural philosophy in its widest meaning was the study of the complete natural world, immaterial as well as material. In practice, it meant the study of the nonliving material world, which corresponds roughly to

[6] Francis Bickley, *The Cavendish Family* (London: Constable, 1911), 202.

[7] Henry Brougham, *Lives of Men of Letters and Science Who Flourished in the Time of George III*, 2 vols. (London, 1845–46); Vol. 1 (Philadelphia, 1845), 250. Henry Brougham, who was first baron Brougham and Vaux, was known as Lord Brougham.

[8] Richard Kirwan to Joseph Banks, 10 January 1789, British Museum (NH), DTC 6: 122–24.

[9] 30 November 1765, Journal Book, Royal Society 25: 656.

[10] 30 November 1757, ibid. 23: 638.

our physical sciences. According to a scientific dictionary of the time, natural philosophy is "that science which considers the powers of nature, the properties of natural bodies, and their actions upon one another." It discovers those general rules that enable us to explain natural effects, both near and far, and to infer past facts and predict future ones; it is humanity's singular claim to "omniscience."[11] Cavendish and most of his colleagues would have accepted this definition or something close to it.

Writing several years after Cavendish's death, when the sciences were becoming more clearly marked out, Thomas Young described him as "a great and justly celebrated Chemist, Natural Philosopher, and Astronomer."[12] Natural philosophy was divided into major branches: electricity, magnetism, optics, pneumatics, acoustics, mechanics, and gravitational astronomy. Cavendish carried out researches in all of the branches, and in addition he pursued several related subjects including, in addition to chemistry and astronomy, geology, meteorology, and mineralogy. Although the mathematical parts of gravitational astronomy were taken up in books on natural philosophy, astronomy was sufficiently distinct in its objectives and instruments to be considered a separate science. Chemistry was considered a separate science partly for the same reason, and partly for conceptual reasons: natural philosophy was concerned with mechanical subjects and their mathematical laws, whereas chemistry was concerned with laws of a different order, not expressible mathematically; alternatively, natural philosophy was concerned with visible motions of bodies, and chemistry with invisible motions. There was general agreement that the boundary between natural philosophy and chemistry could be crossed.[13] Heat, the subject most completely shared between them, was usually grouped with chemistry but increasingly with natural philosophy too.[14] Geology belonged to natural history, but in its study of the origin of strata it "required the most profound researches in every branch of natural philosophy."[15] Meteorology kept a record of the weather, but it also made use of laws and instruments from natural philosophy to study the atmosphere. Mineralogy collected and classified specimens of materials found on earth, but it also subjected them to chemical and physical investigations, and it had

---

[11] Charles Hutton, *A Mathematical and Philosophical Dictionary …*, 2 vols. (London, 1795–96) 2: 139.

[12] Thomas Young, "Life of Cavendish," *Encyclopaedia Britannica*, Supplement (1816–24); reprinted in Henry Cavendish, *The Scientific Papers of the Honourable Henry Cavendish*, 2 vols., ed. E. Thorpe (Cambridge: Cambridge University Press, 1921) 1: 435–47, on 435. Hereafter *Scientific Papers*.

[13] [John Robison], "Physics," *Encyclopaedia Britannica*, 3d ed. (Edinburgh, 1797) 14: 637–59, on 647. William Lewis, *Commercium Philosophico-Technicum; or, The Philosophical Commerce of Arts: Designed as an Attempt to Improve Arts, Trades and Manufactures* (London, 1763), iii–iv. Crosbie Smith, "'Mechanical Philosophy' and the Emergence of Physics in Britain: 1800–1850," *Annals of Science* 33 (1976), 3–29, on 8.

[14] E.g., Thomas Young, *A Syllabus of a Course of Lectures on Natural and Experimental Philosophy* (London, 1802).

[15] J. A. Deluc, *An Elementary Treatise on Geology: Determining Fundamental Points in That Science, and Containing an Examination of Some Modern Geological Systems, and Particularly of the Huttonian Theory of the Earth*, trans. H. De la Fite (London, 1809), 26.

important connections with geology. Distinctions as to what did and did not belong to natural philosophy are of little help in understanding Cavendish's life of science. For our purposes, "natural philosophy" will stand for all of the subjects Cavendish investigated.

By reason of his social position and his wealth, Cavendish stood somewhat apart, but in his work, of course, he did what other natural philosophers did. He served on councils and committees of a scientific society. He read scientific books and journals and met informally with colleagues. He kept records of his researches, sometimes for himself alone, and sometimes for colleagues and for publication. He worked with instruments and apparatus, which he built or had built for him. He performed experiments, made observations, and developed theories in the physical sciences.

Cavendish is best known for his work in chemistry; specifically, for his experimental study of "airs," which we call "gases." His first publication, in 1766, was on "factitious air," gases that are contained in other bodies in an inelastic state and that can be released by chemical means; for this work, he was awarded the Royal Society's Copley Medal. The most important result of his first paper was the determination of the nature and properties of "inflammable air," our "hydrogen"; the larger significance of the paper, according to Sir Edward Thorpe, editor of Cavendish's scientific papers, was its "final and decisive blow to the conception of a universal air, elementary and primordial," and its impetus to the study of gases, which led "eventually to a complete revolution in chemical doctrine."[16] In 1784, he published his best-known paper containing, according to Thorpe, the "first clear and incontestable proof of the compound nature of water, and of the nature and relative proportion of its constituents."[17] In 1771, Cavendish published a masterful mathematical theory of electricity, of which Thomas Thomson wrote in his *History of the Royal Society* 2 years after Cavendish's death that it was "the most rigid and satisfactory explanation of the phenomena of electricity, which has hitherto appeared in any language."[18] The theory predicted a wealth of phenomena, which Cavendish investigated in an extensive series of experiments, including a confirmation of the inverse square-law of electric force. He apparently intended his electrical researches for a book, but this was not to be. He left his experimental work in manuscript, though he later drew on some of it in a second published paper on an electrical

[16] Henry Cavendish, "Three Papers, Containing Experiments on Factitious Air," *Philosophical Transactions of the Royal Society of London* 56 (1766): 141–84. Hereafter *PT*. The article is reprinted in *Scientific Papers* 2: 77–101. Edward Thorpe, "Introduction," in Cavendish, *Scientific Papers* 2: 1–74, on 14–15.

[17] Thorpe, "Introduction," 23. Cavendish gave the experimental "proof," but his interpretation was expressed in the language and concepts of the phlogiston theory. Thorpe noted that it is impossible to tell from Cavendish's statements if he regarded water as a compound substance. Ibid., 35. Wilson differed; he thought that Cavendish interpreted water as a compound of dephlogisticated air and inflammable air. George Wilson, *The Life of the Hon*[ble] *Henry Cavendish* (London: Cavendish Society, 1851; reprint New York: Arno Press, 1975), 369, 435.

[18] Thomas Thomson, *History of the Royal Society from Its Institution to the End of the Eighteenth Century* (London, 1812), 455.

subject.[19] Cavendish's last important published experiment, referred to at the time as "weighing the world," was the determination of the average density of the earth. This experiment is recognized as "the basis of all the most significant experiments on gravitation ever since."[20] Many will agree with the physicist Sir Joseph Larmor, one of Cavendish's editors, that Cavendish was "one of the greatest of scientific discoverers."[21]

---

[19] Henry Cavendish, "An Attempt to Explain Some of the Principal Phaenomena of Electricity, by Means of an Elastic Fluid," *PT* 61 (1771): 584–677; reprinted in Henry Cavendish, *The Electrical Researches of the Honourable Henry Cavendish*, ed. J. C. Maxwell (Cambridge, 1879), 3–63. Maxwell thought that Cavendish planned a book. Ibid., xliii.

[20] Henry Cavendish, "Experiments to Determine the Density of the Earth," *PT* 88 (1798): 469–526; *Scientific Papers* 2: 249–86. A. H. Cook, "Experiments on Gravitation," in *Three Hundred Years of Gravitation*, ed. S. W. Hawking and W. Israel (Cambridge: Cambridge University Press, 1987), 51–79, on 52.

[21] Joseph Larmor, "Preface to Volume 1," in *The Scientific Papers of the Honourable Henry Cavendish, F.R.S.*, vol. 1: *The Electrical Researches*, ed. J. C. Maxwell and J. Lamor (Cambridge: Cambridge University Press, 1921), v–x, on x.

Chapter 2
Normality, Abnormality, and Scientists

2.1 Psychology of Scientists

The psychology of personality is helpful in understanding Cavendish. Taking as its subject the "total person," the psychology of personality studies how persistent patterns of daily activity, thought, and feeling combine to make the "unique, distinctive, individual." It looks for the "integrative quality," or unity, in the behavior of the individual.[1] It is just this, the integrative quality, that is largely missing in accounts of Cavendish's life. Given that his strangeness is what has intrigued people, the psychology of personality by itself is insufficient. Three additional sub-fields of psychology are also helpful: neuroscience, clinical-mental health, and the emerging sub-field of psychology of science.

The psychology of science and the sociology of science are sometimes seen as at odds, as having different ideas about how science works. They are better seen as two approaches to a common subject, science. Temperament, personality, motivation, and talent are properties of individual scientists, who are studied by psychologists. Individual scientists, who differ in these properties, work within social groups and institutional settings, which are studied by sociologists.[2]

Psychologists differ in their definitions of "personality," but they agree that it has at least two parts. One part is individuality and uniqueness of the person, and consistency of the person's behavior over time and from situation to situation; the other part is human nature, or the similarities of individuals. According to a well-known theory, personality can be characterized by basic dimensions, each a continuum. There is a difference of opinion on the number of dimensions, three to five being common. According to a widely used model, the basic dimensions are anxiety or

[1] The concept of the "self" is sometimes called on, though it has its critics. Lawrence A. Pervin, *Personality, Theory and Research* (New York, Chichester, Brisbane: John Wiley & Sons, 1993), 13–15.

[2] Gregory J. Feist, *The Psychology of Science and the Origins of the Scientific Mind* (New Haven and London: Yale University Press, 2006), 30–31.

R. McCormmach, *The Personality of Henry Cavendish - A Great Scientist with Extraordinary Peculiarities*, Archimedes 36, DOI 10.1007/978-3-319-02438-7_2,

neuroticism, introversion, openness, agreeableness, and conscientiousness. They are assumed to be fairly constant over time, to vary between individuals, and to account for behavior.[3]

Studies of twins indicate that dimensions of personality have a genetic linkage, with estimates of their heritability varying between 40 and 50 %. The rest is explained by environment, in which birth order and peer groups are especially significant. Temperament, which depends on genetic makeup and the central nervous system, is the foundation of differences in personality, and like personality, it has dimensions; according to one theory, these are activity levels, sociability, and emotionality. The psychologist of science Gregory Feist argues that personality and temperament act to lower thresholds for "interest, talent, and achievement in science."[4]

Of the various psychological factors, personality is the strongest predictor of an interest in science. A second psychological factor is self-image. A person has first to imagine himself as a scientist. Non-psychological factors of religious background and birth order can also influence a person's choice of science as a career, and they can affect the nature of his work. Protestants and Jews are more likely than Catholics to become scientists, and scientists are more likely than nonscientists to reject organized religion. Firstborns disproportionately choose careers in science, and if they do they are likely to be conservative rather than revolutionary in their scientific work; this agrees with firstborns' tendency to identify with parental authority and to accept the prevailing power structure.[5]

Certain personality traits[6] appear more commonly in scientists than in nonscientists. According to an analysis of 26 studies comparing nearly 5,000 scientists to non-scientists, under the category of "social" personality traits, scientists are found to be more dominant, arrogant, hostile, self-confident, autonomous, and introverted. They are less social than nonscientists, preferring to be left alone or to interact in small groups. Under "cognitive" personality traits, they are found to be more open and flexible. They are open to new experiences, curious about the world, and ready to admit error. Under "motivational" personality traits, they are found to be more driven and ambitious. Upon comparing personality traits of eminent scientists with average scientists, Feist comes up with a slightly altered list. Under "social," eminent scientists tend to be more dominant, self-confident, and deviant. Under "cognitive," they tend to be more open, flexible, intelligent, curious, and imaginative. Under "motivational," as before, they tend to be more driven and ambitious. In addition, they have aesthetic and affective traits. Feist says that researches on scientists' personalities show them to have a high level of conscientiousness,

---

[3] Ibid., 111.

[4] Feist, *Psychology of Science*, 114–15.

[5] Ibid., 72–74.

[6] Traits, a common way of modeling personality, are compatible with a number of psychological theories. They are descriptive, not deterministic; they are dispositions, which result in consistent, enduring patterns of mood, thought, and action. Robert R. McRae and Paul T. Costa, Jr., *Personality in Adulthood* (New York and London: The Guilford press, 1990), 19, 24.

which takes various forms: scientists desire order, organization, and punctuality, and they show caution, discipline, perseverance, and self-control. The most original scientists are an exception, being less conscientious. The psychological understanding of individuals' preferences of scientific fields is not very advanced, though evidence suggests that preferences for the physical sciences are often based on temperament: persons with such preferences are more oriented to inanimate objects than to people.[7] The findings above apply to scientists today, whose structured communities are very different from Cavendish's circle, a reservation we should keep in mind when transferring them to natural philosophers in the eighteenth century. Later in the book, in characterizing Cavendish's personality, we return to traits of scientists discussed here.

2.2 Scientists, Psychology, and Biography

Few would question the usefulness of psychology in biography. The occasional use of psychological insights is common practice among biographers, and it is usually unobjectionable. However, the systematic use of a psychological theory in biography meets with considerable resistance. The result, its critics say, is narrow, reductionist, and destructive of character. It need not be, but there is that danger.

From an early date, biographies written from a psychological perspective were called "psychobiographies." Because the reputation of psychobiography has waxed and waned, its supporters now prefer to use other words to describe it such as "psychological biography." The psychologist Alan Elms, who has written psychological biographies, acknowledges that some psychological biographies merit the derision they receive, doing little more than to label their subjects and to place them in pathological pigeonholes. The qualities needed to write good psychological biography, he says, are an understanding of the relevant parts of psychology and knowledge about and empathy for the subject.[8]

Psychological biography has been dominated by psychoanalytical theory. There are psychoanalytic biographies of great musicians, artists, and writers; for example, of Mozart, van Gogh, and Henry James. Psychoanalytical biographies of great scientists are rare, a notable exception being the historian Frank Manuel's biography of Newton. Manuel warns his readers that he makes "free use" of psychology, if not always its vocabulary. His starting point is notebooks from Newton's youth, to which he applies psychological "theorizing," involving "speculative leaps and analogical thinking." He introduces psychological facts and hypotheses that "are not subject to traditional forms of verification." His object is not the genesis and nature of Newton's scientific inventions but an understanding of Newton's

[7] Ibid., 115–19, 123.

[8] Alan C. Elms, *Uncovering Lives: The Uneasy Alliance of Biography and Psychology* (Oxford: Oxford University Press, 1994), 4–5; "The Psychologist as Biographer," http://www.ulmus.net/aceworks/psycholbiog.html.

Fig. 2.1 Sir Isaac Newton. Portrait by G. Kneller, engraving by A. Tardieu. (Courtesy of Smith Image Collection, Van Pelt-Dietrich Library, University of Pennsylvania)

character, the forces that shaped it, his loves and hates, his way of life, his view of the world, of "Newton the man." By exploring Newton's drives and bringing to life his troubled psyche, by revealing "the humanity of genius," Manuel has written one of the most arresting psychological biographies and, within the limits it concedes, and for those who accept the idea of the unconscious, one of the finest biographies of a scientist.[9]

Cavendish was the outstanding British physical scientist after Newton, and naturally in his time he was compared to his predecessor: "Like his countryman Newton, whom he resembles in so many other respects," Georges Cuvier said, Cavendish died full of honors (Fig. 2.1).[10] The patterns of their lives had similarities. Outwardly, Newton's life was uneventful, studious, secluded, and ascetic. He studied and taught at a university, and later he held a government job and presided over the Royal Society; nothing exciting here. Cavendish was likewise studious, secluded, and ascetic, and his life was similarly uneventful. They shared other traits too, but the

[9] Frank E. Manuel, *A Portrait of Isaac Newton* (New York: Da Capo Press, 1968), 2, 9–10, 18.

[10] Georges Cuvier quoted in Young, "Cavendish," 445.

main comparison has to do their scientific skills. On the workings of their psyches, the similarities end. They could hardly be more different.

The drama of Newton's life, Manuel finds, was largely inward, scripted by childhood traumas.[11] Newton's father died before he was born, depriving him of a mentor and model. His mother remarried when Newton was three, and she moved away to be with her new husband, leaving Newton behind with his grandmother. Newton bore a violent hatred of his stepfather for depriving him of his mother, whom he both loved and resented as a betrayer. This was the beginning of Newton's rage, which he would later unleash on his scientific rivals.[12]

Like Newton, Cavendish lost his mother at an early age, though he had his natural father, and as an adult he gave no sign of having to struggle with rage. He had no open quarrels with his fellows. A chemist who knew Cavendish credited him with "the total absence of all the violent passions."[13] He lived at his father's house until he was in middle age, when his father died, and there is every reason to believe that his father was a mentor and a model to him. We know for certain that he had a high opinion of his father in what mattered most to him, scientific skill at arriving at truths of nature. In a memorandum to his instrument maker, he wrote, "My father's experiments on which what I said concerning the turning of water into vapour are founded seem so convincing as to leave no doubt of the truth of it."[14]

"Truth" is a word Cavendish associated with "father." "Fornicator" is a word Newton associated with "father."[15] Manuel looks to Newton's repressed anxieties and unconscious defenses to explain his passion to understand nature. To address a related question about Cavendish, we call on a different branch of psychology.[16]

2.3 Problem of Cavendish

Cavendish's "habits were so retired" and his "intercourse with society was so small, there is nothing else to relate except his scientific labours," a chemist said. Wilson, the author of the first full-length biography of Cavendish, said that a "more eventless life, according to the ordinary judgment of mankind, than that of Cavendish, could scarcely

[11] Manuel says it was the interplay between his singular personality and what his society repressed and allowed. *Newton*, 5.

[12] Manuel, *Newton*, 24–27, 32–33, 348.

[13] Thomas Thomson, *The History of Chemistry*, 2 vols. (London, 1830–31) 1: 339.

[14] Two-sheet memorandum on the additivity of air pressure and the pressure of water vapor in a receiver. Cavendish Scientific Manuscripts, Devonshire Collections, Chatsworth, IV, 4.

[15] Manuel, *Newton*, 27. The association would have been with his stepfather. His biological father died before he was born.

[16] Geoffrey Cantor, in a stimulating paper "Psychobiography & Science," discusses several psychological biographies of scientists in which critical events or the unconscious play a central role. In the case of Cavendish, whose life was distinguished by its even temper, other factors need to be considered. I thank the author for a copy of his paper.

be conceived."[17] Cavendish indeed offers his biographer little of personal interest to work with. It would seem that conflict and other common elements of the human drama are all but absent. He lived apart, worked at home, did not keep a diary, and so far as we know, with the possible exception of an exchange with his brother, wrote no personal letters (Chap. 13). His personality and activity apparently changed little over time. His whole life, a colleague John Barrow said, was absorbed in science.[18]

Since any biography of Cavendish must lack events that enliven the usual biography, the focus has to be elsewhere. The natural philosopher Thomas Young said that of a benefactor of science so elevated as Cavendish, "we are anxious to learn all the details both of intellectual cultivation and of moral character that the labours of a biographer can discover and record."[19] In evident agreement with Young, George Wilson accepted an invitation to write a book about Cavendish by the Cavendish Society, a chemical printing society named after the scientist. The timing of the invitation was a revival of a priority dispute over the discovery of the composition of water. Cavendish, a principal in the discovery, was accused of behaving dishonorably toward his rival James Watt. Wilson asked the family for permission to examine Cavendish's scientific manuscripts, explaining that he intended to vindicate the "moral character" of Cavendish in his conduct towards other parties of the "water controversy."[20] Because of the form he adopted, a biography, he also attached a wider discussion of Cavendish's character (Fig. 2.2).

To learn about his character, Wilson enlisted the help of a friend in London to track down elderly fellows of the Royal Society and inhabitants of Clapham who had known Cavendish as a neighbor. He received responses from a fair number of them, who it turned out remembered Cavendish very well. Calling on memories of 50 or 60 years, they could repeat his conversations and describe his behaviors in detail. His behaviors had been mentioned in earlier biographical sketches; Wilson supplied more examples from his inquiries, and drew conclusions about the character of his subject.

Wilson's biography is insightful and generous. It is also strangely put together. Cavendish's "life," in the usual sense of the word, is presented in two chapters only, Chaps. 2 and 4, and fills only 50 pages of a book of nearly 500. The remainder of the book is taken up with an episode in the history of science, Cavendish's and others' experiments relating to the composition of water. The episode itself is divided into three parts, one a narrative, one an abstract of the relevant scientific papers, and the other a critical analysis of the several claims to the title of discoverer of the composition of water. Wilson's discussion of the episode is penetrating and judicious.

The imbalance of Wilson's biography had an unintended consequence. The book is called a life of Cavendish, but because it is mainly about the water controversy,

[17] Wilson, *Cavendish*, 165.

[18] John Barrow, *Sketches of the Royal Society and Royal Society Club* (London, 1849), 148.

[19] Young, "Cavendish," 435.

[20] Wilson, *Cavendish*, v, vii, ix. George Wilson to Lord Burlington, 15 March 1850, Lancashire Record Office, Miscellaneous Letters, DDCa 19.

Fig. 2.2 George Wilson (Photograph by James Good Tunny about 1854). Scottish chemist who wrote the first book-length biography of Henry Cavendish (Scottish National Gallery)

readers could come away from it with the idea that Cavendish's life was mainly about it. This reading of Wilson is evident in the biography of Cavendish appearing in the original edition of the *Oxford Dictionary of National Biography*.[21] The entry there reflects the proportions of Wilson's biography, most of it being a recounting of the water controversy. Wilson had not meant to burden Cavendish's memory with the controversy; quite the contrary, he meant to liberate his memory, but he had written the biography when partisan feelings still ran high, and the controversy would not be denied. So far as we know, it was the only controversy in Cavendish's life, and nearly all of it took place long after he was dead, but given the human interest in conflict, after Wilson's biography the identification of Cavendish with the water controversy was assured.

We are interested in a second consequence of Wilson's biography, a depiction of Cavendish's character or personality that has gone unchallenged for a century and a half. Of the two chapters on the life of Cavendish, we are interested in Chap. 4, which contains recollections of Cavendish by his contemporaries. Usually in biographies, sources like these appear as footnotes, with occasional quotations from them worked into the text. Wilson wrote his life of Cavendish not so much

[21] "Cavendish, Henry (1731–1810)," *Dictionary of National Biography*, 22 vols., ed. L. Stephen and S. Lee (New York: Macmillan, 1908–9) 3: 1257–1262. Hereafter *DNB*. This reference is to the first edition. Unless otherwise indicated, later references are to the second edition: 60 vols., eds. H. C. G. Matthew and B. Harrison (Oxford, New York, Aukland: Oxford University Press, 2004).

as a biographer than as a scientist, laying out the evidence in full, inviting his readers to repeat the experiment. "The account I have given of him [Cavendish] has necessarily assumed the character of a Mosaic, made up of fragments furnished by different hands. I have thus supplied each reader with the means of drawing a likeness for himself."[22] Wilson drew his likeness, and we take up his offer to draw our own, should it differ. For this book, which makes limited use of findings of a science, psychology, it is fortunate that Wilson included in his biography a wealth of raw testimony.

The recollections that Wilson collected are one of our sources. Another is obituaries and other published and unpublished accounts of Cavendish. Another is Cavendish's letters and scientific manuscripts. In 1879, under the editorship of James Clark Maxwell, Cavendish's electrical manuscripts and papers were published by Cambridge University Press. The same publisher brought out his other papers together with a selection of manuscripts on other parts of his work in 1921. Most of his manuscripts remain unpublished and are kept in the Cavendish family. Taken together, the sources on Cavendish's work are very considerable, and since his work and his life were so nearly the same, his scientific manuscripts complement accounts of his character in informing us about who he was and what his life was like.

[22] Wilson, *Cavendish*, 185.

Chapter 3
Everyday Life

Manuel discusses scientific genius in connection with Newton to whom the word certainly applies, though he does not presume to "unlock the secret of Newton's genius."[1] When we consider Cavendish, the word "genius" may not come readily to mind, but it was applied to him in his day,[2] and today Sacks uses it. Like Manuel, we can accept the mystery of Cavendish's genius and try to understand his character.

Genius requires a highly uncommon psyche, and for a person to reach the goal for which his genius is remembered, he may have to overcome great obstacles, and call on great resources. Cavendish, to be sure, had an uncommon psyche, and some of his behavior was considered abnormal by his contemporaries. He faced obstacles, perhaps in his youth great obstacles, but he did not let them stand in the way of realizing his goal. His obstacles were social, and his goal was a life of science, which showed the way around them.

Wilson and other biographers of Cavendish wrote principally about his scientific work. What they had to say about his "character" they found in his scientific vocation and his behavior. In this chapter, we look at what they singled out as evidence of his character, drawn mainly from his everyday activities. We begin by identifying the scientific contemporaries who supply us with firsthand accounts.

Banks, Sir Joseph, 1743–1820. Natural historian. President of the Royal Society. He held regular social gatherings of men of science at his house, which Cavendish faithfully attended.

Barrow, Sir John, 1764–1848. Statesman. Fellow of the Royal Society. He became acquainted with Cavendish at dinners of the Royal Society Club.

Biot, Jean-Baptiste, 1774–1862. Physicist. He had no direct knowledge of Cavendish. Though not entirely reliable, his encyclopedia article on Cavendish is highly readable.

[1] Manuel, *Newton*, 2.

[2] Playfair, quoted in Wilson, *Cavendish*, 165.

R. McCormmach, *The Personality of Henry Cavendish - A Great Scientist with Extraordinary Peculiarities*, Archimedes 36, DOI 10.1007/978-3-319-02438-7_3,

Blagden, Sir Charles, 1748–1820. Physician. Secretary of the Royal Society. For several years in the 1780s he was Cavendish's associate, spending much time in his company.

Brougham, Henry Peter, Baron, 1778–1868. Statesman. Fellow of the Royal Society. He knew Cavendish from gatherings at Sir Joseph Banks's house. He included him in a book on the lives of men of science and letters.

Brown, Robert, 1773–1858. Botanist. Fellow of the Royal Society.

Children, John George, 1777–1852. Chemist, natural historian. Fellow of the Royal Society. He knew Cavendish from the Royal Society Club.

Cuvier, Georges, 1769–1832. Naturalist and zoologist. He was a permanent secretary of the Academy of Sciences of the Institute of France, of which Cavendish was a foreign associate. He did not know Cavendish personally. For information for his *éloge* of Cavendish, he went to Blagden. Cuvier's biography is largely accurate and insightful.

Davy, Sir Humphry, 1778–1829. Chemist. Fellow of the Royal Society. Cavendish interviewed him for a position in the Royal Institution as lecturer and director of the chemical laboratory, and Cavendish visited him in his laboratory to observe his experiments. Davy spoke about Cavendish in a lecture he delivered at the Royal Institution following his death.

Davy, John, 1790–1868. Physician. Fellow of the Royal Society. For a time, he assisted his brother Humphry in the Royal Institution, where he saw Cavendish.

Lawson, Henry, 1774–1855. Astronomer, meteorologist. Fellow of the Royal Society.

Michell, John, 1724–93. Natural philosopher. Minister of Thornhill Parish, Yorkshire. Fellow of the Royal Society.

Newman, John. Instrument maker employed by the firm Nairne and Blunt.

Pepys, William Hasledine, 1775–1856. Mineralogist, geologist. Fellow of the Royal Society.

Playfair, John, 1748–1819. Natural philosopher, geologist. When he was in London, he was a frequent guest at the Royal Society Club, where he saw Cavendish.

Thomson, Thomas, 1773–1852. Chemist. Fellow of the Royal Society. He included an account of Cavendish and his work in his history of chemistry.

Young, Thomas, 1773–1829. Natural philosopher. Fellow of the Royal Society. For a time he was professor of natural philosophy in the Royal Institution, where he knew Cavendish. His article on Cavendish for the *Encyclopaedia Britannica* drew on Cuvier's *éloge* and on personal knowledge. It is reprinted in the edition of Cavendish's scientific papers, a good choice.

3.1 Appearance

People noticed Cavendish's appearance. Biot said that his "clothes never changed in form, in color, nor in material."[3] Young said that his "dress was singularly uniform, although sometimes a little neglected."[4] According to a woman who lived in Cavendish's house at Clapham Common after his death, he had no wardrobe, possessing only "one suit of clothes at a time."[5] Her report is likely nearly correct. At the time of Cavendish's death, the value of his wardrobe at his house on Clapham Common was £37 and at his London townhouse nothing.[6]

Thanks to our one image of Cavendish, an ink and wash sketch showing him in full dress, we can be specific about his appearance and relate it to the fashions of the time. To begin with, he wore a wig. This is unsurprising since he lived at a time when men's wigs were in high fashion. In the previous century, to conceal his early baldness the king of France took to wearing a wig, which became a fashion, first in France and then abroad. Later in the century the word "wig," short for "periwig," entered the English language, and in the following century the wig made from hair of various species was a standard item in the dress of middle- and upper-class men in England. By the time Cavendish was born, the loose, full bottom, shoulder-length wigs with cascading curls, the kind of wig worn by Newton, had given way to tied-back wigs, the hair pulled back as in a ponytail or a braid. When Cavendish was in his 40s, in the 1770s, the bob wig came into fashion. Paintings of his colleagues show them wearing wigs of this kind. In the 1780s, fashionable young men wore their own hair lightly powdered, and in the 1790s it was uncommon to see wigs and hair powder on any but older, conservative men. The fashion of wigs effectively came to a close in 1795, when a tax on hair powder was imposed. The sketch of Cavendish late in life, around 1800, shows him wearing a variant of the tied-back wig, described by one person as a "knocker-tailed periwig," and by another person as a "thick clubbed tail," then decades out of fashion.[7]

The three-cornered hat, the "cocked hat" or "tricorne," familiar to Americans as the hat worn by Revolutionary patriots, originated with Spanish soldiers in the seventeenth century. By pinning up the sides and the back of a round, broad brim, the soldiers created a practical headgear for rainy climates; water collecting in the brim was funneled forward or to the side. During a war between France and Spain, French soldiers brought the three cornered hat back to France, where it like the wig was made fashionable by the French king, for civilian as well as military wear. Because

[3] J.-B. Biot, "Cavendish (Henri)," *Biographie Universelle*, vol. 7 (Paris, 1813), 272–73, on 273.

[4] Young, "Cavendish," 444.

[5] Wilson, *Cavendish*, 188.

[6] "Extracts from Valuations of Furniture &c.," Devonshire Collections, Chatsworth.

[7] W. H. Pepys, quoted in Wilson, *Cavendish*, 168. Brougham, *Lives* 1: 259. Anon., "Wig," http://www.en.wikipedia.org/wiki/Wig. Anon., "Colonial Wigs," http://www.stitchnsave.com/wigs/Colonial_Wigs.asp.

wigs were enormous at the time, the popular broad brimmed hat was carried more than worn over wigs, and for carrying, it was convenient if the hat was folded into a tricorne. The hat remained popular in Europe and in England through the eighteenth century, though by the time of Cavendish's sketch, around the end of the century, its use had sharply dropped off, replaced by a tall conical hat with narrow brim and soon by the top hat. We know that Cavendish's tailor paid him an annual visit, at a time predetermined by the almanac,[8] and we suppose that his hatter did the same. We could possibly date this yearly event. At the bottom of a draft letter, Cavendish jotted a reminder of items of business to attend to that day, one of which was "hatter." Three-cornered hats could be elaborate, decorated with gold and silver lace and feathers, and the hats he ordered for his servants were ornate, but from the sketch we see that Cavendish's own hat was simple.[9]

In the sketch, Cavendish is shown wearing breeches. Breeches and high stockings were the usual lower garments for men then, as they had been from the late sixteenth century and would be into the early nineteenth century. The rest of his three-piece suit, knee-length coat and mid-thigh waistcoat were still worn, as was his knee-length dress shirt with ruffles in front and at the wrists and his high stock at the neck. A variety of new styles were appearing in men's fashion, and except for formal dress, the items of Cavendish's wardrobe – powdered wigs, ruffles, lace embroidery, and cocked hats – were falling out of favor.[10] His dated appearance in his later years made an impression on others, as summarized by John Davy, who was in his late teens when he saw him: "His dress was that of the gentleman of the preceding half-century … in brief, almost the court dress of the then and the present time."[11]

Cavendish's "walk was quick and uneasy," Brougham said.[12] A fellow diner at the Royal Society Club said Cavendish "came slouching in, one hand behind his back."[13] To be sure, the sketch of Cavendish shows him walking with one hand behind his back, the other hand inside his coat pressed against his chest, both hands immobile. The latter, one hand inside the jacket, was a common pose then – Cavendish's father posed like that for a painting – but Cavendish was not posing, allowing the possibility that he walked that way.

---

[8] Biot, "Cavendish," 273.

[9] Anon., "Tricorne," http://www.en.wikipedia.org/wiki/Tricorne. Anon., "1750–1795 in Fashion," http://www.en.wikipedia.org/wiki/1750. The hatter's bill for hats for the servants at Cavendish's townhouse on Bedford Square in one year listed four silver laced livery cocked hats and one extra fine beaver hat. 29 February, 29 June, 1 August, 22 September, 14 December, 1809, "Bedford Square. James Fuller's Account with the Exec. of Hen: Cavendish Esq. Balance 37.6.4. Settled 30 August 1810," Devonshire Collections, Chatsworth.

[10] Anon., "1795–1820 in Fashion," http://www.en.wikipedia.org/wiki/1795. Colonial Williamsburg Foundation, "A Colonial Gentleman's Clothing: A Glossary of Terms," http://www.history.org/history/clothing/men/mglossary.cfm.

[11] John Davy, quoted in Wilson, *Cavendish*, 168.

[12] Brougham, *Lives* 1: 259.

[13] A chemist, quoted in Wilson, *Cavendish*, 168.

Like the days of his life, Cavendish's features held no particular interest. Thomson said that "his countenance was not strongly marked," and it "did not much prepossess strangers in his favour."[14] Playfair said that he had an "unpromising exterior, "that "he is of an awkward appearance, and has certainly not much the look of a man of rank."[15] John Davy gave a little more detail: apart from his dress, his "appearance was ... nowise distinguished: of fair complexion, small, and not marked features."[16] There was agreement on his height but not on his build. By one account, "he was somewhat above the middle size, his body rather thick, and his neck rather short."[17] By another, "his person was tall, and rather thin."[18] If we combine these verbal descriptions with the ink sketch, we have as complete a picture as we ever will of the man who cut so strange a figure in the world.

3.2 Speech and Silence

People apparently had no difficulty understanding Cavendish, but they found his voice and speech strange. He had a "thin shrill voice," Brougham said; Barrow said he had a "shrill disagreeable voice."[19] The rhythm of his speech was odd too. He had "a hesitation or difficulty of speech," "an imperfection in his speech, a difficulty in bringing out his words," a slight stutter that "gave him an air of awkwardness."[20]

Cavendish rarely joined in general conversation even at occasions where conversation was the point, as it was at the Royal Society Club or at Joseph Banks's house. Barrow thought that Cavendish was "averse from conversation" because he was self-conscious about the sound of his voice.[21] He entered "diffidently into any conversation," and then only when it was on a scientific subject that interested him.[22] If the conversation veered from science, "he turned aside, and all the cold indifference of his nature returned."[23]

There is a single recollection of Henry with his brother, Frederick. It seems that they crossed the English Channel, and upon landing at Calais they stayed the night in a hotel. On the way to their room they passed an open door exposing a body ready

[14] Thomson, *History of Chemistry* 1: 339.

[15] John Playfair, *The Works of John Playfair*, ed. J. G. Playfair, 4 vols. (Edinburgh, 1822) 1: appendix, lxxxiv.

[16] John Davy, quoted in Wilson, *Cavendish*, 168.

[17] Thomson, *History of Chemistry* 1: 339.

[18] Young, "Cavendish," 444.

[19] Brougham, *Lives* 1: 258. Barrow, *Royal Society*, 144.

[20] Pepys, quoted in Wilson, *Cavendish*, 168. Brougham, *Lives* 1: 258. Barrow, *Royal Society*, 144. Thomson, *History of Chemistry* 1: 339. Barrow, *Royal Society*, 144. Playfair, *Works* 1: appendix, lxxxiv.

[21] Barrow, *Royal Society*, 144–45.

[22] Brougham, *Lives* 1: 258.

[23] A fellow of the Royal Society, quoted in Wilson, *Cavendish*, 182.

for burial. The brothers said nothing, but on the road to Paris the next day Frederick said to Henry, "Did you see the corpse?" Henry answered, "I did." The anecdote was preserved, it is thought, "to show that taciturnity was a family characteristic."[24] Henry Holland, who knew Henry Cavendish and his family, spoke of his "hereditary taciturnity."[25] His was only the most notable instance of it in the family.

At social gatherings, Cavendish tended to be a spectral presence. Although he might approach "to hear what was passing among others," he "seemed to dislike being spoken to" and "be annoyed if looked at."[26] When he did speak, he did so without looking at the other person. If he was already in conversation, any eye contact could put an end to it. When J. G. Children, a new member of the Royal Society Club and unknown to Cavendish, spotted him talking with several other members, he joined the group. "My eye caught that of Cavendish, and he instantly became silent: he did not say a word."[27] According to another member of the Club, if someone tried to draw Cavendish into conversation, "he always fought shy." The best way to engage him was "never to look at him, but to talk as it were into vacancy."[28] Banks advised visitors to his salon "to avoid speaking to him … [but] if he speaks to you, continue the conversation."[29] People used stratagems to get Cavendish to talk to them, but their rate of success may not have been high. Brougham said that Cavendish "uttered fewer words in the course of his life than any man who ever lived to fourscore years, not at all excepting the monks of La Trappe."[30] Cavendish was known as much for his silences as for his words.[31]

In company, Cavendish made involuntary sounds. When someone spoke to him the wrong way, he might leave "abruptly, with a kind of cry or ejaculation, as if scared and disturbed." Brougham recalled the "shrill cry he uttered as he shuffled quickly from room to room."[32] This rather startling behavior occurred regularly, it seems. Naturally people remembered him.

Two overheard conversations illustrate Cavendish's tendency to repeat words and phrases. When Cavendish's former librarian was mentioned at the Royal Society

[24] Bickley, *Cavendish Family*, 198–99.

[25] Henry Holland, quoted in Gwendy Caroe, *The Royal Institution: An Informal History* (London: John Murray, 1985), 39.

[26] Brougham, *Lives* 1: 258–59.

[27] Children, quoted in Wilson, *Cavendish*, 169.

[28] A chemist, quoted, ibid.

[29] Pepys, quoted, ibid., 168.

[30] Brougham, *Lives* 1: 259.

[31] Lidbetter suggests that Cavendish had no control over his silences because he was autistic. Elsewhere in his article says that intelligence has great importance for an autistic person's ability to adapt. "Henry Cavendish," 786. Cavendish was intelligent, and speech can be adapted. As mentioned before, Cavendish's colleague John Barrow thought that Cavendish controlled his speech: he attributed Cavendish's "aversion from conversation" to his self-consciousness about his unpleasant voice. *Royal Society*, 144–45.

[32] Ibid., 258–59.

Club, he said, "'Ah! poor fellow: how does he do? How does he get on?' 'I fear very indifferently,' said this person. 'I am sorry for it,' said Mr. C. 'We had hopes you would have done something for him, sir.' 'Me, me, me, what could I do?' 'A little annuity for his life; he is not in the best of health.' 'Well, well, well, a check for ten thousand pounds, would that do?'"[33] A banker announced himself to Cavendish's servant, who passed the information to his master. "'Sir, there is a person below, who wants to speak to you.' 'Who is he? Who is he? What does he want with me?' 'He says he is your banker, and must speak to you.' Mr. Cavendish, in great agitation, desires he may be sent up, and before he entered the room, cries, 'What do you come here for? What do you want with me?'" The banker told him he had a large balance and he might like some of it invested. "'Well! Well! What do you want to do?' 'Perhaps you would like to have forty thousand pounds invested.' 'Do so! Do so, and don't come here to trouble me, or I will remove it.'"[34] As we would expect, repetitions of words and sentences do not appear in Cavendish's writing. The presence of another person was needed.

Cavendish was worth hearing out. Banks told visitors that if they managed to get Cavendish to talk, they would find him "full of information, particularly as to chemistry."[35] Pepys said that "he was very quick of comprehension."[36] John Davy, who came to know Cavendish when he was about 70 and physically infirm, said that "his conversation and queries denoted quickness and acuteness, and undiminished vigour of mind."[37] Humphry Davy said that when Cavendish was familiar with a person, his "conversation was lively, and full of varied information," and that "upon all subjects of science he was luminous and profound; and in discussion wonderfully acute."[38] Playfair said that "gleams of genius break often through this unpromising exterior" and that "he never speaks at all but that it is exceedingly to the purpose, and either brings some excellent information, or draws some important conclusion."[39] On Cavendish's clarity of speech, there was unanimity of opinion.[40]

[33] Pepys, quoted in Wilson, *Cavendish*, 174.

[34] Ibid., 175–76.

[35] Banks's advice to Pepys, quoted in Wilson, *Cavendish*, 168.

[36] Pepys, quoted in Wilson, *Cavendish*, 168.

[37] John Davy, quoted in Wilson, *Cavendish*, 168.

[38] Humphry Davy, quoted in John Davy, *Memoirs of the Life of Sir Humphry Davy, Bart.*, 2 vols. (London, 1836), 222.

[39] Playfair, quoted in Wilson, *Cavendish*, 165–66.

[40] In our biography of Cavendish, Christa Jungnickel and I wrote that "on the subjects he cared to speak about, Cavendish spoke precisely and sparingly as a point of conscience." Lidbetter calls this statement "hyperbole." I do not know what he means. We use the word "conscience," if that is his point, because of the high value Cavendish placed on truth, accuracy, and simplicity; these values are discussed in our biography, as they are in this book. Persons from Cavendish's time who wrote about him said in various ways much the same thing about his speech.

About Cavendish's use of language, we have only his writings and a few remembered fragments of conversation to go on. His writings were about factual matters, and so was his language. His letters contain no examples of a sense of irony, a sense of humor, or figurative language, but neither do the letters he received. There may be an exception. In a letter to John Michell, he wrote, "If your health does not allow you to go on with that [completing his 'great telescope'] I hope it may at least permit the easier & less laborious employment of weighing the world."[41] The experiment of weighing the world, or determining the average density of the earth, was itself laborious, as Cavendish learned when he carried out the experiment himself, but even so it was less laborious than completing the telescope, a goal Michell never accomplished. So Cavendish was correct in what he said to Michell, but his wording would seem to be a turn of phrase, and if so it is evidence he did not lack a sense of humor.

Silence not only spared Cavendish painful social encounters, it confirmed a marked scientific trait of his, circumspection. He was sensitive to words as potential instruments of deception, as shown by his criticism of new words in the language of chemistry, discussed below. He was more comfortable with the language of numbers and mathematics, instruments of "strict reasoning," a favorite expression of his and his ideal in scientific work. Cavendish's inclination to remain silent was reinforced by his preoccupation, science.

3.3 Feelings

Wilson's sources found Cavendish emotionally stunted. According to one of them, he was the "coldest and most indifferent of mortals."[42] After a meeting of a scientific club, Blagden and the astronomer Alexander Aubert discussed Cavendish and agreed that "he had no affections, but always meant well."[43] By denying that Cavendish had affections, his colleagues did not mean that he had no feelings; they meant that he lacked personal warmth. An eighteenth-century meaning of "affection" applies to Cavendish's defect: passion as opposed to reason.[44]

Cavendish had strong feelings on subjects that were tied to his sense of who he was and what he did: he belonged to a proud aristocratic family, and he worked in natural philosophy. In the following two examples, Cavendish's words convey his strong feelings. The first example comes from natural philosophy. In the course of the Chemical Revolution, the leader of the new chemistry A. L. Lavoisier sent Cavendish a copy of a book laying out a corresponding new nomenclature. This prompted an angry outburst from Cavendish in a letter to Blagden. What Lavoisier says in the preface "has only served the more to convince me of the impropriety of systematic names in Chemistry & the great mischief which will follow from his

[41] Henry Cavendish to John Michell, 27 May 1783, draft; in Jungnickel and McCormmach, *Cavendish* (1999), 567–69, on 567.

[42] Quote from one of his informants, Wilson, *Cavendish*, 173.

[43] 15 September 1794, Charles Blagden Diary, Royal Society 3: 16(back).

[44] *Oxford Universal Dictionary on Historical Principles*, 3d ed. (Oxford: Clarendon Press, 1955), 31.

scheme if it should come into use." He closed the letter: "As I think this attempt a very mischievous one it has provoked me to go out of my usual way & give you a long sermon. I do not imagine indeed that their nomenclature will ever come into use but I am much afraid it will do mischief by setting peoples minds afloat & increasing the present rage of name making."[45] Cavendish used a variety of strong words of disapproval to express the harm it would do to chemistry: "great inconvenience," "confusion," "ill consequence," "improper idea," "more difficult to understand," and "the fashion which so much prevails among philosophers of giving new names." In his "sermon," Cavendish expressed his feelings about a threat to chemistry. Systematic naming was not a form of systematizing he welcomed because it was based on physical theories, which were impermanent. Systematic naming was helpful in natural history – Cavendish bought two volumes of Linnaeus in or after 1785, adding to the seven volumes he already owned –but he considered it a dangerous trend in natural philosophy, in which we include chemistry.

The second example comes from Cavendish's activity as a land developer, a typical outlet for the energy and leisure of a gentleman. Shortly before he moved into his house on Clapham Common, a suburb of London, he bought 15 acres fronting the Common, which he leased to builders. He soon found himself in conflict with the original owner of the 15 acres, a principal developer of the Common, Christopher Baldwin. Baldwin owned a small piece of land, about one half acre, lying between Cavendish's 15 acres and Balam Lane, which opened onto the Common, and Cavendish wanted to buy it. Baldwin had extended the small piece of the land part way into the lane. This slip of land extending onto the lane was the problem. Baldwin evidently had incurred legal expenses because of it, and he asked Cavendish to pay him £60 to cover them. Cavendish's lawyer advised him that that £60 was too much, that £40 was closer to it. Letters passed between Baldwin, Cavendish, and Cavendish's lawyer, expressing growing impatience and exasperation on both sides. The lawyer threatened to take Baldwin to court. Baldwin brought up an independent issue: Cavendish's fences were ruined, allowing cattle into Baldwin's garden. Cavendish said he would have fixed the fences long ago if it had not been for the business of the slip of land. "I shall observe my agreement about the fence but will not be prescribed to about it nor bear your delays or cavils." In the midst of the wrangling, there appeared a rumor that the people of the Common were going to pull down all enclosures of the Common, in which case Cavendish should not pay Baldwin a farthing, his lawyer told him.[46] Cavendish attempted to cut through the claims and complaints, writing to Baldwin:

---

[45] Henry Cavendish to Charles Blagden, n.d. [after 16 September 1787], draft; in Jungnickel and McCormmach, *Cavendish* (1996), 638–39.

[46] Christopher Baldwin to Henry Cavendish, 7 July, 19 September 1785; 8, 22, 27 February, 3 March, Saturday 1786. Henry Cavendish to Christopher Baldwin, six undated draft letters. Thomas Dunn to Henry Cavendish, 6 September 1785; 6 February 1786, Devonshire Collections, Chatsworth, 86/comp. 1. The slip of land was part of a larger piece of land, of roughly one half acre. See Fig. 14.1 showing the slip of land. Soon after the end of the exchange of letters, Baldwin deeded to Cavendish the one half acre of land "abutting or bounding" Balam Lane. On the same day, Baldwin released all claim on the 15 acres he sold to Cavendish, for a consideration of £80. 5 April 1786, Christopher Baldwin to Henry Cavendish, "Lease" for the one half acre. 6 April 1786,

I only desire a direct answer to this question whether you agree to the offer made you on my account by Mr. Dunn namely that if you would make over to me in proper form whatever right you may have to the ground taken in by you from Balam Lane, I would pay you 40£ deducting only the composition which may be required by the parish. If I do not receive a direct answer to this question in 2 or 3 days, I must consider it as a refusal & act accordingly.[47]

Baldwin replied that he was prepared to release the claim he had made on the encroachment some years before. He said he never thought he had any right to it but possession, and he was ready to sign a deed once Cavendish agreed to put up fencing. Their conflict came to a head with four letters passing between them on a single day. Cavendish wrote:

I can not at all conceive what is the cause of this behavior whether you have any private reason for wishing to delay the agreement or whether you distrust my honour about the tailings & wish to make some further conditions about it. If the latter is the true cause you may assure yourself that I will never submit to make any such conditions or explanation with a person who distrusts my honour.[48]

Baldwin told Cavendish that his lawyer did not mind harming him to save "a *little* matter to *you*." It is true that £40 and £60 were trivial sums to Cavendish, a rich man, but the issue was no longer a disagreement about money but an aspersion on Cavendish's honor, which was not a little matter. On the subject of honor, this scion of the House of Cavendish had strong feelings, as shown by the wording of his last letter to Baldwin.

There is evidence that Cavendish experienced feelings of anxiety in varying degrees. Reports of his agitated speech point in this direction, as does the following report. He was repeatedly observed to pause for a long time on the landing at Sir Joseph Banks's house, "evidently wanting courage to open the door and face the people assembled, nor would he open the door until he heard someone coming up the stairs, and then he was forced to go in."[49] Perhaps anticipating embarrassment, he was fearful to join the only company he sought out, that of scientific men.

Cavendish may also have experienced depression, the evidence for which is a state of low energy. "Mr Cavendish did not seem to be very busy," Blagden wrote to a friend.[50] When the natural philosopher and chemist Joseph Priestley invited an exchange of letters, Cavendish wrote back that he would be glad to send him reports of his experiments, "but I am so far from possessing any of your activity that I am afraid I shall not make any very soon." When Priestley's second letter went unanswered for 6 months, Cavendish apologized: "as I make not a tenth part of the exper

Christopher Baldwin to Henry Cavendish, "Release of a Piece of Land on Clapham Common." 6 April 1786, Christopher Baldwin to Henry Cavendish, "General Release," for a consideration of £80, of all claims by Baldwin on the fifteen acres he sold to Cavendish. Devonshire Collections, Chatsworth, 38/78.

[47] Henry Cavendish to Christopher Baldwin, n.d., draft, ibid.

[48] Henry Cavendish to Christopher Baldwin, n.d., draft, ibid.

[49] A fellow of the Royal Society, quoted in Wilson, *Cavendish*, 169.

[50] Charles Blagden to Richard Kirwan, 20 March 1790, draft, Blagden Letters, Royal Society 7: 322.

that you do & as my facility in writing falls short of yours in a still greater proportion I am afraid you will think me a bad correspondent & that the advantage lies intirely on my side."[51] In the middle of a political revolt in the Royal Society, in which Cavendish was caught up, his reason for not assuming leadership was his "unfitness for active exertion."[52] Blagden wrote in his diary, Cavendish "looked out of spirits."[53] Cavendish was not supercharged, and he may have had episodes of low energy brought about by depression.

3.4 Strangers, Shyness, and Solitude

Two encounters illustrate Cavendish's aversion to strangers. While Cavendish was talking with several men at the Royal Society Club, a new member joined them. Cavendish immediately fell silent. "The fact is he saw in me a strange face, and of a strange face he had a perfect horror," the new member said, but once he had been introduced, Cavendish behaved "very courteously" to him.[54] The second encounter was with a new guest at one of Banks's conversational gatherings. While he was talking with Banks about an experiment, he noticed that an old gentleman was "very attentive" to what he was saying. "When I caught his eye he retired in great haste, but I soon found he was again listening near me." He inquired and was told that the old gentleman was Cavendish. Later he met Cavendish at the Royal Society Club, and Cavendish picked up the conversation, showing he had been listening.[55] Thomson made a similar observation: Cavendish was normally silent, but when he was seated near persons he liked, he frequently talked a "great deal."[56]

The words "shy" and "shyness" reappear in accounts of Cavendish, as do related words such as "diffident," "reserved," "bashful," and "embarrassed." His shyness was seen as an underlying trait of character. In the family obituary, Blagden wrote of a "shyness and diffidence natural to his disposition."[57] Brougham said he had "a most reserved disposition, and peculiarly shy habits," which accounted for his "singularity of manner."[58] Barrow called his shyness "extreme."[59] Thomson said he was "shy and bashful to a degree bordering on disease."[60]

---

[51] Henry Cavendish to Joseph Priestley, n.d. [May or June 1784] and 20 December 1784, drafts; in Jungnickel and McCormmach, *Cavendish* (1999), 594, 598–99.

[52] Charles Blagden to Joseph Banks, 5 April 1784, British Museum (NH), DTC 3: 20–21.

[53] 12 March 1804, Charles Blagden Diary 5: 210 (back).

[54] Children, quoted in Wilson, *Cavendish*, 169.

[55] Pepys, quoted in Wilson, *Cavendish*, 168.

[56] Thomson, *History of Chemistry* 1: 337.

[57] Family obituary of Henry Cavendish.

[58] Brougham, *Lives* 1: 258.

[59] Barrow, *Royal Society*, 144.

[60] Thomson, *History of Chemistry* 1: 337.

Cavendish went to lengths to avoid the gaze of strangers. Two women observed him on his daily punctual walk, and the next day they brought a man with them to point out the famous philosopher. As Cavendish was climbing over a stile, he noticed "to his horror that he was being watched." He changed the route and the time of his walk as a result.[61] Another example shows his reaction to being singled out. He was introduced to a foreign visitor as a celebrated natural philosopher, and he was then subjected to a flattering speech. "Mr. Cavendish answered not a word, but stood with his eyes cast down quite abashed and confounded. At last, spying an opening in the crowd, he darted through it with all the speed of which he was master; nor did he stop till he reached his carriage, which drove him directly home." Unable to bear being pointed out as remarkable in any way, any attention to his person caused him "embarrassment."[62] The prospect of a portrait was repugnant to him, and the ink and wash sketch of him was made surreptitiously.[63] Young characterized Cavendish's situation in the world as one of "painful pre-eminence."[64]

In his contribution to the family obituary of Cavendish, Blagden wrote, "His habits had, from early life, been retired." Having second thoughts about it, he submitted a correction, replacing "retired" with "secluded."[65] The two words mean much the same thing, but not exactly. "Retired" suggests withdrawal from an activity; "secluded" allows for the activity, suggesting only that it is carried on away from others. "Shut up apart" is an eighteenth-century meaning of "seclude." Blagden's second choice of words better describes Cavendish.[66] The obituary was sent to the newspapers before Blagden's correction was received, but his preferred word appeared in its reprinting in *Gentleman's Magazine*.

The world had "little intercourse" with Cavendish.[67] The only society he entered willingly was "that of his scientific friends,"[68] but even among them he remained apart. When he joined diners at the Royal Society Club, "he sat down without taking notice of anybody."[69] At home, he rarely received visitors. On walks, he never spoke to anyone or tipped his hat. Blagden told a prospective borrower that Cavendish's library was accessible year-round, but that Cavendish did not want people even to sit in his library. They were to "borrow such books as they wish & take them home," and they did not even need to come to his house, as books would be sent to them.[70]

---

[61] Dr. Sylvester, quoted in Wilson, *Cavendish*, 170.

[62] Thomson, *History of Chemistry* 1: 337–38.

[63] Wilson, *Cavendish*, 171.

[64] Young, "Cavendish," 445.

[65] Lord George [Augustus Henry] Cavendish to Charles Blagden, 8, 9, and 10 March 1810. Charles Blagden to Lord George Cavendish, [8 March 1810], draft, Blagden Letters, Royal Society, C.17 and C.19. Blagden wrote a draft of the word change on the bottom of Lord George's letter to him on 10 March.

[66] It is a fine distinction. Both Young and Thomson used Blagden's first choice of word, and they may have borrowed from it, referring to Cavendish's "retired habits of life" and "whose habits were so retired." Young, "Cavendish," 444. Thomson, *History of Chemistry* 1: 338.

[67] Young, "Cavendish," 445.

[68] Thomson, *History of Chemistry* 1: 337.

[69] Chemist, quoted in Wilson, *Cavendish*, 169.

[70] Charles Blagden to Thomas Beddoes, 12 March 1788, draft, Blagden Letters, Royal Society 7: 129.

His habits confirmed his view of himself as "a solitary being in the world," "unfit for society." "His desire was to be alone and to be left alone, Barrow said."[71] Wilson concluded from reports like these that the "most striking" peculiarity of Cavendish was "a singular love for solitariness, and the reluctance to mix with his fellows." He called him "one of the most ungregarious of beings" and his desire for solitude "*passive* selfishness."[72]

From Cavendish's shyness and craving for solitude, traits which his contemporaries judged to be defining features of his character, we could easily mistake our man. The family's obituary of Cavendish contains an observation that rounds out the portrait: he was "firm of mind."[73] Within the world that interested him, science, Cavendish moved with supreme confidence.

3.5 Women

With the rare exception, Cavendish did not interact with women, and his reaction to women was seen as a distinct trait of character. We might think of women as a variety of strangers, but this would overlook the intensity of his aversion, as Wilson described it. At the Royal Society Club, Cavendish gave the impression he despised men who liked female company.[74] At a meeting, members noticed a pretty girl watching them from a window across the street, and they gathered around their window to admire her. Thinking they were looking at the moon, he joined them at the window. When he saw what they were about, he turned away in "intense disgust."[75] In his neighborhood, he was regarded as a woman hater, and so it was with surprise that the inhabitants learned he had saved a woman from a mad cow.[76] At home, he took precautions to avoid encountering his female servants, discussed below.

Cavendish's reputation as woman hater needs qualifying. To judge from the household accounts, it would seem that Wilson's testimonials give an exaggerated idea of Cavendish's severity toward female servants. At his death, he had three female servants at his house on Clapham Common and two female servants at his house on Bedford Square. His housemaid at Clapham Common had worked for him for twenty-five and a half years, the entire time Cavendish owned the house. His cook at Bedford Square had worked for him for 20 years, nearly the entire time Cavendish owned that house. Only one male servant, a footman, had worked for

71 Barrow, *Royal Society*, 144. Dr. Sylvester, quoted in Wilson, *Cavendish*, 170.

72 Wilson, *Cavendish*, 165, 173.

73 Family obituary of Henry Cavendish.

74 Barrow, *Royal Society*, 145.

75 A fellow of the Royal Society, quoted in Wilson, *Cavendish*, 170. John Timbs regarded this anecdote as apocryphal, though he used it all the same. It well may be apocryphal, but we have no way of knowing, and it is consistent with less colorful reports of Cavendish's aversion to women. *English Eccentrics and Eccentricities*, 2 vols. (London, 1866) 1: 143.

76 Mrs. Herbert, quoted in Wilson, *Cavendish*, 178.

him that long.[77] In public, Cavendish could be civil to women, as Blagden recorded in his diary.[78]

Two examples put the other side of Cavendish in a clear light. During time Blagden served as Cavendish's associate, he considered marriage. He asked his brother to inquire if his prospective bride would enjoy his kind of company and "particularly could so far enter into the pursuits of my friend Mr. C[avendish] as not to think some portion of time spent in his company tedious," and "should not find it disagreeable to be present when such matters [his and Cavendish's studies] were the subject of conversation, or when any experiment which had nothing offensive in it, was going on."[79] This marriage did not take place, but we learn that Blagden contemplated the three of them together, he, his wife, and Cavendish, in harmonious company, his concern being her reaction, not Cavendish's. Cavendish did not think the poorer of his associate for contemplating marriage. On the contrary, he encouraged the marriage, giving Blagden to believe that he would make ample settlement on him should he carry through with his plan.[80] The other example is from a marriage within the Cavendish family. Georgiana (Spencer) Cavendish, duchess of Devonshire, was on friendly terms with Henry Cavendish. Known as the Queen of London fashion, she had a wide-ranging curiosity about the world beyond fashion. Cavendish rarely attended events at Devonshire House, but he went there frequently to see the duchess. When Blagden called, he found "Mr. Cav. there; saw none had notes"; the duchess asked Cavendish to "shew extracts from Js de Physique." When Blagden and the duchess happened to be abroad at the same time, they engaged in "much talk about chemistry & mineralogy," and the duchess "said she was quite wild with studies of that nature: asked much about Mr Cavendish & his pursuits," and she asked him to give her compliments to Cavendish.[81] Cavendish, it would seem, was not shy with women when he knew them and the talk kept to his interest, science.

3.6 Regularity and Routine

We have noted the uniformity of Cavendish's dress. At home, his boots were always placed in one location, his walking stick in one of the boots, always the same one. The few persons who visited Cavendish and stayed for dinner were treated to the

[77] "Wages Due to the Servants at Clapham and in Bedford Square," 5 April 1810, Devonshire Collections, Chatsworth.

[78] "Cav. civil then to L^y A[---]sts." 25 May 1794, Charles Blagden Diary, Royal Society, 3: 2(back).

[79] Charles Blagden to John Blagden Hale, 13 November 1789, draft, Blagden Papers, Beinecke Rare Book & Manuscript Library, Yale University, box 5, folder 49. Hereafter Yale.

[80] Joseph Banks to Charles Blagden, 27 March 1790; Charles Blagden to Joseph Banks, 8 April 1790, British Library, Add Mss 33272, pp. 74, 81.

[81] 1 September 1794, Charles Blagden Diary, Royal Society 3: 14. Charles Blagden, "The Diary of Sir Charles Blagden," ed. G. De Beer, *Notes and Records of the Royal Society of London* 8 (1950): 65–89, on 76, 80, 83. Georgiana, duchess of Devonshire to Charles Blagden, 4 March 1794, Blagden Letters, Royal Society, D.61.

same fare, "a leg of mutton, and nothing else." On one occasion, Cavendish agreed with his housekeeper that a leg of mutton would not be enough for guests, and he ordered two legs of mutton.[82] For his table, it would seem, he thought in terms of quantity, not of variety. Cavendish always took a walk at dusk and always down the middle of the lane.[83] At the Crown and Anchor, where the Royal Society Club met, he always hung his hat on a particular peg.[84] Over decades, he took regular readings of the earth's magnetic inclination and dip at his house. He was "constantly present" at meetings of the Royal Society, the Royal Society Club, and Joseph Banks's salons.[85] "The current of his life passed on with the utmost regularity," Thomson said. The "description of a single day would convey a correct idea of his whole existence."[86]

It might be thought that a person who prizes clockwork regularity and routine would have his affairs in fine order, but that does not seem to have been Cavendish's way. Before he hired a librarian, the books in his large library were in a "bad state of arrangement."[87] Friends and family on occasion looked to him as a banker in-need. When he did not remember if he had received interest on money he had loaned to a family member, he said "I am not very regular in my accounts."[88] After Cavendish died his scientific papers were found to be in "great disorder."[89] It may be that Cavendish knew where every book and every paper was, but to an outsider his possessions looked to be a mess.

3.7 House and Servants

When Cavendish was among people, he was also apart. The same can be said of the places where he lived. His apartment behind his father's house was separated from the house by a garden. The other three sides of the apartment were bounded by a side street, an alley, and probably stables. After leaving his apartment, he moved to an end-of-row terraced house at Hampstead. He next bought a terraced house on Bedford Square, which was also end-of-row. The last house he bought

[82] A Fellow of the Royal Society, quoted in Wilson, *Cavendish*, 164. Like other anecdotes about Cavendish, this one may need to be qualified. From butcher and fishmonger bills at the end of his life, we find that although leg of mutton looks to be the most common meat course, his house-keeper also ordered beef, loin pork, cod, and oysters. Perhaps the variety was only for the servants, but we cannot know. "Vouchers to Mrs Stewarts household," at Clapham Common, Devonshire Collections, Chatsworth, box 31.

[83] Dr. Sylvester, quoted in Wilson, *Cavendish*, 170.

[84] A chemist, quoted in Wilson, *Cavendish*, 169.

[85] Young, "Cavendish," 445.

[86] Thomson, *History of Chemistry* 1: 338.

[87] Pepys, quoted in Wilson, *Cavendish*, 174.

[88] Henry Cavendish to John Dalrymple, n.d., draft, Devonshire Collections, Chatsworth, L/34/64.

[89] Charles Blagden to "My Lord" [George Augustus Henry Cavendish], n.d., draft, Blagden Collection, Royal Society, Miscellaneous matter – Unclassified.

was freestanding, a villa on Clapham Common. The pattern suggests that he preferred what apartness was possible in living quarters that were at the same time convenient for his scientific activities.

The alterations Cavendish made to his house on Clapham Common show how completely he subordinated his domestic life to the needs of his science. According to a scientific instrument maker who saw it, Cavendish's whole house was "occupied as workshops and laboratory."[90] The drawing room was a laboratory, the room next to it was a forge, the upstairs held an observatory, and instruments were stuck all over the outside. Only a small part of the house was set aside for personal use.[91] The few persons who entered it recounted "its desolate appearance, and its scanty and mean-looking furniture." On the morning Cavendish died, his physician toured the "forlorn abode."[92] The rich aristocrat who chose to live there seemed strange to people who lived normal lives in normal houses.

Servants were necessary in Cavendish's time. At his Clapham Common house, he employed seven, and at some point he included a scientific instrument maker.[93] One of the responsibilities of a servant, presumably the butler, was to shield him from interruptions of his routine of research and study. The banker's call on Cavendish, referred to earlier, shows how it worked. The servant asked the banker what he wanted with his master. The banker did not say. The servant did not inform Cavendish he had a caller, and the banker had to wait until the master rang the bell. In about a quarter of an hour the bell rang and the servant went upstairs to announce the banker, who was then allowed to follow. Upstairs, he found Cavendish upset with him for coming to his house. In ill temper, Cavendish gave the banker the instructions he came for.[94]

Cavendish was a strict employer. If a female servant showed herself against the rules, he immediately dismissed her. When one day he encountered a maid with broom and bucket on the stairs, he had a back staircase built.[95] To avoid having to speak to a female servant, he left a written note on the hall table at a certain hour containing instructions for dinner.[96] At his townhouse in London, which he visited once or twice a week, he did not speak to his servants, and they did not speak to him without invitation.[97] When he wished to communicate with his butler, he rang a bell. Cavendish's silence in public was repeated in the silence of his houses (Figs. 3.1, 3.2, 3.3, 3.4, 3.5, 3.6, 3.7, 3.8, 3.9, 3.10, 3.11, 3.12, and 3.13).

---

[90] Newman, quoted in Wilson, *Cavendish*, 164.

[91] Wilson, *Cavendish*, 164.

[92] Barrow, *Royal Society*, 150, 154.

[93] "Account of the Number of Persons Residing in the Parish of Clapham … Dtd 18 Feb. 1788," Greater London Record Office, P 95/TRI 1/72. "Wages Due to the Servants at Clapham and in Bedford Square," 1810, Devonshire Collections, Chatsworth.

[94] Wilson, *Cavendish*, 175.

[95] Ibid., 169–70.

[96] Brougham, *Lives*, 258–59.

[97] Thomson, *History of Chemistry* 1: 337.

Cavendish's Places

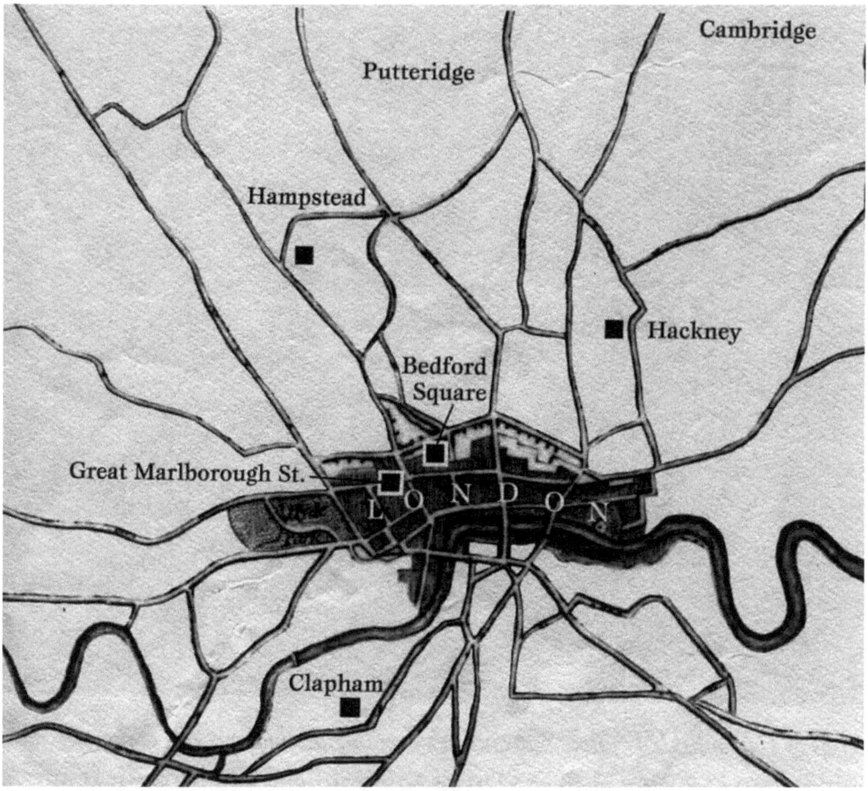

Fig. 3.1 Places Where Henry Cavendish Lived. All of the places on this map are mentioned in the book with the exception of Putteridge, one of three manors and several farms north of London that Lord Charles Cavendish bought after his marriage; he also kept a residence in the city. In 1738, he sold the Putteridge estate and bought a house on Great Marlborough Street. The map shows that although Henry Cavendish did not always live in London, London was never far away

Fig. 3.2 Westminster Bridge. By Canaletto (Courtesy of the Yale Center for British Art, Paul Mellon Collection). The Lord Mayor's procession on the River Thames, 1747. This second bridge in London over the river is nearly finished in this painting; final construction can be seen at the far right. Lord Charles Cavendish did extensive work on Westminster Bridge construction committees

Fig. 3.3 "Panoramic View of London." By T. Bowles in 1751 (Wikimedia Commons)

Fig. 3.4 Great Marlborough Street House. No. 13. Demolished. View of the back premises of the house on Blenheim Street (Courtesy of the Westminster City Archives). This was Lord Charles Cavendish's house from 1738 to the end of his life. Henry Cavendish lived there too

Fig. 3.5 Plan of Great Marlborough Street. Detail from Horwood's survey circa 1792–99, updated to 1813. The plan of the house No. 13, on the corner of Great Marlborough and Blenheim, shows a building at the end of the property, designated No. 1 Blenheim Street. There is a divided garden between it and the main house. It would seem that Henry Cavendish lived in the rear building while Lord Charles Cavendish lived in the main house

Fig. 3.6 Hampstead House, No. 34 Church Row. Between 1782 and 1785, Cavendish lived in a house at the end of this row next to the church. But for the automobiles, this street with its terraced houses and church looks much the same as it did then (Photograph by the author)

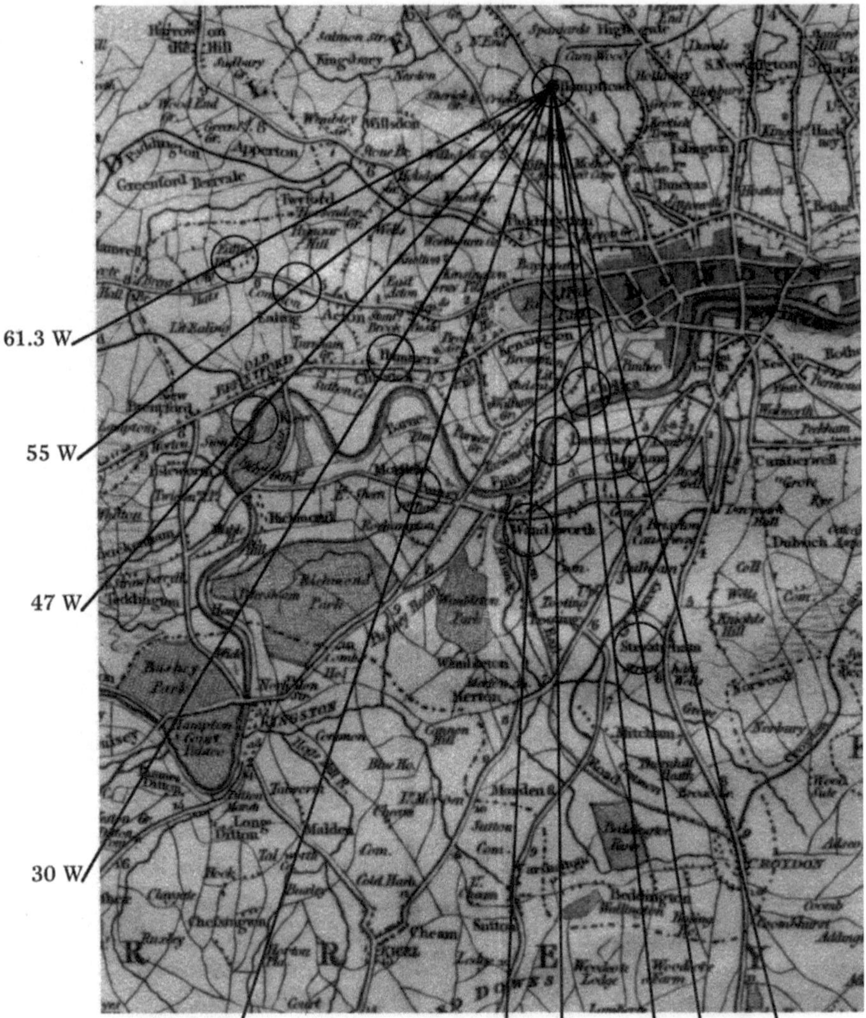

Fig. 3.7 Hampstead Bearings. From his country house in Hampstead, Cavendish took bearings in the direction of London. With a theodolite, he recorded the angular position of objects such as trees, houses, hills, and church steeples through an arc of about 60°. Prominent among the objects he selected were steeples, as we would expect from the picture of Westminster Bridge and the panoramic picture of the city above; the London skyline was marked by steeples. On the not-too-accurate map of London and environs, published by R. Phillips in 1808, I have drawn Cavendish's lines of sight for a number of steeples, labeling them with the angles he measured. From *right* to *left*: 1. New houses on the road to Clapham. 2. Streatham steeple. 3. Chelsea steeple. 4. Battersea steeple. 5. Wandsworth steeple. 6. Putney steeple. 7. Hammersmith steeple. 8. Kew Chapel. 9. Acton steeple. 10. Ealing steeple. "Bearings," (Cavendish Scientific Manuscripts, Devonshire Collections, Chatsworth, Miscellaneous)

Fig. 3.8 Bedford Square House, No. 11. This was Cavendish's townhouse from 1784 to the end of his life (Photograph by the author)

Fig. 3.9 Cavendish's familiar destinations in London are identified by numbers superposed on *Plan of London, with Its Modern Improvements*, published by Richard Phillips, on 2 May 1808 or 1809. 1. Royal Institution. 2. Great Marlborough Street house. 3. Sir Joseph Banks's house. 4. Bedford Square house. 5. British Museum. 6. Royal Society. 7. Crown & Anchor. 8. Edward Nairne's instrument shop. 9. George & Valture

Fig. 3.9 (continued)

Fig. 3.10 Clapham Village. This is a view of the village from Clapham Common (William Thornton, *The New, Complete, and Universal History, Description, and Survey of the Cities of London and Westminster… Likewise the Towns, Villages, Palaces, Seats, and Country, to the Extent of Above Twenty Miles Round*, rev. ed. [London, 1784], facing p. 490)

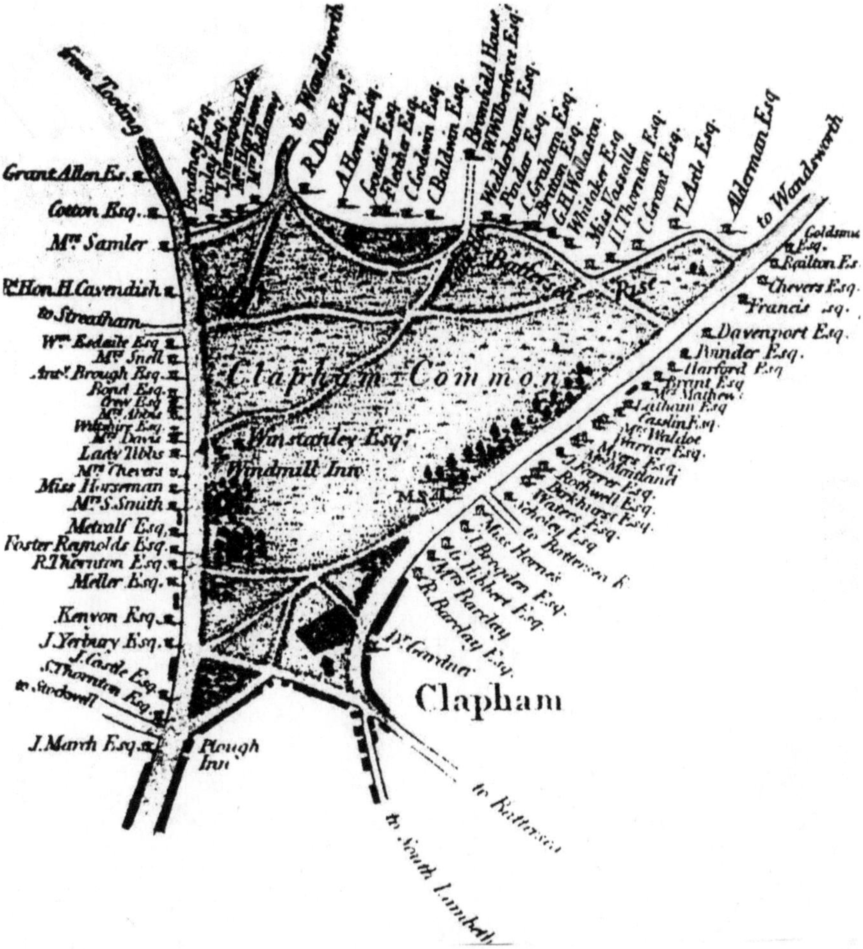

Fig. 3.11 Plan of Clapham Common. Cavendish's house is on the left, fourth house from the *top*. ("Perambulation of Clapham Common 1800. From C. Smith's 'Actual Survey of the Road from London to Brighthelmston,'" *The Chronicles of Clapham [Clapham Common]. Being a Selection from the Reminiscences of Thomas Parsons. Sometime Member of the Clapham Antiquarian Society* [London: The Ramsdan press, 1929], opposite p. 112. Reproduced by permission of the Bodleian Library)

Fig. 3.12 Clapham Common House. This was Cavendish's country house from 1785 to the end of his life. We see the back of the house, much altered since Cavendish lived there, and now demolished (Frontispiece to *The Scientific Papers of the Honourable Henry Cavendish*, 2 vols., ed. J. C. Maxwell, Sir J. Larmor, and Sir E. Thorpe [Cambridge: Cambridge University Press, 1921].

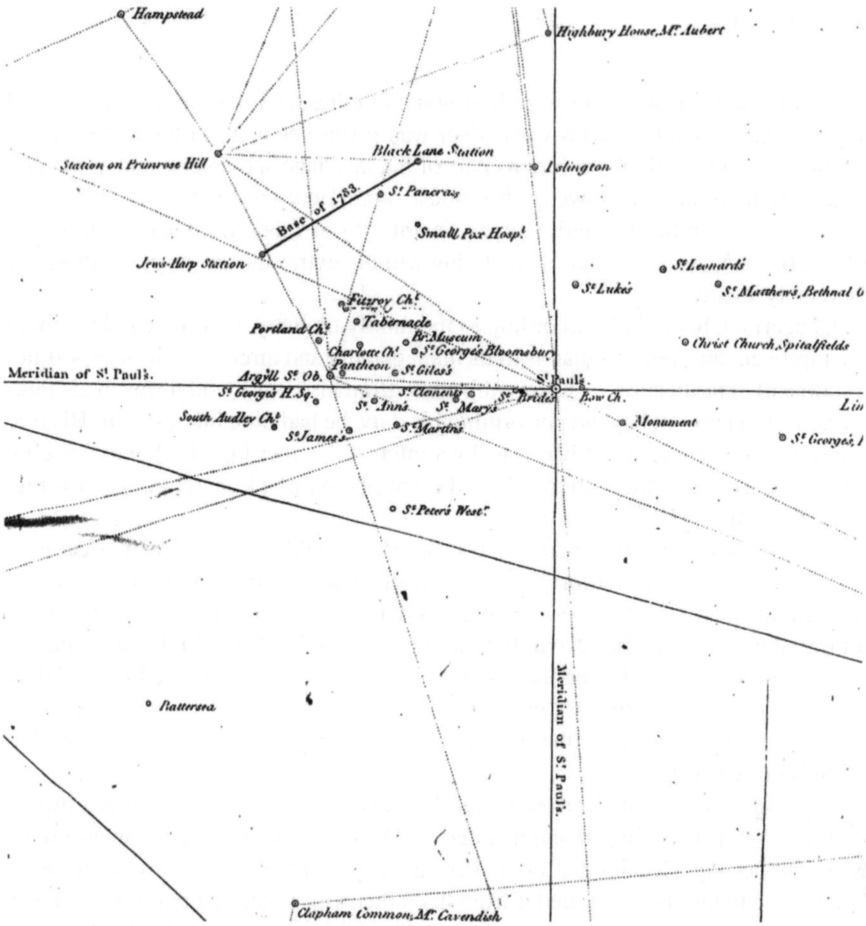

Fig. 3.13 Triangulation. This detail is of a map originating in an Anglo-French project to determine the relative locations of the Greenwich and Paris observatories. The English half began with the baseline of 1783 on Hounslow Heath. The object of laying down secondary triangles was to improve maps of London and the country. One of the stations was Cavendish's observatory at Clapham Common, shown at the bottom (From a map by William Roy, "An Account of the Trigonometric Operation, Whereby the Distance between the Meridians of the Royal Observatories of Greenwich and Paris Has Been Determined," *Philosophical Transactions* 80[1790]: 111–270)

3.8 Wealth

Land was the main source of wealth of noble families, and the Cavendishes owned a great deal of land. Charles Cavendish was given about 20 farms and tithes in Derbyshire and Nottinghamshire as part of his marriage settlement, and at the end of his life he turned them over to his eldest son.[98] Henry received rent from them, but it was insignificant relative to his income from funds he inherited from his father. By reinvesting dividends and living without extravagance, he gradually grew extremely wealthy.

Concerning his wealth, according to the family obituary, he was "careless about its improvement," but "regular as to its management and direction"; its size was not a source of "gratification."[99] His wealth came from his family, and it was to his family he would leave it; beyond performing his duty, he had little interest in it. His true interest was science, and it was a jealous interest. Neither high birth, Cuvier said, "nor great riches, which offered him the temptation of pleasure, could turn him away from his goal."[100]

Wilson said that with regard to Cavendish's wealth, there were two cardinal facts: he "was, for the first forty years of his life, a poor man, and for the last thirty-nine an unusually wealthy one."[101] The first half of Wilson's statement is probably an overstatement, but the second half is unquestionable. Biot said that as a man of science with a fortune, Cavendish was without a peer. Newton and Leibniz had died wealthy, but their wealth was "incomparably less" than Cavendish's. Biot's characterization of Cavendish as "the wisest of the rich and the richest of the wise" has often been repeated.[102]

In a conversation with Barrow, Cavendish talked about various properties owned by the Cavendish family, among them Holker Hall, a grand manor in Lancashire on Morecambe Bay. He said that his father had acquired the property by inheritance, and that he in turn had inherited it from his father. Upon learning that Barrow knew the place well, he spoke "somewhat eagerly" of his wish to extend his land into the bay, as the ironmaster John Wilkinson had extended his property on the other side of it. Cavendish said he would make a trip there soon to look into the possibility, but he died before he could.[103] Cavendish showed a typical wealthy aristocrat's desire to improve his properties (Chap. 15).

Cavendish's colleagues were curious to know what he did with his wealth besides increase it. Davy said that Cavendish's income at the end was above £40,000, and he

---

[98] Henry Cavendish to William Gould, 30 December, draft, Devonshire Collections, Chatsworth, L/34/7.

[99] Family obituary of Henry Cavendish.

[100] Cuvier, "Cavendish," 229.

[101] Wilson, *Cavendish*, 160.

[102] Biot, "Cavendish," 273.

[103] Barrow, *Royal Society*, 146–47. The ownership of Holker Hall is discussed in Chap. 15.

"was not the person who gave the £40,000 to hospitals."[104] Cuvier said that he was very generous with his wealth,[105] but that is not exactly how others saw it. The family obituary described Cavendish as "liberal, without being profuse; and charitable, without ostentation."[106] Whenever someone called on Cavendish for charity, he looked at the list of subscriptions and wrote a check to match the largest, "neither more nor less," according to a physician who lived at Clapham.[107] At a christening, Cavendish gave the nurse an expected gift, putting his hand in his pocket and presenting her "with a handful of guineas without counting them."[108] We have seen that when he was told that his former librarian was in need and poor health, he was sorry and he did not know how he could have helped, but when an annuity was mentioned he promptly wrote a check for £10,000.[109] Pepys came across Cavendish at the Royal Institution in conversation with Davy, who feared he was not going to have enough money for a worthy scientific object. Cavendish "joined most truly in deploring the want of liberality in the patrons of science to carry it into effect. He did not think he was called on to forward the desired object." Wilson supposed that if Davy had asked Cavendish to write a check for £500, he would have done so at once.[110]

In his will, Cavendish distributed his estate according to custom, leaving the bulk of it to another Cavendish and his male offspring. Young said that some people, including himself, regretted that Cavendish had not apportioned "a small share of his immense and neglected wealth" for the establishment of scientific fellowships or professorships. He also recognized that "the most successful cultivators of science are not always the most strenuous promoters of it in others."[111] Cavendish did intend to leave a legacy to science: he wanted his extensive library to remain accessible to borrowers after his death.[112] Serious readers were indebted to private libraries like his. Blagden, a serious reader, explained why: "Books … are now so dear, that none but *rich* persons can afford to buy them."[113]

Some years before his death, Cavendish was asked to donate money to establish Bowdoin College together with an observatory in Maine. To a very long letter explaining the plan, which included "the prospect of a Cavendish observatory and

[104] 1 March 1810, Charles Blagden Diary, Royal Society 5: 428(back).

[105] Ibid., 237.

[106] Family obituary of Henry Cavendish.

[107] Dr. Sylvester, quoted in Wilson, *Cavendish*, 189. Probably this was Cavendish's usual practice, but if so he frequently gave the same amount. The porter's account for 1806 lists charity donations: Magdalen £5.5.0. African Association £5.5.0. Asylum £5.5.0. St. George's Hospital £5.5.0. Society of harts (sic) £2.2.0. The porter's account for 1807 adds St. Giles Charity School £2.2.0, insolvent debtors £1.1.0. "Bedford Square. James Fuller's Account with the Exec. of Hen: Cavendish Esq. Balance 37.6.4. Settled 30 August 1810."

[108] Lord Burlington, quoted in Wilson, *Cavendish*, 181.

[109] Pepys, quoted in Wilson, *Cavendish*, 174.

[110] Wilson, *Cavendish*, 174.

[111] Young, "Cavendish," 446.

[112] 3 and 4 March 1810, Charles Blagden Diary, Royal Society 5: 429, 429(back), 430.

[113] Charles Blagden to Lady Palmerston, 7 November 1804, draft, Blagden Letters, Yale.

Cavendish professors," perhaps as inducement, Cavendish replied in a letter of two sentences. In the first sentence, he declined the proposal, for "many reasons," which he did not specify. The second sentence read, "As my mind is made up on the subject it is unnecessary to give yourself any further trouble on the subject."[114] In the matter of wealth, his thinking was firmly in line with his family's. Wealth was to be concentrated, not scattered.[115]

3.9 Religion

To explain why Cavendish left Cambridge University without taking a degree, it has been suggested that he wished to avoid having to subscribe to the articles of the Anglican Church, a requirement of graduates. That may be, but he was also following custom. Students who entered Cambridge as "fellow commoners," as he did, rarely took degrees. Biot said that Cavendish was "religious in the manner of Locke and Newton." Wilson, who was very interested in this side of Cavendish, questioned Biot's statement. Cavendish showed none of the earnestness of Newton and Locke on the subject of religion, and the reference was ambiguous in any case given the uncertainty then of Newton's opinion on the Trinity. Wilson thought that Biot probably meant that Cavendish's views were, like Newton's, unorthodox. Wilson learned that at Cavendish's college in Cambridge, there was "a sort of hereditary belief" that Cavendish was a Unitarian, but there was no foundation for it.[116]

If Cavendish had a spiritual side, he kept it to himself. He turned up at one or more christenings in the family, but it is unclear if he did more than attend the dinner.[117] Sundays for him were like any workday, as we know from his laboratory and field notes.[118] An inhabitant of Clapham said that so far as he knew, Cavendish "never attended a place of worship."[119] A fellow of the Royal Society

---

[114] Benjamin Vaughan to Henry Cavendish, 21 December 1799 [enclosed in a letter from his brother William]; Henry Cavendish to William Vaughan, draft, [after 27 October 1803]; in Jungnickel and McCormmach, *Cavendish* (1999), 721–23, 729.

[115] He left legacies to Blagden, £15,000; to another colleague, by then deceased, Alexander Dalrymple. £5,000; and to his one-time physician, also deceased, John Hunter,£5,000. Servants received £1,000. The remainder of his personal property was in the form of annuities, stocks, cash, and mortgages, excluding securities held in trust, totaling £892,254. This was the actual value at the time of Cavendish's death, not the face value, which was considerably higher. Less 4 % legacy duty and various expenses, it together with his real property was left to various members of the Cavendish family. "The Personal Property of the Late Hon. Henry Cavendish 24 Feb. 1810," Devonshire Collections, Chatsworth, 114/74.

[116] Biot, "Cavendish," 273. Wilson, *Cavendish*, 180–81. Privately, Newton held a Unitarian view.

[117] From Lord Burlington, in Wilson, *Cavendish*, 181.

[118] Wilson, *Cavendish*, 182.

[119] From Mr. Allnutt, in Wilson, *Cavendish*, 181.

said, “As to Cavendish’s religion, he was nothing at all.”[120] His one reference to a church, other than to a church as a landmark in taking bearings, occurred during a dinner of the Royal Society Club: he said that some wood at Clapham Church was eaten “thro’ by the insects… working their way out.”[121] In the absence of any outward display of interest in religion on Cavendish’s part, Wilson concluded “that the World to come did not engross his thoughts.” Ever hopeful, Wilson wondered if on his deathbed Cavendish may have turned his thoughts in a spiritual direction.[122]

Newton wrote in the *Principia* that the discussion of God “does certainly belong to Natural Philosophy.” In the biography of Cavendish I wrote with Christa Jungnickel, we said, correctly, that he did not record any thoughts on religion in his writings on natural philosophy. We did not mention in this connection a poem Cavendish wrote, which I now give more weight to. It came about this way. In 1748, the year before Cavendish entered the University of Cambridge, the crown prince Frederick made a bid to become chancellor of the University, but his father George II disapproved, and the University followed the safe course. Frederick died soon after. As if to make amends for what it had denied him in life, the University produced a handsome volume of academic exercises, poetic lamentations in Latin in honor of the crown prince. Cavendish contributed a “Lament” to this volume. It was his first publication, and because it is his only publication on a subject other than science, it holds an interest for us. The poem follows form for memorials of this kind, but it is also revealing of its author’s thinking. Cavendish writes that by understanding nature we can understand the occasion of the lament, the prince’s death. Nature has nothing to do with human comforts and desires, dread thistles and sweet roses both flourish and die, but to the “intimate” of nature, by which I take him to mean the student of nature, “natural truth” is disclosed.[123] What is disclosed to the student of nature is the destination of the royal prince, a dwelling in heaven. Cavendish would seem to have had in mind natural religion or a version of religion close to it, certainly a religion without the notion of a personal God. The fragility of life and other references in the poem are conventional, but the goal of the author of the poem is foreshadowed, a life of natural philosophy. “However so much lovely Nature may mock, She does lay bare hidden causes/And the wandering paths of the stars.” Cavendish was only 18 when his poem was published, and as he matured his thoughts on religion may have changed, or never returned. He may have rejected religion altogether, an impression he gave the world, or he may have rejected only its social forms. Because after his youthful poem, he wrote nothing again on the subject, we cannot know his subsequent religious leanings, if he had any.

---

[120] A fellow of the Royal Society, quoted in Wilson, *Cavendish*, 181–82.

[121] 19 February 1807, Charles Blagden Diary, Royal Society 5: 39.

[122] Wilson, *Cavendish*, 182, 185.

[123] Henry Cavendish, “Luctus,” in Cambridge University, *Academiae Cantabrigiensis Luctus in Obitum Frederici celsissimi Walliae Principis* (Cambridge, 1751).

3.10 Death

Cavendish suffered occasionally from "calculous complaints" and other minor ailments, but otherwise he enjoyed remarkably good health.[124] To the last week of his long life, Davy said, he "retained his activity of body, and all his energy and sagacity of intellect."[125] Cuvier thought that Cavendish's "good health and longevity" owed to the regularity of his habits.[126] No doubt it did, but inheritance played the larger part. His father and brother too had healthy lives almost identical in length to his.

In his "Lament," Cavendish treated death in an unemotional way. When the crown prince died, the prince's family grieved, and that was understandable, but when the citizens grieved, it was for "naught." "Why is a fruitless tear shed over a gloomy grave? Why complaints?" Indifferent to beauty, youth, and faith, death plays no favorites. It is part of the natural course of things. "Sweet roses flourish at dawn; frail, transitory – they fly apart at dusk."[127] There are four accounts of Cavendish's death, which vary, but all agree that he was resigned to an imminent end, and that he was serene. The accounts also agree that Cavendish's desire for solitude lasted to his dying breath. The most detailed account has Cavendish returning from the Royal Society one evening with blood on his shirt, and as always proceeding silently to his study. His servant saw the blood but dared not ask why. Cavendish lived only a few days longer. On his last day, he called his servant to his bedside and told him he was going to die. The servant was to leave and not return until the breath was out of his body, upon which time he was to inform his principal heir. After a while, having received no further summons, the servant returned to find him a corpse. By another account, the servant returned too soon, finding Cavendish still alive and much annoyed with the intrusion. Another account supplies a new detail. Cavendish told his servant to leave and not return until a certain hour when he expected to be dead, for he wanted to pass his final moments "in the tranquility of perfect solitude." To his physician Everard Home, who attended him, he said "it is all over, with unusual cheerfulness." He wished Home "good by with uncommon mildness," "quite composed & firm." There was a "shortness of questionings"; he "seemed to have nothing to say, nor to think of anyone with request."[128] What was right for Cavendish never changed, and to the end he ensured that nothing did.

---

124 Young, "Cavendish," 444.

125 Humphry Davy lecture in 1810, in John Davy, *Memoirs* 1: 222.

126 Cuvier, "Cavendish," 237.

127 Cavendish, "Luctus."

128 25 February 1810, Charles Blagden Diary, Royal Society 5: 426 (back), 427. Young, "Cavendish," 445–46. Wilson, *Cavendish*, 182–84. In Cavendish's time, it was thought that a person's behavior and words at the approach of death were revealing. We find similarities in the accounts of persons' deaths. Peter Woulfe, an accomplished chemist and member of the Royal Society, in his last hours, told his servant to leave him and shut the door. The servant returned at midnight to find him still alive. The servant came again in the morning and found him dead, in the same seat, bearing a calm and serene countenance. Timbs, *English Eccentrics*, 137.

Cavendish's burial was elaborate and expensive, following the family tradition. The funeral procession of carriages for the family and of men on horses set out from Clapham Common with the body, then passed through London and on to Derby, the location of the family vault. At the gates of Derby, the procession was met by 24 burghers, 24 constables, and various city officials all dressed in black.[129] The display of pomp and ceremony befitted a powerful family, who came together to claim one of their own in death, however distant in thought he was in life. Cavendish died as he was born, an aristocrat.

[129] Lord Bessborough to Charles Blagden, 7 March 1810, Blagden Letters, Royal Society, B.149. "Mr. Swift's Bill for Expenses Attg the Funeral of Hen: Cavendish Esq.," 29 August 1810, Devonshire Collections, Chatsworth, L/114/74.

Chapter 4
Life of Natural Philosophy

4.1 Physical World

Although Cavendish's researches went far beyond his father's, it is clear that he got his start at home. The physician John Walker, a neighbor of Cavendish's on Bedford Square, said that Cavendish had been "educated and trained by his father from very early youth to scientific pursuits."[1] Examples of his father's help can be is seen in his work. Lord Charles Cavendish had instruments for observing magnetic declination and dip, with which he seems to have made systematic observations in the garden of his house on Great Marlborough Street, and his son Henry assisted him.[2] In his physical and meteorological studies, Charles took special interest in heat. Henry's earliest known experiments included an extended study of specific and latent heats. Charles measured the pressure of water vapor over a range of temperatures. Henry measured the pressure of water vapor and of common air over a range of temperatures, and he later measured the pressure of several other kinds of air, or gases.[3] Charles received the Royal Society's Copley Medal for his thermometers; Henry was recognized by the Royal Society as its expert on thermometers and on meteorological instruments in general (Fig. 4.1). Charles measured the depression of mercury in tubes of different diameters. Henry included these measurements in his examination of the Royal Society's barometers.[4] In the Royal Society, Charles's skills were called on when precision of measurement was at issue; for example, he served on a committee to compare the Royal Society's standards of weights and measures with

[1] John Walker to James Edward Smith, 16 Mar. 1810, in *Memoir and Correspondence of the Late Sir James Edward Smith, M.D.*, 2 vols., ed. Lady Smith (London, 1832), 170–71.

[2] Thorpe, ibid., 326, 438.

[3] Cavendish, *Scientific Papers* 2: 355, 362–80. The pages contain manuscripts of Cavendish's experiments on the pressure of water vapor and gases together with comments by the editor.

[4] Henry Cavendish, "An Account of the Meteorological Instruments Used at the Royal Society's House," *PT* 66 (1776): 375–401; in Cavendish, *Scientific Papers* 2: 112–26, on 116.

R. McCormmach, *The Personality of Henry Cavendish - A Great Scientist with Extraordinary Peculiarities*, Archimedes 36, DOI 10.1007/978-3-319-02438-7_4,

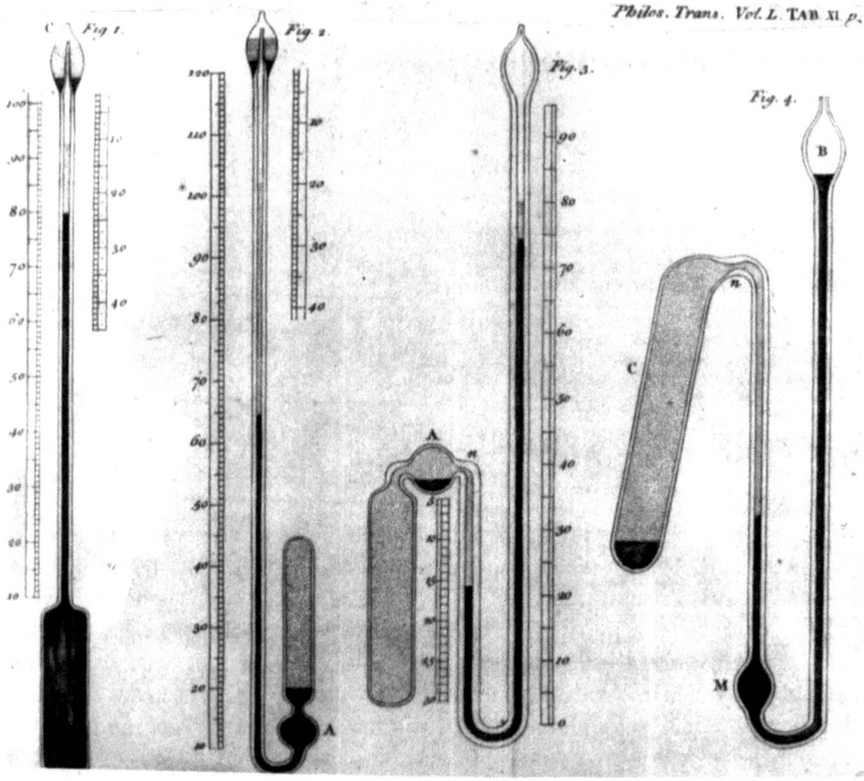

Fig. 4.1 Charles Cavendish's Thermometers. These thermometers of Lord Charles's contriving measured the greatest degree of heat or cold. The thermometer on the *right* is folded to make it more compact, desirable if it were sunk to the bottom of the sea or raised in the atmosphere. The drawing is from a paper he communicated to the Royal Society, for which he was awarded the Copley Medal for that year. ("A Description of Some Thermometers for Particular Uses," *Philosophical Transactions* 50[1757]: 300)

those of the Paris Academy of Sciences. Henry served on committees of a similar kind, and he made standards of measurement a special concern of his.

Thorpe said that "almost every department of the physical science of his time appealed to him with equal force and he pursued all with equal zeal and success."[5] That could not be said of any other natural philosopher. Distinctive for the sweep of his learning, he was also clear about its limits. His interest was, as Thorpe said, "physical science." He was interested in the living world insofar as it exhibited the same phenomena as the physical world. To determine that the shock of an electric fish is caused by the same electricity as the shock of a battery of Leyden jars, he

[5] Thorpe, "Introduction," 6.

performed experiments on a model of the fish made from leather.[6] With Blagden, a physician, he experimented on the eye as an optical instrument.[7] To assist a chemist, he experimented on the part of a certain plant that is most like a rock.[8] Near the end of his life he was made an honorary member of a new society, the Society for the Improvement of Animal Chemistry.[9] The science naturally appealed to him, living processes understood as chemical reactions.

Cavendish's library tells us more about his interests. Other than for living expenses, his biggest expense was books, which he added to the already substantial library his father left him. The catalog of his library[10] lists approximately 9,000 titles, representing about 12,000 volumes. This was a large private library for the time, though there were much larger ones. In the catalog, books are listed under headings. Natural philosophy accounts for the largest number of books, nearly 2,000. Chemistry is included under natural philosophy, as are most of Cavendish's other major interests such as mechanics, instruments, and meteorology. Also included are some books that lay outside of his known interests such as medicine, anatomy, and animal economy, very few of which were published after his father's death. Exceptions, which fit the pattern, are *Diseases/Apollo Mathematicus, or Art of Curing by the Mathematics*, 1795, and *Treatise on Materia Medica*, 1789, by William Cullen, professor of medicine in Edinburgh, who lectured on chemistry, a subject he regarded as belonging to natural philosophy. Another major category is mathematics, which includes in addition to books on pure mathematics books on natural philosophy with mathematical content such as Newton's *Principia* and *Opticks*. Astronomy is another major category. The category of natural history contains books on the sciences of life, subjects in which Cavendish showed little interest, as seen by the near total absence of books on them published after his father's death. Natural history also covers geology and mineralogy, subjects which did interest Cavendish; in addition to buying many books on them, he bought books belonging to the related categories of geography, maps, and voyages and travels. Roughly half of the books in Cavendish's catalog are scientific. The categories listed first in the catalog, astronomy, natural philosophy, and mathematics, are subjects that came first in Cavendish's life.[11]

---

[6] Henry Cavendish, "An Account of Some Attempts to Imitate the Effects of the Torpedo by Electricity," *PT* 66 (1776): 196–225.

[7] In response to a paper by his physician Everard Home, Cavendish with Blagden's assistance carried out an experiment to detect changes in the convexity of the cornea accompanying changes in the focus. 8, 11, and 16 November 1795, Charles Blagden Diary, Royal Society 3: 75(back), 76, and 77(back).

[8] Cavendish's young assistant James Lewis Macie presented him with the problem of determining the density of tabasheer, a hard substance located in the joints of tropical bamboo. Macie identified it correctly as siliceous earth. James Lewis Macie, "An Account of Experiments on Tabasheer," *PT* 81 (1791): 368–88, on 370, 384–85, 388.

[9] 27 April 1809, Minutes of Council, Royal Society 7: 527–31.

[10] The catalog is incomplete, extending only to the early 1790s. Cavendish continued to buy books after that.

[11] R. A. Harvey, "The Private Library of Henry Cavendish," *The Library* 2 (1980): 281–84.

The non-scientific half of the books are arranged under a number of headings, poetry and plays accounting for the largest number. These were clearly his father's books, for practically none was published after the year his father died. An exception is an Indian drama, which may relate to Cavendish's interest in Indian calendars. There are numerous books under the headings of history and antiquities; Cavendish added some books about India and other faraway lands, but otherwise he took little interest in these subjects. There are numerous books under the heading of classical authors, but since none of the editions was published after his father's death, his father probably bought them. There are only a few books published after his father's death under the headings architecture, language and education, and works of authors. Under the heading of novels, the number of books is small, and they were published before Cavendish took over the library. His disinterest in the feelings of imagined persons is consistent with what we know about him. There is no category of biographies, though Cavendish bought *The Life of Samuel Johnson*, written by James Boswell, who came as a guest of the Royal Society Club to dinners at which Cavendish was present; Cavendish may have seen Johnson too, for he often dined at the Crown & Anchor, where the Club met. There are a number of books on the behavior of people in the aggregate, dealing with regularities in populations and commerce. We conclude that Charles Cavendish's library, the nucleus of Henry's, was that of a rounded gentleman of the eighteenth century, revealing a keen interest in science but not to the neglect of literature and varied branches of learning. The library that Henry built was that of a man whose interest was a comprehensive knowledge of the physical world.

4.2 Mathematical and Musical Mind

Numbers and their laws have been called the language of science. It was a language in which Cavendish was fluent, and his scientific work was a model of its use. "He carried his mathematical habits into the laboratory," Brougham said.[12] Davy said, "Of all the philosophers of the present age, Mr. Cavendish was the one who combined, in the highest degree, a depth and extent of mathematical knowledge with a delicacy and precision in the methods of experimental research"; "he reasoned with the caution of a geometer upon the results of his experiments." With reference to his own field of chemistry, Davy said that Cavendish "was the first who carried the mind and methods of a mathematician into the field from which the alchemist had not long retired, and in which the speculator still remained."[13] He worked with mathematics extensively, though primarily as a means of studying nature, not as an end in itself. Mathematics was important to him as a means of achieving his goal of

[12] Brougham, *Lives* 1: 251.

[13] Humphry Davy, quoted from a chemical lecture he gave in 1810. John Davy, *Memoirs* 1: 221. Humphry Davy, *Elements of Chemical Philosophy* (London, 1812) 1: 37. Humphry Davy, quoted in George Godfrey Cunningham, ed., *Lives of Eminent and Illustrious Englishmen,* vol. 8 (Glasgow, 1837), 169.

strict reasoning in natural philosophy. He had a mathematical mind, but he was not a mathematician (Figs. 4.2, 4.3, 4.4, 4.5, and 4.6).

Wilson said that Cavendish subjected nature "to a canon which knew of no limitations, and required that every phenomenon and every physical force should be held to be governed by law, and admit of expression in mathematical or arithmetical symbols." It seemed that it was "impossible for Cavendish to investigate any question otherwise than quantitatively," and to have experimented otherwise would have been "painful" to him.[14] Thorpe said that quantitative methods enabled Cavendish to grasp the "real nature of the phenomena," and they enabled later scientists us to "see how frequently he was on the verge of discoveries which are accredited to later workers."[15]

Cavendish may have been drawn to music as well as to mathematics. His maternal family was musical, and if he had shown an interest as a child, he would certainly have been encouraged. His name appears on the list of subscribers in 1758 or 1759 to concerts at a London academy run by the Italian musician Felice Girardini.[16] The name does not prove he was our subject – Sir Henry Cavendish, a distant Irish relative of the same age, was in London around this time[17] – but the evidence, sketchy as it is, points to him. There were 17 subscribers, young singers and players of instruments, who took part in performances at the academy. While we may have difficulty imagining Cavendish performing musically before an audience, it is not out of the question. Playing to his skill, he was capable of performing experiments before a scientific audience. Shyness is distinct from stage fright.[18] At Cavendish's death, the auction catalog of his house at Clapham Common listed a grand piano-forte. Because apart from servants, Cavendish was the only person who lived in the house, the piano must have been there because he wanted it.[19]

Music in the eighteenth century was thought to be the art that speaks most directly to the emotions. For all his solitary ways, Cavendish was not isolated from

---

[14] Wilson, *Cavendish*, 187–88. In another place, Wilson seemed to have backed off from his statement that Cavendish evidently could not investigate other than quantitatively. "Many of his scientific researches were avowedly *quantitative*." Ibid., 187. Here he said "many," not "all." With this wording, Wilson is correct.

[15] Thorpe in Cavendish, *Scientific Papers* 2: 317.

[16] Great Britain, Historical Manuscripts Commission, *Report on Manuscripts in Various Collections*, vol. 8*: The Manuscripts of the Hon. Frederick Lindley Wood; M. L. S. Clements, Esq.; S. Philip Unwin, Esq.* (London: His Majesty's Stationary Office, 1913), 188–89.

[17] He descended from an illegitimate branch of Henry Cavendish's family. An English and Irish politician, he is best known as a parliamentary diarist. Peter D. G. Thomas, "Cavendish, Sir Henry, second baronet (1732–1804)," *DNB* 10: 627–28.

[18] Arnold H. Buss, "A Conception of Shyness" in *Avoiding Communication: Shyness, Reticence and Communication Apprehension*, eds., J. A. Daley and J. C. McCrosky (Beverley Hills: Sage, 1984), 39–49, on 43.

[19] *A Catalog of an Assortment of Modern Household Furniture ... the Genuine Property of a Professional Gentleman; Which Will Be Sold by Auction by Mr. Squibb, at His Great Room Saville Passage, Saville Row, on Wednesday, December 5, 1810, and Two Following Days, at Twelve O'clock.* Item 45 is a grand pianoforte, by Longman and Broderip, in a mahogany case. Devonshire Collections, Chatsworth.

Cavendish's Exactitude

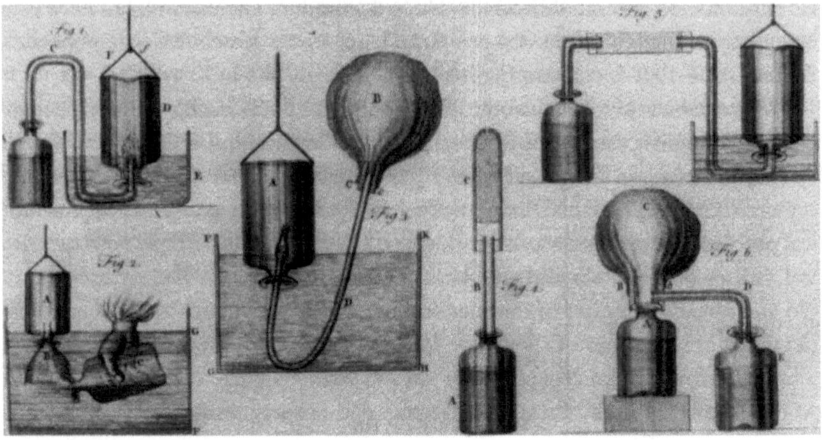

Fig. 4.2 Factitious Air Apparatus. The numbered figures are from Cavendish's first publication, for which he received the Royal Society's Copley Medal. He carried out exacting quantitative work in the chemistry of gases using apparatus that appears to us very simple. *Figure 1* shows his technique for filling a bottle D with air. The bottle, containing water, is inverted in the vessel of water E; the air to be captured is generated by dissolving metals and acids and by other means in bottle A. His measure of quantity of air is the weight of the water it displaces in D. *Figure 2* shows how he transfers air from one bottle to another. *Figure 3* shows how he withdraws air from a bottle by means of a bladder. The speckled substance in *Figures 4* and *5* is dry pearl ash, through which air is passed to free it from water and acid. ("Three Papers, Containing Experiments on Factitious Air," *Philosophical Transactions* 56[1766]: 141)

Fig. 4.3 Standard Gas Volume Measures. These measuring glass vessels with brass caps on the bottom, indicative of Cavendish's laboratory technique, are described and drawn in his paper on an important instrument in the chemistry of gases. Any inaccuracies incurred in using these volume measures, he said, were insignificant because he determined the "exact quantity of air used by weight." ("An Account of a New Eudiometer," *Philosophical Transactions* 73[1783]: 128 and plate III. The standard volume measures are kept in the Royal Institution of Great Britain. Photograph by the author)

Fig. 4.4 Chemical Balance. Built by "Harrison," probably Cavendish's private instrument-maker William Harrison, this instrument is the earliest of the great precision balances of the eighteenth century (Reproduced by permission of the Royal Institution of Great Britain)

Fig. 4.5 Mathematical Instruments. These two drawers, which fit into a cabinet belonging to Henry Cavendish, contain many scales and rulers made of brass and wood, a brass globe map projection, an ivory triangle, and other aids to exact drawing. The top drawer contains the regular solids, made of boxwood (Photograph by the author). The instruments belong to the Devonshire Collections, Chatsworth (Reproduced by permission of the Chatsworth Settlement Trustees)

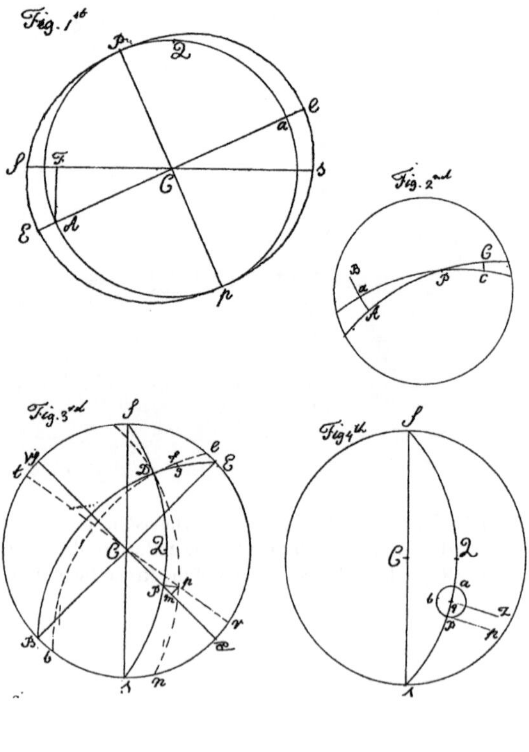

Fig. 4.6 Mathematical Drawings. The figures are part of a carefully drafted but unpublished manuscript, ("Precession of Equinoxes," Cavendish Scientific Manuscripts VIII, 9. Reproduced by permission of the Chatsworth Settlement Trustees)

currents of ideas and feelings of the day. Insofar as we know him, his inner life was directed outwardly, to nature and to the tools for understanding it, but if we are right about his interest in music, it possible that music gave him access to feelings and a means of expressing them, an alternative to the spoken and otherwise conventional expressions of feelings, to which he was without access. Cavendish does not neatly fit most categories, and we are at liberty to consider his possibilities.

4.3 Senses

Sensations originating in the eyes, the fingertips, and other parts of the body are translated in the mind into meaningful experiences. Their record is the raw material of science, and because Cavendish's life was about science, his sensory experiences in the laboratory, the observatory, and the field were his most important. The mass of scientific manuscripts he left behind are a faithful record of his life. Looked at this way, his life is one of the best documented lives of the eighteenth century. In this section, we look at the senses as Cavendish employed them in his work. We take up them up one by one, though acknowledging that "our brains derive meaning from the world" by blending the kinds of sensory perception.[20]

In 1749, the year Cavendish entered the University of Cambridge, an English translation of a book by the French physiologist Nicolas Claude Le Cat, *A Physical Essay on the Senses,* was published. Reviewed in the *Philosophical Transactions* and in the *Monthly Review,* Cavendish may have seen it. The author wrote that the senses are "a Matter the most interesting of Physics," for they "are our means of Correspondence with the rest of the Universe." "This Part of Physiology is so linked with all the other Parts of natural Philosophy, that it is almost impossible to treat of the Senses, without touching at the same Time, slightly at least, on the other Branches of Physics."[21] The connection Le Cat made between the senses and natural philosophy introduces our subject here.

Cavendish compared the sensitivity of the sense of smell to the sensitivity of the eudiometer, a principal instrument in his favored field of chemistry. He said that if the air contains impurities, our sense of smell may detect them even when the instrument cannot. He gave an example: if 10 oz of nitrous air (nitric oxide) are released in a room measuring 12 by 12 by 12 ft, a strong odor is detectable, but to the eudiometer it is "utterly insensible." The reason is that the sample of nitrous air is only 140,000th part of the air of the room, and it can affect the eudiometer reading by only 1 part in 47,000, far too little to register. This shows that "our sense of smell can, in many cases, perceive infinitely smaller alterations in the purity of the air than can be perceived by the nitrous test [eudiometer]." Cavendish's point

[20] Lawrence D. Rosenblum, "A Confederacy of Senses," *Scientific American* 308:1 (2013): 73–75, on 75.

[21] Nicolas Claude Le Cat, *A Physical Essay on the Senses* (London, 1750), 29, 284. Transcribed and summarized in *Monthly Review*, 2nd ed., 1 (May 1749): 20–29, 2 (March 1750): 364–68.

was to show the limitations of the eudiometer, but also to clarify what it does that the sense of smell does not: it measures, and what it measures is a known physical property of the air.[22] (What it was thought to measure is the degree of phlogistication of common air, and the nitrous air in Cavendish's example strongly phlogisticates common air. To explain this would require a detour through the reigning phlogiston theory of chemistry.) Indirectly, Cavendish's discussion of the eudiometer shows that what is most important in quantitative work generally is the sensitivity of the instrument rather than the limits of sensitivity and the acuity of sense organs. Investigators had some control over this kind of sensitivity.

Cavendish's calculation shows what an extraordinarily small quantity in terms of ounces of an air-borne substance is needed to produce a sensation of smell, though in numbers of molecules it is a very large quantity. The sense of smell was rarely relied on as the primary sense in experiments in natural philosophy. An exception was experiments on passing odors through electrified glass, reported from abroad, and repeated by skeptical fellows of the Royal Society meeting in one of their houses.[23] In his experiments in chemistry, Cavendish regularly made use of the sense of smell, though together with other senses. Compared with the sense of smell, the sense of taste is relatively crude, requiring 25,000 times the number of molecules for a sensation to register.[24] On occasion, Cavendish made use of this sense too in his chemical experiments.

Touch, hearing, and sight, the remaining three of the standard five senses, are capable of fine discrimination. Touch is one of the main senses called upon throughout natural philosophy. In experimental work, the hands are constantly in touch with instruments, apparatus, and objects. Touch can detect a difference between a smooth surface and one with grooves only 1/2,500th of an inch deep, and a fingertip is extremely sensitive to vibrations, a form of touch. Yet compared with our favored senses, hearing and sight, touch is crude, requiring of the order of a billion times more energy to register.[25]

Our range of hearing is great, spanning nearly 10 octaves, each octave having double the number of vibrations of the previous. The sensitivity of our hearing is great too: at frequencies where sensitivity is greatest, if it were any greater our ears would pick up the random motion of molecules of air and hear it as background noise.[26] Hearing is the primary sense in experiments in a branch of natural philosophy, acoustics, and it finds applications in other branches too. An example of its use in precision measurement is a pyrometer, which measures the expansion of metals

---

[22] Henry Cavendish, "An Account of a New Eudiometer," *PT* 73 (1783): 106–35; in Cavendish, *Scientific Papers* 2: 127–44, quotation on 144.

[23] William Watson, "A Letter… Declaring That He as Well as Many Others Have Not Been Able to Make Odours Pass thro' Glass by Means of Electricity …," *PT* 46 (1750): 348–56.

[24] Lors and Margery Milne, *The Senses of Animals and Men* (New York: Atheneum, 1962), 147.

[25] Ibid., 18–19.

[26] Ibid., 39, 41.

with heat, designed by John Smeaton, a colleague of Cavendish's. Instead of by touch or sight, as one might expect, the point of contact of the heated metal object with the point of the micrometer screw was determined by the more discriminating sense of hearing. Capable of measuring an expansion to an accuracy of 1/4,000 part of an inch, and upon repeated measurements of 1/20,000 part of an inch, his pyrometer, Smeaton said, exceeds in sensibility "any thing I have met with."[27] In his chemical work, Cavendish frequently made use of the sense of hearing. He tested the explosibility of "inflammable air" mixed with common air in different proportions by listening to the loudness of the sound, the "bounce," when a flame was held to the mouth of a bottle containing the mixture.[28]

Our sense of sight is responsive to a great range of intensities, of the order of a billion to one. Like hearing, seeing is the primary sense of a branch of natural philosophy, optics. The eye, Cavendish's biographer Wilson wrote in a book on the five senses, is "Queen of the Senses … its empire is far wider than those ruled over by its sisters." The ear, nostril, tongue, and hand are bounded in space and time as the eye is not.[29] There was very little in Cavendish's work that did not make use of vision. He relied on it when handling apparatus and instruments and when reading numbers off scales. He needed sight to make qualitative distinctions. In his investigation of neutral arsenical salt, he made numerous visual distinctions: milky, cloudy, yellow, pale straw, reddish yellow, pale madeira, red, reddish brown, dirty red, green, blue, bluish green, pearl colored, transparent, turgid, and muddy. To interpret what was going on in his experiments on arsenic, he combined visual impressions with impressions drawn from other senses. He felt as well as observed textures: dry, hard, thin jelly, gluey, thick, stiff mud, and lump. By smell and taste, he distinguished between acids and their products. He produced a new substance that had "all the properties of an acid," a claim which he qualified with the implicit acknowledgment of the fatal reputation of arsenic, "unless perhaps it should fail in respect of taste which I have not thought proper to try."[30] The senses were Cavendish's window onto the physical world, his special interest, and he employed them with great skill in advancing natural philosophy.

There are many more senses than the standard five, each sense organ responding to a different kind of stimulus. We have, for example, organs for sensing heat, a branch of natural philosophy and a major field of research for Cavendish. For

[27] John Smeaton, "Description of a New Pyrometer, with a Table of Experiments Made Therewith," *PT* 48 (1754): 598–613, on 600.

[28] He reasoned that "the loudness of the explosion should be in proportion to the quantity of phlogiston containd in the mixture." Henry Cavendish, "Experiments on Factitious Air. Part IV," in Cavendish, *Scientific Papers* 2: 307–16, on 309–10. He also designed a mechanical apparatus to measure the force, rather than the noise, of explosions of inflammable air. Henry Cavendish, "Measurer of Explosions of Inflam. Air," ibid., 318–19. Thorpe, "Introduction," 10.

[29] George Wilson, *The Five Senses; or, Gateways to Knowledge* (Philadelphia, 1860), 35–36.

[30] Henry Cavendish, "Arsenic," Cavendish Scientific Manuscripts, Devonshire Collections, Chatsworth, 1(b), 10.

electricity, another branch of natural philosophy and a major field for Cavendish, we lack a sense organ, but we can feel electrical shocks. High-voltages applied to different parts of the skin affect various sense organs: in addition to pain, there can be touch, pressure, cold, warmth, even taste. Cavendish measured electrical resistances of conducting materials before there existed a means for generating a continuous current and the instrument for measuring it, the galvanometer. He relied instead on transient currents produced by the discharge of Leyden jars (capacitors), using his body as an instrument. He passed the discharges through solutions of salt contained in calibrated tubes a yard long and bent at one end. Corks were placed at both ends of the tubes and wires were run through them into the solutions. Inserting himself in the circuit, with one hand touching the top wire of a tube and the other hand the knob of an electrified Leyden jar, he felt a shock in his wrists and elbows. By varying the solutions and the length of the wires in the solutions, he compared the degrees of shock when Leyden jars were discharged through different tubes. The solution that allowed the larger shock had the smaller resistance. When Cavendish's results were compared with later results using galvanometers and steady currents, the accuracy he achieved by his method of sensing shocks was found to be, Maxwell said, "truly marvelous."[31]

4.4 Objectivity

The desire for objectivity is a complement of other traits we treat in this book such as a desire for perfection or for truth. In practical terms, objectivity meant the repeatability of experiments, eliminating or reducing the personal element in research. William Watson, who recommended Cavendish for membership in the Royal Society, repeated an experiment that had been carried out at a number of locations in Europe, with different outcomes. He explained the need for repeating it: to establish the outcome, it was necessary for the experiment to succeed in Venice and Leipzig, as it did, and also in Paris, Geneva, Turin, and Wittemberg, where it did not. Watson reported that it did not succeed in London either.[32]

Cavendish responded in print twice to criticisms. What he said in each case reveals the value he placed on objectivity. Several foreign chemists tried to repeat one of his experiments "without success," an implicit criticism. Cavendish could not imagine why they failed "except want of patience." He arranged for the clerk of the Royal Society to repeat the experiment, using the same method and apparatus, and

[31] Maxwell, in Cavendish, *Electrical Researches,* lvii–lviii. Experiments on electrical resistance, on 321–61.

[32] William Watson, "An Account of Professor Winkler's Experiments Relating to Odours Passing through Electrified Globes and Tubes …," *PT* 47 (1751): 231–41, on 237–38; "A Letter … Declaring That He as Well as Many Others Have Not Been Able to Make Odours Pass thro' Glass by Means of Electricity," 349.

for several fellows of the Royal Society to witness it. In this way, by removing himself from the experiment and by bringing in competent observers, he established the conditions for an objective judgment. The repeated experiment reproduced his result, establishing, he said, "the truth of my position."[33]

In his reply to the second criticism, Cavendish again defended the conclusion he had drawn in his original paper. In that place he had given his reasons "without any arguments of a personal nature, or which related to any one person who espouses the contrary doctrine more than to another." His critic brought up several experiments that seemed to disagree with his conclusion there. Taking "no notice of the arguments used by him," Cavendish examined the experiments, showing that in each case the disagreement did not hold up. His object was not to "point out any inconsistencies or false reasonings" in his critic, but to show what the experiments found and did not find. At the end of his paper, Cavendish summarized the "state of the argument," which "sufficiently justified" the conclusion he had drawn from his own experiments. He disliked entering into "controversy," and by allowing experiment the final say, he did not engage in one.[34] Cavendish took objectivity to another level. Not only was his conclusion objectively arrived at and confirmed, but his relations with colleagues were objective; he did not engage with them in a personal way.

The original controversy over the discovery of the composition of water, which took place around the same time, was driven by passion, but there is no indication that Cavendish shared in the passion. His one public action was to write a reasonable letter to a French journal that had made an error in the dating the translation of a paper of his, asking that the correction be published in the next issue; Blagden, who was Cavendish's associate at the time, took several actions on the part of Cavendish. As we see in Part II of this book, in the year following the water controversy Cavendish paid the principal rival in the controversy James Watt a visit, which went off well. Later that year, Watt came to London, where he "was received very kindly by Mr. Cavendish and Dr. Blagden."[35]

Cavendish's objectivity played out in another setting. In the early 1780s a body of mathematical members of the Royal Society charged their president Banks with giving preference to natural historians and to persons of rank, to the detriment of the Society, and they wanted to replace him. The regular proceedings of the Society were suspended while intemperate speeches were made, and Cavendish was drawn in. He discounted the display of emotion in the opposition and looked only for "proof" of wrong, the same standard he used in defending his scientific

---

[33] Henry Cavendish, "On the Conversion of a Mixture of Dephlogisticated and Phlogisticated Air into Nitrous Acid, by the Electric Spark," *PT* 78 (1788), 261–76; in Cavendish, *Scientific Papers* 2: 224–32, on 227, 230.

[34] Henry Cavendish, "Answer to Mr. Kirwan's Remarks upon the Experiments on Air," *PT* 74 (1784), 170–77; in Cavendish, *Scientific Papers* 2: 182–186, on 182, 186.

[35] Watt, quoted in Samuel Smiles, *Lives of the Engineers. Harbours – Bridges. Smeaton and Rennie*, rev. ed. (London, 1874), 169.

work. In the only faithfully recorded conversational fragment of Cavendish's we have, he corrected Blagden:[36]

Cavendish: I did not expect any success from the D^{rs} negotiations [Dr William Heberden and no doubt Dr Sammuel Horsley, a mathematician and opponent of Banks]. But whatever violence *they* may express, that is no reason against proceeding with all moderation, as by such conduct the sense of the Society will be insured against them.

Blagden: I wish you would see D^{r} H[orsley] & learn from him itself the implacable temper expressed; as I think you would then change the opinion to which you seemed inclined when we conversed last, that those gentlemen might have nothing in view but the good of the Society.

Cavendish: I did not say they had nothing else in view, but only that no proof yet appeared of other motives.

When the conversation ended, Cavendish said he would approve a vote of confidence in Banks, but only so long as its wording did not give offense. Critical of Banks, but not hostile, Cavendish gave his support to the existing order, Banks's presidency, even though he himself was one of the Society's mathematical members. His goal was to restore the meetings to their intended purpose, the reading of scientific papers and other business. He thought that any abuses should be corrected by the proper choice of Council, not by taking over the meetings. What was at stake was important enough to him that he willingly played a behind-the-scenes role in shaping the outcome. His participation in the "dissensions," as they were called, was entirely impersonal, entirely objective.

Objectivity was a family trait. Cavendish's first cousin William Cavendish, fourth duke of Devonshire, was the "supremely objective man, never led away by passion, completely reliable," according to the editors of his state papers. His "detachment was natural to him and inevitably confirmed his exalted station," the key to his "usefulness, recognized by everyone."[37] From all we know about Henry Cavendish, his detachment was natural too, and in his scientific setting it confirmed his usefulness. In the Royal Society, he was the completely objective observer of the physical world and of the society of scientific men.

4.5 Accuracy, Precision, and Perfection

The words "accuracy" and "precision" were repeatedly used by Cavendish in his work. Accuracy in measurement means that the measured values are close to the truth; precision means that the values are close to one another. Cavendish distinguished the two words. In one experiment, he measured the whole quantity of air

[36] Charles Blagden to Joseph Banks, 24 December 1783, copy, Fitzwilliam Museum Library, Perceval, 178–79, on 178.

[37] William Cavendish, *Devonshire Diary*, 19–21.

that was absorbed "with considerable precision," but he could not determine "with any accuracy" the quantities of each kind of air that were absorbed because of the uncertainty of the composition of the air remaining at the end of the experiment.[38] Cavendish was generally both accurate and precise in his work.

In this respect, Henry Cavendish took after his father, whose experiments were made with an "accuracy and exactness which are peculiar to his Lordship," according to the president of the Royal Society.[39] Charles Cavendish in his turn had had a good mentor. His first recorded scientific observations were made with the astronomer James Bradley, who was renowned for the exactness of his measurements and of his instruments. Charles's observations were taken with a telescope Bradley used to look for parallax of the fixed stars, a minute measurement then beyond the capabilities of the best instruments.[40] In 1748, Bradley discovered a small annual change in the position of stars, nutation, which was perceptible to him, he said, only because "of the exactness of my instrument." He made the point that instruments as well as theory lead to discovery, and that the "more exact instruments are … and the more regular the series of observations is … the sooner we are enabled to discover the cause of any new phaenomenon."[41] Bradley urged astronomers to begin their investigations with an examination of the accuracy of their instruments, advice Henry Cavendish would follow in every branch of physical science. It is noteworthy that Bradley recommended him for membership in the Royal Society.

The single surviving laboratory record of Charles Cavendish's researches shows the standard he set for his son. The occasion was the repeatability of experiments, the mark of objectivity. In 1762, the London schoolmaster John Canton published experiments demonstrating that water is compressible, contradicting the finding of a prestigious scientific academy that it is not. Canton's apparatus was a glass ball a little over an inch across attached to a very narrow capillary tube. The ball together with a few inches of the tube was filled with mercury and placed in a water bath. The bath was heated until the mercury rose to the top of the tube, at which time the tube was hermetically sealed. When the mercury had cooled to its original temperature, Canton observed that it stood about a half-inch higher than before. He found the same when water was used in place of mercury, only the water rose a little higher than the mercury. The only difference before and after the expansion

---

[38] Cavendish, "On the Conversion of a Mixture of Dephlogisticated and Phlogisticated Air into Nitrous Acid, by the Electric Spark," 229. The two words together with a third "exactness" were sometimes used interchangeably, and sometimes the words were applied to reasoning rather than to measurement. Young, in characterizing Cavendish's work, spoke of the "precision of the experimental demonstration" and the "accuracy of the argumentative induction." Young, "Cavendish," 436. Dugald Stewart said that a natural philosopher was distinguished from other people by the "accuracy" of his observations and the "precision" of his judgments. *The Collected Works of Dugald Stewart, Esq.*, F.R.SS., ed. W. Hamilton, 11 vols. (Edinburgh, 1854–60) 3: 242–44.

[39] 30 November 1757, Journal Book, Royal Society 23: 639.

[40] James Bradley, *Miscellaneous Works and Correspondence of the Rev. James Bradley, D.D., F.R.S.*, ed. S. P. Rigaud (Oxford, 1832), 237.

[41] James Bradley, "A Letter … Concerning an Apparent Motion Observed in Some of the Fixed Stars," *PT* 45 (1748): 1–43, on 1–4.

of the water was that the pressure of the atmosphere over it had been removed. Water, Canton concluded, is compressible. He showed it another way: by placing the apparatus in the receiver of an air pump and reducing the air pressure, he observed the water rise in the capillary tube. The Council of the Royal Society moved to reward Canton with the Copley Medal for this work, but doubts were raised, and for the honor of the Royal Society the Council ordered that Canton's experiments be repeated. To this end, a committee of instrument makers and experimenters including Cavendish was appointed in 1764.[42] It seems that Cavendish performed most of the experiments himself. He drew up a paper informing the Council that the new experiments confirmed Canton's.[43] He said that the art of experiment had improved since the earlier finding of the incompressibility of water, proof of which was Canton's skill (and his own) in demonstrating "so small a degree of compressibility."[44] In his report, Cavendish discussed the "causes of Error" in his experiments. The main cause was heat, accounting for one fifth of the whole compression of the water. He tested for the effect of different sizes, shapes, and thicknesses of the glass. He compared the alternative instruments of air pump and condenser, and he analyzed and improved the gauge. He took the means of the measurements of repeated experiments.[45] His concern with accuracy would reappear throughout Henry's career as an exacting experimenter. Charles Cavendish annotated the record of his experiments throughout with "by my measure,"[46] an expression his son would take to heart; most of Henry's experiments were carried out with measuring instruments. It is a meaningful coincidence that the year 1765 when Charles repeated Canton's experiment was the year Henry published his first paper, a quantitative experimental study of factitious air.[47]

Henry Cavendish's skill with instruments and his concern for accuracy were recognized early in the Royal Society. His first assignment, in 1766, was to improve the accuracy of thermometers. Even the best-made thermometers varied among themselves by as much as two or three degrees in their reading of the boiling point of water. Cavendish set about to determine if the boiling point of a thermometer was affected by its immersion in the boiling water or in the steam above the water and by the rapidity of the boiling. He found that immersion of the stem and the ball of the thermometer in steam offered a "considerably more exact method" for adjusting the

---

[42] 28 November 1764, Minutes of Council, Royal Society 4: 57.

[43] The Council resolved that Canton's experiment was verified. 21 November 1765, ibid. 4: 141.

[44] Charles Cavendish, "Observations on Mr. Blake's Objections to Mr. Canton's Experiments," Canton Papers, Royal Society.

[45] Two papers drawn up by Charles Cavendish were read to the Society, and a long summary of them was entered in the Journal Book, Royal Society 25: 668–79, on 670–71 (5 December 1765).

[46] Charles Cavendish sent Canton his computations and measures to review. This material is preserved in John Canton's papers in the Royal Society.

[47] Henry Cavendish, "Three Papers, Containing Experiments on Factitious Air," *PT* 56 (1766): 141–84; in *Scientific Papers* 2: 77–101.

boiling point than the customary method.[48] His recommendation that the Royal Society standardize the method for determining the fixed points of mercury thermometers led several years later to his appointment as head of a committee to decide the best method for this purpose. The published report of the committee, most or all of which Cavendish wrote, was influential, affecting the construction of mercury thermometers and with it the science of heat.[49] In 1773, the Society called on Cavendish to draw up a plan for regular meteorological readings to be taken by the clerk of the Society at its house on Crane Court and to be entered in a meteorological journal. To carry out the plan, the Society acquired instruments for recording temperature, pressure, humidity, rainfall, wind, and magnetic variation and inclination.[50] Two years later, Cavendish was asked to write up an account of these instruments. In his discussion of the accuracy of the several kinds of instruments, he considered both the "error of observation" and the "error of the instrument." To determine the error of the Society's earth-magnetic variation compass due to the attraction of the Society's house, he removed it to the garden of his father's house at Great Marlborough Street, out of reach of the influence of any iron. For the same reason, and also to compare differently constructed needles with one another, he removed the Society's earth-magnetic dipping needle to his father's garden. He showed how to achieve "accuracy" with the Society's meteorological instruments by attending to their indoor and outdoor placement; by applying corrections such as his corrections for the thermometer if the stem is cooler than the bulb and J. A. Deluc's corrections of the barometer for heat; by taking into account the capillary depression of mercury in the barometer tube as determined by Cavendish's father; and by taking a mean of observations. Thorpe said that Cavendish's report on the Society's meteorological instruments showed him "at his best," revealing the "range of his knowledge, his painstaking care, his sense of accuracy, his perspicacity and the thoroughness with which he studied any problem he attacked."[51]

---

[48] Henry Cavendish, "Boiling Point of Water. At the Royal Society, April 18, 1766," in Cavendish, *Scientific Papers* 2: 351–53.

[49] "The Report of the Committee Appointed by the Royal Society to Consider of the Best Method of Adjusting the Fixed Points of Thermometers; and of the Precautions Necessary to Be Used in Making Experiments with Those Instruments," *PT* 67 (1777): 816–57; the report is signed by the committee, Cavendish's name coming first. Thorpe, "Introduction," 56.

[50] The Council of the Society ordered its clerk to make daily observations of the weather "with the instruments to be procured for that purpose, & proper accommodations under the inspection of the Hon. Henry Cavendish," 22 November 1773, Minutes of Council, Royal Society 6: 194. "The following scheme drawn up by the Hon. Henry Cavendish for the regulating the manner of making daily meteorological observations by the Clerk of the Royal Society …," 9 December 1773, Minutes of Council, Royal Society 6: 197–200.

[51] Henry Cavendish, "An Account of the Meteorological Instruments Used at the Royal Society's House," *PT* 66 (1776): 375–401; in Cavendish, *Scientific Papers* 2: 112–26. The Society's variation compass incorporated a contrivance taken from Charles Cavendish's variation compass. Ibid., 120. Thorpe, "Introduction," 53–56.

The attention he gave to accuracy is illustrated by a visit he paid to the firm Nairne and Blunt to pick up a meteorological instrument. An employee having made a wind-gauge on order from Cavendish was required to be in attendance when Cavendish appeared. Cavendish "insisted upon his taking the whole apparatus to pieces. Then by means of a file and a magnifying glass, he tested the pinions to see that they were properly hardened and polished, and of the right shape, according to his written directions."[52] The accuracy of the pinions was important for the accuracy of the instrument over a wide range of wind velocities; he insured that the instrument was made correctly so he could trust to its accuracy. The clearest example of Cavendish's attention to precision and accuracy is the instrument he had built for weighing small quantities. It was the first of the great precision balances of the eighteenth century, capable of an accuracy of 5 mg.[53]

Cavendish was a perfectionist. Davy said of him, "what can, perhaps, hardly be said of any other person, that whatever he has done has been perfect at the moment of its production."[54] His precise ways were more appreciated abroad than at home.[55] Cuvier judged Cavendish's works as "so many master-pieces of sagacity and methodical reasoning; each perfect as a whole and in its parts, and leaving nothing for any other hand to correct."[56] The mathematical astronomer P. S. Laplace, who had recently taken up chemistry, said of Cavendish's chemical experiments that they were made with the "precision and finesse that distinguish that excellent physicist."[57] His colleague the physicist Biot spoke of Cavendish's "spirit of precision." Cuvier thought that Cavendish's "spirit of exactness ... benefited science as much as his individual discoveries." At the end of his life, Cavendish pursued "his passion for exactitude even in the one science where this quality has been sought most vigilantly," showing how to make astronomical instruments more accurate.[58]

In the Cavendish family, caution was a family trait, captured in the family motto *cavendo tutus*, "safe by being cautious." Thomson said that Cavendish "never advances a single opinion, which he has not put to the test of experiment." He never allows himself to "go any further than his experiment will warrant," and whatever is not "accurately determined" he states as a "conjecture," laying little stress on it.[59] According to Young, he exhibited "a degree of *caution* almost unparalleled in the annals of science, for there is scarcely a single instance in which he occasion to retrace his steps or to recall his opinions."[60]

---

[52] Mr. Newman, son of the instrument maker, quoted in Wilson, *Cavendish*, 179.

[53] In this meaning, "precision" is the limit of accuracy in decimals. Ernest Child, *The Tools of the Chemist* (New York: Reinhold, 1940), 79. Maurice Daumas, *Scientific Instruments of the Seventeenth and Eighteenth Centuries*, trans. M. Holbrook (New York: Praeger, 1972), 134–35, 221–23.

[54] Humphry Davy, quoted in John Davy, *Memoirs* 1: 221.

[55] Young, "Cavendish," 444.

[56] Cuvier, quoted in Young, "Cavendish," 444.

[57] Pierre Simon Laplace to Charles Blagden, 7 May 1785, Blagden Letters, Royal Society, L.181.

[58] Cuvier, "Cavendish," 234, 236.

[59] Thomson, *History of Chemistry* 1: 339, 348.

[60] Young, "Cavendish," 435–36.

4.6 Knowledge and Truth

Colleagues were impressed by Cavendish's breadth of knowledge. In casting a wide net, he again followed the example of his father. We do not know the full range of Charles's researches, since he largely kept them private or communicated them privately, but we know that he carried out research in nearly all parts of physical science. The president of the Royal Society Lord Morton said that Lord Charles was "eminent for his great abilities, and deep knowledge in all the branches of science that come before him."[61]

Henry Cavendish's education was "very complete," Thomson said.[62] He was "thoroughly educated in all branches of the Mathematics and Natural Philosophy," Brougham said.[63] Playfair said he "joins together the knowledge of mathematics, chemistry, and experimental philosophy."[64] These impressions are supported by what he surrounded himself with: his townhouse on Bedford Square, which he converted into a library containing books on all the subjects mentioned, and his country house at Clapham Common, which he converted into a physical and chemical laboratory and observatory. He lived among books and instruments.

In a letter initiating a short-lived correspondence, Priestley wrote to Cavendish, "Nothing I well know is so acceptable [to] yo[u] [as] truth."[65] "The love of truth was sufficient to fill his mind," the family obituary said. A common motive for entering into scientific study, a desire for distinction, glory, or power, did not apply to Cavendish, Davy said: "love of truth and of knowledge" was his sole motive.[66] Cuvier, who believed that great truths were humanity's most important legacy to history, agreed with Davy: the "disinterested love of truth was his [Cavendish's] motivating force."[67] Cavendish's colleagues agreed about what he was on earth to do: seek truth and knowledge.

Wilson said that Cavendish hated error "as a transgression of law."[68] He probably did, but more important than his hatred of errors was his understanding of them. His researches were advanced for their time for his treatment of errors. Mathematical conclusions follow necessarily from premises, but observations are true only within certain limits, or margins of error. Errors are part of the data of observations. They depend on the skill of the observer and the quality of the instrument, and their measure is accuracy and precision. Cavendish's knowledge of limits brought him as close to certainty as a natural philosopher was allowed. He stated the result of his celebrated experiment on the law of electric force within limits: the force varies

[61] 30 November 1765, Journal Book, Royal Society 25: 656.

[62] Thomson, *History of Chemistry* 1: 339.

[63] Brougham, *Lives*, 250.

[64] Playfair, *Works* 1: lxxxiv–lxxxv.

[65] Joseph Priestley to Henry Cavendish, 13 May 1784, Cavendish Scientific Manuscripts, Devonshire Collections, Chatsworth; in Jungnickel and McCormmach, *Cavendish* (1999), 593.

[66] Humphry Davy, quoted in John Davy, *Memoirs* 1: 221.

[67] Cuvier, "Cavendish," 229.

[68] Wilson, *Cavendish*, 190.

inversely as some power of the distance between 2 + 1/50 and 2 – 1/50, and because of the narrowness of the limits he concluded that there was no reason to believe that the force varies inversely as any other power than 2. He calculated the limits by analyzing how far the law could differ from 2 without the difference being perceived in the experiment. The limits were the measure of confidence he placed in the truth of his conclusion.[69] A second example is Cavendish's experiment of weighing the world, or measuring the average density of the earth. It was known that errors of the observer were greatly diminished by making repeated observations and taking their mean. Cavendish expressed the mean of his observations in the experiment together with a measure of its precision: the extreme values of the observations do not differ from their mean value by more than one fourth of the whole, and therefore "the density should seem to be determined hereby, to great exactness."[70] Cavendish incorporated error in his work by making it an object of investigation.

The number of decimal places of a measurement is part of the truth of the measurement. In Cavendish's time, the importance of significant figures was not widely appreciated. The instrument maker and astronomer James Short explained why it should be. Newton had calculated the departure of the shape of the earth from a perfect sphere as 229 to 230, the ratio of the diameters. If he had calculated the diameters to more than three decimal places, he would have claimed for the measurements on which his calculation was based a precision they did not have, a show rather than the substance of precision. It was absurd to compute a length to an inch or an angle to a second if the instrument could only measure to the foot or to the degree.[71] In stating the final results of his investigations, Cavendish did not use more decimal places than were warranted. This detail fits a pattern of accuracy, precision, and truth-seeking in his work.

Scientific investigators naturally strove to avoid making large errors. Cavendish was unusual only in the degree of precaution he took to minimize errors. In this objective, he was completely successful in the eyes of his contemporaries. He earned a reputation of never making a careless investigation or an avoidable error.[72] Priestley was more tolerant. He thought that science benefited most from the investigator who hazards "his own reputation so far as to risk even *mistakes* in things of less moment. Among a multiplicity of new objects, and new relations, some will necessarily pass without sufficient attention; but if a man be not mistaken in the principal objects of his pursuits, he has no occasion to distress himself about lesser things."[73] Cavendish could not have written that. Inseparable from truth, error was not a lesser thing with him.

---

[69] Henry Cavendish, "Experimental Determination of the Law of Electric Force," in *Electrical Researches*, 104–13.

[70] Cavendish, "Experiments to Determine the Density of the Earth," 284.

[71] James Short, "An Account of a Book, Intitled, P. D. Pauli Frisii Mediolanensis, & c. Disquisitio mathematica ..., *PT* 48 (1753): 5–17, on 5–7.

[72] It makes no difference that he interpreted his chemical researches with the help of phlogiston. Within the phlogiston theory, which was the accepted theory at the time, he made no mistakes. Like all theories, the phlogiston theory came to be replaced.

[73] Joseph Priestley, *Experiments and Observations on Different Kinds of Air*, 2d ed. (London, 1775), ix.

4.7 Publication and Writing

People have long wondered why Cavendish did not publish more of his researches. What he did publish was a small fraction of the total, and what he did not publish contained much that was important. To address the question, we begin as usual with his father, who set the precedent. Benjamin Franklin wrote of Charles Cavendish, "it were to be wished, that this noble philosopher would communicate more of his experiments to the world, as he makes many, and with great accuracy."[74] The president of the Royal Society made the same point when he presented him with the Copley Medal: because of Charles's modesty and unawareness of his great talent, the public had been "deprived" of his discoveries and inventions.[75] His only publication was the Copley Award paper, which described his self-registering maximum and minimum thermometers, useful in tracking the weather.

Among the leading members of the Royal Society to whom Charles Cavendish introduced his son, there was wide disparity in their habits of publication. We see this in the guest list of a dinner hosted by Charles at his home. On one night, there were eight guests, all professional men in middle age, sometime members of the Council of the Royal Society and one of them currently secretary. They were good friends with common interests, including a serious interest in science. Half of them were primarily interested in antiquities, as was the membership of the Royal Society at large. Only one of the guests had published extensively in the Society's journal, though another guest was about to begin publishing a series of papers there, and two of the guests had published four papers each there. The remainder of the guests were, like Charles himself, one-paper men, though some had published extensively on nonscientific subjects outside the journal.[76] Henry did not happen to be at this dinner, but he was at other dinners with the same company. Their scientific activity was familiar to Henry when he set out on a lifelong quest to understand the regularities of nature. Publication was a valued contribution, but it was an individual choice, not the driven need that was later to attach to scientific activity. What he made of the example, we can only surmise from his own habits of publication.

In some cases, the answer to why Cavendish did not publish more of his researches is obvious: the papers he left contained preliminary results that he later incorporated in publications. In other cases, the papers dealt with side issues, questions that came up in the course of his main investigation. Sometimes these papers had original ideas, which if Cavendish had developed them would have been important contributions. In still other cases, the papers contained complete studies that wanted only to be written in final copy to be ready for publication.[77]

---

[74] Benjamin Franklin to Ebenezer Kinnersley, 20 February 1762, *The Papers of Benjamin Franklin*, vol. 10, ed. L. W. Labaree (New Haven: Yale University Press, 1966), 37–53, on 42.

[75] 30 November 1757, Journal Book, Royal Society 23: 638–39.

[76] 21 October 1758, Thomas Birch Diary, British Library, Add Mss 4478C.

[77] Thorpe, in Cavendish, *Scientific Papers* 2: 297.

Curiosity about Cavendish's reluctance to publish is closely tied to what has been seen as his anticipation of ideas and discoveries credited to other scientists. On this count, he has exerted a perpetual fascination over scientists and others who know about his work. We see it in his biographies, in the commentaries by his editors, and in historical writings. Here are some examples. Around the same time as Joseph Black, possibly a little later, but in any case independently, Cavendish arrived at the laws of specific and latent heats. He wrote a major paper about his experiments and conclusions on the subject, but he did not complete it for publication. Wilson speculated that the reason was fear of rivalry with Black, but in truth, he said, "with Cavendish, publication was the exception, not the rule," and so no "special hypothesis" is needed to explain why he did not publish his early work on heat. Thorpe agreed: "it is probably useless to surmise," for "Cavendish was no ordinary man, and ordinary conventions and rules of conduct are inapplicable to him."[78] Thorpe did sometimes surmise, but more often he wondered. He wondered, for example, why Cavendish did not publish his early paper on arsenic acid, since he surely knew it was original. If he had, he would have anticipated Carl Wilhelm Scheele's memoir in 1775, and moreover his method was superior and more accurate.[79] Thorpe said that Cavendish's early paper on tartar was remarkably accurate, and that by clarifying the subject it would have made a notable contribution to chemistry, whether or not he had been anticipated by Scheele, and Thorpe thought he had not. But Cavendish was "never in a hurry to publish," Thorpe said; his only need was to "satisfy the questioning of his own intelligence."[80] Cavendish omitted from his publication on factitious air a part about air distilled from vegetable and animal matter. In it he said he had not "made sufficient experiments to be quite positive" about a certain point, a possible reason he withheld it. The air he distilled was in reality a mix of several gases, not all of them known. Still he showed that the inflammable airs from the solution of metals in acids and from the distillation of charcoal are different substances, an important conclusion. If he had published it, Thorpe said, Priestley might not have made the mistake that led indirectly to the water controversy.[81] Cavendish carried out a very thorough investigation of the elastic force of water vapor above and below 212°. The work was entirely original, but again Cavendish did not write it up for publication. Later John Dalton published a book of much less accurate tables, which were long regarded as the authority on the subject. Thorpe speculated that Cavendish did not publish this work because did not want to come into conflict with Deluc, whose theory of the thermometric scale his experiments contradicted.[82] At around the same time, Cavendish carried out analogous experiments on the expansion of common air with heat. It had been studied by others, who had achieved widely varying results, owing in part to the unsuspected effect

[78] Thorpe, ibid., 2: 327. Thorpe quotes Wilson.

[79] Ibid., 301.

[80] Ibid., 304–5.

[81] ibid., 316.

[82] Ibid., 371.

of moisture in the air or in the apparatus. Cavendish recognized that the pressure of the water vapor needed to be added to the pressure of the air, anticipating Dalton's later discovery. In a carefully written paper on mechanics, "Remarks on the Theory of Motion," Cavendish derived a law of central importance in physics in the next century and beyond. Larmor, editor of Cavendish's papers on physics, said of it, "This surely is the earliest precise enunciation of the principle of the conservation of energy, kinetic and potential, including enumeration of the causes that lead to its degradation, which on the principles of [a corollary in this paper] would be into heat of precisely equivalent amount."[83] Frank Watson Dyson, an astronomer and editor of Cavendish's astronomical manuscripts, wrote to Thorpe that he found the papers "extremely interesting, and one cannot help regretting that he did not publish more."[84] The above and many more anticipations by Cavendish are singled out by other commentators.[85]

The main literary occupation of Maxwell for last 5 years of his life was his edition of Cavendish's electrical researches. This edition is "without parallel," there being no other instance of a major scientific investigator devoting "such profound and extensive efforts to the elucidation of the unpublished work of a predecessor in his own subject."[86] In the course of preparing the edition, Maxwell wrote to a friend in 1874:

> In the MS. he appears to be familiar with the theory of divided currents and also of conductors in series.... He made a most extensive series of experiments on the conductivity of saline solutions in tubes, compared with wires of different metals, and it seems as if more marks were wanted for him if he cut out G. S. Ohm long before constant currents were invented. His measures of capacity will give us some work at the Cavendish Laboratory, before we work up to the point where he left it.[87]

Maxwell wondered why Cavendish did not publish his experiments on the charges of bodies, even though he had drawn up an account in almost fair copy. He believed the reason might be that Cavendish had returned to chemistry, but he also had a general thought about it. Like Wilson and Thorpe, Maxwell saw Cavendish not as a normal man but as one who lived outside the conventions of society, and lacking the desire of normal scientists to communicate their work, "Cavendish cared more for investigation than for publication. He would undertake the most laborious researches in order to clear up a difficulty which no one but himself could appreciate, or was even aware of, and we cannot doubt that the result of his inquiries, when successful, gave him a certain degree of satisfaction."[88]

---

[83] Larmor in Cavendish, *Scientific Papers* 2: 428.

[84] Frank Watson Dyson, ibid., 433.

[85] Examples are the chapter "Cavendish and Modern Science," in Berry, *Cavendish*, 175–91; and a summary of a number of anticipations in James Gerald Crowther, *Scientists in the Industrial Revolution: Joseph Black, James Watt, Joseph Priestley, Henry Cavendish* (London: Crescent Press, 1962), in Pearson, *Serpent and Stag*, 120.

[86] Crowther, *Scientists in the Industrial Revolution*, 304.

[87] Maxwell's letter to W. Garnett, in July 1874, quoted by Larmor, "Preface to Volume 1," in *Cavendish Scientific* Papers 1: vii.

[88] Maxwell, *Electrical Researches of the Honourable Henry Cavendish*, xlv.

Maxwell's explanation joins a number of other general explanations of Cavendish's habits of publication. One of these we owe to Blagden, and because of his closeness to Cavendish, we give it weight. After Cavendish died, Blagden and several others examined his manuscripts to see if he had prepared any of them for printing. Finding none, Blagden was not surprised, for "Mr Cavendish was always ready to publish whatever he had made out to his full satisfaction." He published only such work as met his high standards.[89] Another explanation is that because he disliked controversy he was reluctant to expose himself. True, he did not like controversy, and on occasion he may have chosen not to risk it, but there were times when there were no rivals and he did not publish. "The real reason," according to a related explanation, "must be attributed to Cavendish's morbid shyness and timidity."[90] This is a possibility, but he was shy in company, not in his writing. Another reason is that he did not care what others thought.[91] He may have been indifferent to what people thought of the way he lived, but he cared what colleagues thought; he answered criticism, and he was concerned when others could not repeat his experiments. Another reason is that he looked at his discoveries the way he looked at his wealth, as his to do with as he saw fit, expressing a social outlook; and because of his class and wealth, he lacked the common inducements to bring his work before the world, and that for the same reason he was set apart from the scientific men of the Industrial Revolution who might have encouraged him.[92] His social outlook was probably a factor, but not an overriding one. There were a number of aristocrats in the Royal Society who, like his father, published very little or nothing at all in science but who devoted time and energy to the Society. Cavendish too devoted time and energy but he differed from them in that he published works repeatedly, though not all he might have. Another explanation is that Cavendish was "wholly indifferent to scientific fame"; "he disliked notoriety; he was, as it were, fearful of the voice of fame."[93] It is correct that any attention to his person made him uncomfortable, and he was not motivated by the desire for fame, but this does not explain why he published some things and not others. Another explanation is that he sometimes lacked the drive to complete the writing of a final draft. Preliminary drafts of a number of his unpublished researches look as though he originally had publication in mind but did not follow through, as though "literary composition was irksome to him."[94] This is a plausible explanation in some instances. Although not

[89] Charles Blagden to Lord George Augustus Henry Cavendish, n.d., Blagden Collection, Royal Society, Misc. Matter – Unclassified.

[90] Berry, *Cavendish*, 21.

[91] Thorpe said Cavendish "cared little for the judgment and opinion of his fellows." "Introduction," 6. A similar observation was made by Pierre Lépine and Jacques Nicoll, *Sir Henry Cavendish: l'homme qui a pesé la terre* (Paris: Seghers, 1964). This is pointed out by John Heilbron, who disagrees with it. *Electricity in the 17th and 18th Centuries: A Study of Early Modern Physics* (Berkeley: University of California Press, 1979), 489.

[92] Crowther, *Scientists in the Industrial Revolution,* 318.

[93] Thorpe in Cavendish, *Scientific Papers* 2: 6. Humphry Davy, quoted, ibid. 2: 5. Wilson, *Cavendish*, 173.

[94] Thorpe, ibid.

expressly directed at Cavendish, Priestley gave a general explanation for withholding publication: "for the sake of a little more reputation," he said, some investigators hold back their work, which may have questionable merit, under the illusion that when it is ready, it will "astonish the world with a system as complete as it is new," and they are justly punished when their work is anticipated by other investigators who properly "immediately communicate to others whatever occurs to them in their inquiries."[95] The reason Priestley proposed, vanity, was foreign to Cavendish's character, but Priestley was right about the consequences of silence. If Cavendish had communicated his findings on the composition of water sooner, science would have advanced a little sooner, and he might have avoided a nasty priority dispute, the water controversy.[96]

Secrecy is not a plausible explanation. Cavendish understood perfectly the harm that it brought on persons who practiced it. When a colleague asked him to keep a new method in astronomy secret for a time, he replied that he was "sorry" he wished him to do this. "The surest way if securing merit to the author is to let it be known as soon as possible & those who act otherwise commonly find themselves forestalled by others."[97] His colleague agreed and allowed him to make the method public right away; "indeed the more people see it the better."[98] When the government considered imposing secrecy on a colleague's experiments on alloys for coins, Cavendish persuaded the government to allow his colleague to publish his experiments.[99] When asked his opinion on the author of a pamphlet about dissensions in the Royal Society who wished to remain anonymous, Cavendish "answered at once & decisively that the only way to make it produce any useful effect was for the author to sign his name";[100] the author agreed with Cavendish, and released his name. When a foreign colleague accused Cavendish of withholding information he asked for to repeat an experiment of his, Cavendish published the letter he had sent the colleague so that readers could judge for themselves the fairness of the complaint, saying he "should be sorry to be thought to have refused any necessary information."[101] The Royal Society, to which he gave his services unstintingly, stood for the public exchange of

---

[95] Joseph Priestley, Preface to *Experiments and Observations on Different Kinds of Air*, 2d ed., 1775; quoted by Thorpe, "Introduction," 6.

[96] Thorpe, ibid. But Wilson made the case that Cavendish delayed publication for the reason that he had not finished with his research, and as soon as he was finished he published. *Cavendish*, 442–44.

[97] Henry Cavendish to John Michell, 27 May 1783, draft; in Jungnickel and McCormmach, *Cavendish* (1999), 567–69, on 567.

[98] John Michell to Henry Cavendish, 2 July 1783; ibid., 570–78, on 570.

[99] Lord Liverpool, head of a select committee on coins, was satisfied with Cavendish's opinion that the material could be published. Henry Cavendish to Charles Hatchett, 15 October 1802; Charles Hatchett to Joseph Banks, 24 October 1802, British Library, Add Mss 38424, f. 160.

[100] Charles Blagden to Joseph Banks, 24 and 26 October 1784, British Museum (NH), DTC 3: 83–86.

[101] Letter from Henry Cavendish to Martin van Marum, included in Henry Cavendish, "On the Conversion of a Mixture of Dephlogisticated and Phlogisticated Air into Nitrous Acid," *PT* 78 (1788), 261–76; in Cavendish, *Scientific Papers* 2: 224–32, on 231–32.

knowledge, and he offered his very large collection of books as a lending library. Clearly Cavendish was a warm advocate of openness.

Of the proposed general explanations, the most satisfactory is that the first object of Cavendish's researches was, as Thorpe put it, to "satisfy his own spirit of inquiry."[102] This was Maxwell's view above. When Cavendish brought himself to publish, he seemed to perform "not what was a duty to himself, but what was a duty to the public," Davy said.[103] But Cavendish's researches were not "aimed at his personal gratification alone," Larmor said, otherwise "they could not have been recovered so completely." Larmor continued this thought: Cavendish devoted his life to research "as the one means of ensuring an adequate foundation for sciences then being born."[104] It is significant that Cavendish devoted himself equally to the work of the Royal Society as to his researches, and the object of the Royal Society was to advance the understanding of nature. Once Cavendish convinced himself of a truth, he might put off its time-consuming preparation, diverting him from the search for further truth, but he was a member of a scientific body whose objectives he fully shared.

In Cavendish's way with publications, we see a tension between the private and the public man, the same tension we see in his appearances in society. In the absence of a fully satisfactory general explanation, the question of why he decided to publish or not to publish certain researches can be looked at on a case-by-case basis. He may have found an imperfection or incompleteness in the work, or another subject may have diverted his attention and he did not return to the work.

William Watson took to task an experimenter who allowed that by his "style and expressions" he had "embellished a little" his account of an experiment. "The language of philosophers should not be tainted with the license of the poets," Watson said; "their aim in the communicating their discoveries to the world, should be simple truth without desiring to exaggerate," for nature herself was cause enough for admiration.[105] This admonition echoes the original statutes of the Royal Society: "in all reports of experiments to be brought into the Society, the matter of fact shall

---

[102] Thorpe, "Introduction," 6.

[103] Quoted in John Davy, *Memoirs* 1: 221–22. Lidbetter gives a psychological explanation: because Cavendish was autistic, he did not spontaneously share his enjoyment, interests, and achievements with others. If Cavendish was autistic, this is a credible general reason for Cavendish's habits of publication. Lidbetter misunderstands what Christa Jungnickel and I say in our Cavendish biography where he says we explain why Cavendish held back from publication by Cavendish's" views on the inadequacy of language." That is not what we say. In a discussion of Cavendish's taciturnity we say that words, as used in normal *speech*, do not adequately represent Cavendish's world; for that mathematics and quantities are essential. Publications are, of course, exactly where quantities and mathematical reasoning are proper and necessary. My co-author and I offered a suggestion arising from his work bearing on his habits of speech, not of publication. Lidbetter, "Henry Cavendish," 786. Jungnickel and McCormmach, *Cavendish* (1996), 370–71.

[104] Larmor, "Preface to Volume 1," ix.

[105] Watson, "A Letter… Declaring He as Well as Many Others Have Not Been Able to Make Odours Pass thro' Glass," 355–56.

Fig. 4.7 Cavendish's Handwriting. This close of a letter could hardly be simpler, in writing and wording (Reproduced by permission of the Chatsworth Settlement Trustees)

be barely stated, without any prefaces, apologies, or rhetorical flourishes."[106] Cavendish's writing followed the rule: it was spare, there was not a word too many, only the "simple truth." "Every sentence will bear the microscope," Brougham said of it.[107] Davy said that Cavendish wrote with the "greatest dignity and simplicity and in the fewest possible words, without parade or apology."[108] Young said that Cavendish's scientific publications were "expressed in language which affords a model of concise simplicity and unaffected modesty."[109] The way Cavendish formed letters of the alphabet on the page agreed with the way he formed sentences: the letters were clear and free of all embellishment. Cavendish's handwriting belongs with his traits of accuracy and precision (Fig. 4.7).

Young's "concise simplicity" applied to Cavendish's ways of proceeding in general. Cavendish did not desire any special form of address. His library stamp was typically simple and plain, his name only. He cared nothing for the appearance of instruments and apparatus, never using hardwood unless it was required, preferring plain fir.[110] Blagden told a colleague that Cavendish not only had great scientific ability but as well "a truly philosophical simplicity of manners."[111]

Newton, whose writings Cavendish studied closely, wrote in his optical treatise that "Nature will be very conformable to herself and very simple."[112] Newton associated simplicity with truth: "Truth is ever to be found in the simplicity, and not in the multiplicity and confusion of things."[113] Cavendish, in a plan of a treatise on mechanics, wrote that Newton's second law of motion is "the most simple & therefore the most likely to be true of any law one can invent."[114] Truth is simple, nature is simple; simplicity was Cavendish's preference in all things.

---

[106] Quoted by Thorpe, "Introduction," 6.

[107] Brougham, *Lives* 1: 257.

[108] Humphry Davy, quoted in Cavendish, *Scientific Papers* 2: 5–6.

[109] Young, "Cavendish," 436.

[110] Wilson, *Cavendish,* 178.

[111] Charles Blagden to William Cullen, 7 June 1784, draft, Blagden Letterbook, Yale.

[112] Isaac Newton, *Opticks; or A Treatise of the Reflexions, Refractions, Inflexions and Colours of Light*, 4th ed., 1730, reprint (New York: Dover Publications, 1952), 372.

[113] Quoted in Frank Edward Manuel, *The Religion of Isaac Newton* (Oxford: Clarendon Press, 1974), 120.

[114] Henry Cavendish, "Plan of a Treatise on Mechanicks," Cavendish Scientific Manuscripts, Devonshire Collections, Chatsworth, VI(b), 45: 17.

4.8 Scientific Society

However attractive a life of solitude, it could not be a complete life for Cavendish. To practice natural philosophy, he needed to engage with men of science, to enter into scientific society. Given his painful shyness, the question arises of how he managed it. Part of the answer is his father, who included science in his private life. One evening, for example, Charles Cavendish invited seven "Brethren of the Royal Society" to a "small Party," during which he supplied a "philosophical Entertainment of an artificial Frost by a Solution of Sal Ammoniac in common Water," followed by "a very good Dinner."[115] Soon after Henry ended his studies at Cambridge, his father began including him in dinners like this, where he became familiar with influential members of the Royal Society. Henry took his baby steps in scientific society in the least threatening setting, his home, and in the homes of persons he met there.

It was common for fellows of the Royal Society to bring their sons as guests and to introduce them to other members before they were proposed and balloted for membership. Charles brought Henry as his guest for the first time in 1758, by which time he had introduced him to many of the more active members of the Society at private dinners. He brought his son as a guest a total of 15 times, and his friends brought him 3 more times. At about half of these meetings, he presided. The shy Henry entered scientific society with the support of a highly-respected member, his father.[116]

In 1760, Henry was recommended for membership on a certificate drawn up by the prominent physician William Heberden and signed by the astronomer royal James Bradley, the president of the Royal Society George Parker, earl of Macclesfield, and Hugh Willoughby, baron Willoughby of Parham, described as an "ingenious and learned Nobleman," who like his good friend Charles Cavendish served as vice-president of the Royal Society. These were senior members who represented both aristocratic rank and scientific judgment, an impressive endorsement. His certificate was posted, and Heberden and six other members added their signatures. All of them were members of Charles's dining circle, which now included Henry. As for Cavendish's qualifications, the certificate read, "Having a great regard for Natural Knowledge & being studious of its improvement …."[117] It is a good description of his life, complete in its way.

Around the time Henry began attending meetings of the Royal Society as a guest of his father's, he was recommended for membership in the Royal Society Club by its president Lord Macclesfield. This took place at a dinner of the Club attended by his father. There was no need for him to make himself known to the members, since they knew him through his father. He was elected, but because the Club had a fixed membership, he had to wait until there was an opening. In the meantime, he was invited to

[115] Thomas Birch to Philip Yorke, 18 August 1750, British Library, Add Mss 35397.

[116] Guests and members who brought them are recorded in the Journal Books kept at the Royal Society Library.

[117] Cavendish was proposed 31 January 1760. Certificates, Royal Society, vol. 2, no. 10, f. 198.

dine with the Club repeatedly as a guest of his father's, and he was treated as though he were a regular member. In 1760, 2 months after he was elected to the Royal Society, he formally became a member of the Club. He was 29. After that, he sometimes accompanied his father to dinners of the Club, but his father did not come regularly any longer. Charles had accomplished what he had set out to do: to ensure that his son was accepted into an association of men who shared his interest. As a member of the Royal Society and Royal Society Club, Henry could move with a degree of assurance in the public world of science (Figs. 4.8, 4.9, 4.10, 4.11, 4.12, 4.13, and 4.14).[118]

Henry's father was his mentor as a scientific administrator as well as a scientific investigator. Charles Cavendish was regularly elected 1 of the 21 members of the Council of the Royal Society. In 15 of the 16 years he served on the Council between 1748 in 1761, he attended 106 meetings, missing only 9, several of which absences were excused by a life-threatening accident of his younger son. His last year as Council member was 1769, when he served together with his son Henry.[119] His commitment was uncommon, given that the usual number in attendance was between a half-dozen and a dozen. Only the two secretaries of the Society attended more frequently, and they had no choice short of dereliction of duty. Charles served on the committee that accepted papers for publication in the Society's journal. From its beginnings in 1752, with the exception of 1 year, he attended every meeting of the committee in the years he was active in the Society. In addition, as a vice president of the Society, he chaired meetings when the president was absent, as he frequently was. When a president resigned or died, Charles was the obvious candidate to replace him, but he made it known he would not accept the honor. In 1768, for example, when the presidency was vacated by Lord Morton, the Royal Society's expert on magnetism Gowin Knight wrote to a colleague, "L[d] Charles declares absolutely against accepting it; tho' much sollicited to it by his Friends."[120] During his long service in the Society, he preferred to second a motion rather than propose one, to support rather than to lead. He was the unassuming, expertly capable servant of the Society who could be depended upon without fail. Following his father's example, Henry played the same role in the Society.

It was traditional for an English aristocrat to recognize a duty of public service, normally taking the form of a career in politics or the military or occasionally in the church. Charles Cavendish performed his duty in politics as a Member of Parliament. He also performed it in learned institutions. The remarkable feature of his service in the Royal Society was the year-in year-out regularity with which he assisted in its affairs.

---

[118] The record of the dinners and business of the Club at this time is kept in the Minute Book, Royal Society Club, Royal Society, 4 (1760–64). Archibald Geikie, *Annals of the Royal Society Club: The Record of a London Dining-Club in the Eighteenth and Nineteenth Centuries* (London: Macmillan, 1917), 63.

[119] The record of attendance is given in the Minutes of Council, Royal Society.

[120] Gowin Knight to Benjamin Wilson, 28 October 1768, Benjamin Wilson Papers, British Library, Add Mss, 3094. The importance the Royal Society placed on social rank is brought out in this letter. Knight says that the duke of Northumberland declined to be considered for the presidency, and he asked Wilson to urge his friend Sir George Savile to accept it. Savile was an important Whig politician who was interested in science, but he was not a scientific investigator.

Fig. 4.8 Royal Society. Painting by Frederick William Fairholt, engraving by H. Melville. This is the meeting room of the Royal Society at Somerset House 1780–1857. Over the last 30 years of his life, Cavendish came regularly to meetings here. The president of the Society is at the center and elevated. The two secretaries of the Society sit below him at either side. The paintings on the wall are of distinguished members. It would seem that Cavendish did not want to join them, for he resisted appeals to sit for a painting (Reproduced by permission of the President and Council of the Royal Society)

Fig. 4.9 Crown and Anchor. 1743. Every Thursday Cavendish dined with the Royal Society Club at this tavern on the Strand at the top of Arundel Street (Courtesy of Westminster City Archives)

Fig. 4.10 Sir Charles Blagden. Etching from the portrait by Thomas Phillips. Secretary of the Royal Society and Cavendish's associate in the 1780s (Courtesy of the President and Council of the Royal Society)

Fig. 4.11 Sir Joseph Banks. Painting by Sir Thomas Lawrence, engraving by H. Robinson. President of the Royal Society (Courtesy of the Trustees of the British Museum)

Fig. 4.12 Thomas Young. Painting by Sir Thomas Lawrence, engraving by G. Adcock. Natural philosopher and Cavendish's colleague at the Royal Institution (Courtesy of Smith Image Collection, Van Pelt-Dietrich Library, University of Pennsylvania)

Fig. 4.13 Sir Humphry Davy. Painted by Lonsdale, engraved by Worthington. Chemist and Cavendish's colleague at the Royal Institution (Courtesy of Smith Image Collection, Van Pelt-Dietrich Library, University of Pennsylvania)

Fig. 4.14 Distinguished Men of Science. Engraving by William Walker around 1862, from a drawing by Sir John Gilbert. The full title is "Distinguished Men of Science Living in Great Britain in 1807–8." The setting is the library of the Royal Institution, but the men shown in the print never gathered in this room. Rather the artist created the group from individual portraits. Henry Cavendish is placed in the front, sitting apart, his eyes downcast; perhaps this is the artist's interpretation, and if so it is true to life, Cavendish's solitude in company. Cavendish's profile and dress are based on William Alexander's sketch, with obvious differences: Cavendish's hat is removed; he is seated instead of walking; he faces the other direction; and he is made to appear 30 years younger. Cavendish was a manager of the Royal Institution from 1800 (Wikimedia Commons)

Henry likewise recognized a duty of public service. He was elected to the Council of the Royal Society for the first time in 1765, and for the next 20 years he was on the Council about half the time. In the last 25 years of his life, he was on the Council every year. Like his father again, he regularly served on committees appointed by the Council, including the committee of papers. He rarely missed a Council or committee meeting. He was unlike his father only in that he did not become a vice president, a departure we would expect.[121] His record of attendance was repeated at the Royal Society Club, which met every week of the year, without recess. In 1760, his first year as a member, he came to half the dinners. After that, he came to nearly every one of them, a 50-year record without parallel. In 1809, the last full year of his life, he attended 51 dinners, and in January and February of 1810, the last months of his life, he came every time except for the last 2 weeks.

Charles and Henry Cavendish were both elected trustees of the British Museum, a national library and natural history collection, with close connections to the Royal Society. The library was an adjunct to the collection, the purpose of which was the advancement of "Natural Philosophy and Other Branches of Speculative Knowledge." Charles was appointed to the standing committee of the trustees, which was responsible for the management of the Museum, and after Henry became a trustee he likewise was appointed a member. For 10 years he came with his father to its biweekly meetings, as well as to meetings of the trustees. Their record of attendance was exemplary. Rarely as many as six members turned up at committee

[121] From Minutes of Council, Royal Society, 5–7 (1763–1810).

meetings, and often not enough members came to meetings of the trustees to form a quorum, but usually the Cavendishes were among the few, Charles frequently presiding. After his father died, Henry continued to serve until the end of his life.[122]

Cavendish and his father were members of the Royal Society of Arts, which offered prizes to promote general progress, and Henry was a member of the Society of Antiquaries. Henry took no active part in these societies. It was otherwise with the Royal Institution, an organization dedicated to teaching the applications of science and to spreading knowledge of inventions. Henry was one of its nine managers, and by far the most conscientious. At their frequent meetings, commonly only three of four managers attended, Cavendish invariably one of them. He was also a conscientious member of the scientific committee of council, which oversaw the syllabus and experiments. The membership of the Royal Institution had considerable overlap with that of the Royal Society, which for him was the base membership and his main connection with scientific society.

Cavendish belonged to several clubs that met in coffeehouses in the city. The most formal of them was the Royal Society Club, which because of its close association with the Royal Society was first among Cavendish's clubs. What all the clubs had in common was regularity – they met at a fixed time and place, and their members were known to one another – and interest, which was scientific.

4.9 Talent and Recognition

Cavendish had a talent for natural philosophy. In a letter to a colleague in 1785, another talented natural philosopher John Michell gave his opinion of him:

> I never saw any thing yet, that came from his [Cavendish's] pen, or heard any thing from him in conversation, that appeared to me to have a weak place in it, that I recollect; & however I may sometimes differ from him in opinion, which I think has very seldom happen'd, when I have heard his reasons for his opinion, and he has always appear'd to me to possess so much sound judgment, & good reasoning, as well as general knowledge in philosophical subjects, that I know no one, on whose authority, I should be so ready to receive any opinion implicitly, where I had no means of judging it myself.[123]

Playfair said that most fellows of the Royal Society regarded Cavendish as "one possessed of talents confessedly superior."[124] J. G. Children, who knew him from the Royal Society Club, said that he "was regarded by all as a great authority."[125] Likewise,

---

[122] Cavendish was elected trustee of the British Museum on 8 December 1773. The record of his attendance is in the minutes of the British Museum: Committee, volumes 5–9; General Meeting, volumes 3–5.

[123] John Michell to Charles Blagden, 27 July 1785, American Philosophical Society, Misc. MS Collection; in Russell McCormmach, *Weighing the World: The Reverend John Michell of Thornhill* (Dordrecht, Heidelberg, London: Springer, 2012), 386–94, on 391.

[124] Playfair, *Works* 1: lxxxiv.

[125] Children, quoted in Wilson, *Cavendish*, 169.

Young recalled the "profound respect" in which he was held, owing to the "decided superiority of his faculties."[126] On the occasion of Cavendish's death, Davy gave a memorable evaluation: "He was acute, sagacious, and profound, and, I think, the most accomplished British philosopher of his time."[127] Blagden agreed: writing to Banks from Paris, he compared Cavendish with "Laplace, who is as much superior among them here as Mr Cavendish is with us." Several years later, at Cavendish's death, Blagden wrote to a correspondent in Paris that Cavendish was "by much the best philosopher in my opinion that we have, or have had, in my time, at the R[oyal] S[ociety]."[128] Davy compared Cavendish with the greatest of them: "Since the death of Newton, if I may be permitted to give an opinion, England has sustained no scientific loss so great as that of Cavendish."[129] Upon learning of Cavendish's death, John Walker described him to a colleague as "a man of wonderful mind, more nearly approaching that of Newton than perhaps any individual in this country since his time."[130]

Young speculated on what the benefactor of science received from science in return: Cavendish "must have enjoyed the highest and purest of all intellectual pleasures, arising from the consciousness of his own excellence," compounded by the certainty that posterity would remember and venerate him.[131] Cavendish's talent in science brought him to the attention of the world, and it eased his way, but from what we know of him it is likely that his talent was its own reward.

Talent, a natural aptitude, requires intelligence, but over and above intelligence, personality lies behind high achievement. Introversion and intensity are of the "essence," according to the psychologist Raymond Cattell, directing the individual away from human relations to enter "an orderly, impersonal world, such as science." The epitome of this type of individual is "Cavendish hiding from society in a remote wing of his mansion, Newton forever wandering on 'strange seas of thought alone', Einstein remote in the patent office library, taking his solitary walks in the woods."[132]

4.10 Way of Life

For all intents and purposes, Cavendish's life was his work, and all of his work was related to natural philosophy. The several institutions and clubs he belonged to were all related to natural philosophy, as were all of the people he had meaningful

126 Young, "Cavendish," 444.

127 Humphry Davy, quoted in Wilson, *Cavendish*, 167.

128 Charles Blagden to Joseph Banks, 1 April 1802, British Library, Add Mss 33272, pp. 172–73. Charles Blagden to B. Delessert, 20 March 1810, Blagden Letters, Royal Society, D.44g.

129 Humphry Davy, quoted in John Davy, *Memoirs* 1: 222.

130 John Walker to James Edward Smth, 16 Mar. 1810.

131 Young, "Cavendish," 445.

132 Raymond B. Cattell, *Abilities: Their Structure, Growth, and Action* (Boston: Houghton Mifflin, 1971), 410–11.

exchanges with. His father provided an example and support, but he himself had to want to become a natural philosopher. Instead of entering one of the professions, for which he was not suited, he chose natural philosophy, for which he obviously was. Scientific research was not then a profession; no career was laid out for him, and instead he invented an equivalent that was right for him. In place of "career," the expression "a life of natural philosophy" better describes what Cavendish had.

The philosopher John M. Cooper has done some of our work for us. With him as our guide, we take a brief detour into the history of philosophy. Since antiquity, philosophy has taken two main directions. One is theoretical, going back at least to the Ionian philosophers, who explained the world by natural causes. The origin of natural science is traced to these philosophers. The other direction originated with Socrates, who was uninterested in philosophy as science, and who instead looked to philosophy as a way of life. He spent his time in philosophical discussion in pursuit of wisdom about human nature and human good, and he followed philosophical principles in all his daily activities. The theoretical direction of philosophy predominates today, and because of its historical and ongoing connections to natural science, it has a modest bearing on our understanding of Cavendish. The other direction, philosophy as a way of life, is out of fashion today, though a part of it continues as a subfield of philosophy, moral philosophy; it turns out that philosophy as a way of life may be of help to us.[133]

Cooper would partly revive the Socratic direction, and in his 2011 Locke Lectures and in his subsequent book *Pursuits of Wisdom* in 2012 he tells how. The Stoics and Epicureans, for example, had very different ideas, but they agreed on what it meant to be a Stoic or an Epicurean: it was to live according to his "philosophical worldview." They and other Greek philosophers shared three basic ideas. (1) Philosophical reasoning and investigation are the "constant focus" of one's life and an inherently good activity, available to us because of our natural capacities. (2) Philosophy is the authority on how to learn about human nature and virtue, on how to live best in all of one's activities, choices, and actions. (3) Thorough knowledge of one's philosophy provides the "psychological motivation" for living the "life of philosophy." Reason has within itself the power to motivate by informing the person that a chosen action will bring about something good. A philosophical knowledge of one's own good is a "sufficient and invincibly powerful psychological motivation" to live according to one's philosophy.

Let us consider Cavendish in light of ideas shared by Greek philosophers and see how they line up. (1) "Philosophy" for Cavendish was "natural philosophy." (At Cambridge when he was a student, philosophy was interpreted as natural philosophy in the curriculum.) Natural philosophy, with its reasoning and enquiries, was the "constant focus" of Cavendish's life. (2) Natural philosophy was the "authority" for Cavendish on all questions having to do with the physical world. It was the authority in his private and social life only by analogy: observation and

[133] John M. Cooper, "Ancient Philosophies as Ways of Life," *Oxford Philosophy* (summer 2012), 16; *Pursuits of Wisdom: Six Ways of Life in Ancient Philosophy from Socrates to Plotinus* (Princeton, Oxford: Princeton University Press, 2012), 2–4.

reason, the mental tools of natural philosophy, guided him in all aspects of life. Because he kept his distance from people, he would have escaped some of the moral complexity normal people encounter in living. Wilson put it bluntly: morally his character "was a blank." This judgment is too absolute. We have seen that Cavendish had a sense of justice, and he acted on it. Blagden said that Cavendish had the "most amiable candor" and the "strictest integrity."[134] When Blagden considered a journey abroad and wanted to know if it would inconvenience Cavendish in any way, he said, "Now I trust to the strict principles of <openness> sincerity by which I know you are always guided" to tell him if he had any objection to his going.[135] Justice, candor, integrity, and sincerity are words that enter our moral vocabulary. The practice of natural philosophy would be impossible without these same virtues. (3) Natural philosophy contained the "psychological motivation" to act in certain ways. To use a common word of the time, natural philosophy was open to "improvement." Improvement was the object of scientific societies. The natural philosopher differed from a person who took an interest in science but did nothing more; the natural philosopher acted to improve it. On the first and third points, natural philosophy would seem to meet the criteria of a philosophical way of life. It is deficient in an obvious way on the second point – it does not contain rules on how to live – but let us continue.

Ancient philosophers believed that reason and the truths that reason alone could discover were the basis of philosophy, and that philosophy was the basis for all of life, not just of moral actions.[136] More influence over our actions was granted to reason in Cavendish's day than in ours, but it was known that the discovery of truths could not be the sole motivation to act. There were other well-known motives. However dominant a role reason played in Cavendish's life, the philosophical life in its pure form, as the ancients understood it, was not open to him.

Cooper introduces a limited idea of philosophy as a way of life into the present. It is meaningful to speak of a "distinctive way of life of physics," the practice of physics constituting the "central complement of a person's life," but this is not a "philosophical" way of life in the Greek sense. It holds reason in high regard, but it is little help in matters of daily life and moral decisions. What a way of life of physics shares with the ancient way is its offer of the good life. We have every reason to think that natural philosophy offered Cavendish the good life. With due qualifications, we may say that Cavendish lived a way of life of natural philosophy.

Natural philosophy provided Cavendish with his alternative to the world in which most people spend most of their time. It presented challenges of a kind that were right for him, ones that fully engaged his mind and steered clear of emotional traps. He built his way of life from elements of that alternative, insuring that his way was at once intellectually fulfilling and protective of his psyche.

---

[134] Charles Blagden to William Cullen, 17 June 1784, draft, Blagden Letterbook, Yale.

[135] Charles Blagden to Henry Cavendish, August 1789, draft; in Jungnickel and McCormmach, *Cavendish* (1799), 666–67.

[136] Cooper, *Pursuits of Wisdom*, 6.

Cavendish showed “a strongly developed will,” Wilson said. His observation bears on the present subject. Cavendish came by his interest in the physical world by several converging paths: his father, his university, his native intelligence, talent, and temperament, and ultimately his desire. What makes him memorable is what he made of his interest in the physical world, a life of natural philosophy. He arrived at it by acts of will. This is the central meaning of Blagden’s observation that Cavendish always knew what was right for him.

Chapter 5
Aspects of Personality

5.1 Complexity

The standard view of Cavendish, based upon Wilson's biography and his sources, needs qualifying on several points, as we have seen. In this section, we draw on a source Wilson may not have know about, Blagden's diary, which offers another view of Cavendish, this one immediate and personal. It suggests further qualifications.

At the time of Cavendish's death, Blagden recalled in his diary the "time I was intimate with him,"[1] a closeness no one else could claim. Unfortunately, Blagden's diary in the Royal Society Library is incomplete, missing those years we are most interested in. Between 1781 and 1789, the interval covering Blagden's association with Cavendish, the diary contains entries only for scattered short periods, adding up to less than a year and a half, though even this scant sample has useful information for us. From the early 1790s, after their association ended, the diary is complete to the end of Cavendish's life, and because Blagden continued to meet Cavendish on Thursday, the day the Royal Society and its Club met, and also on Sunday at Banks's house and Monday at the George & Vulture, the later diary contains further useful information.

Following an evening in Cavendish's company, Blagden often noted Cavendish's behavior, which normally signaled a desire to be left alone: "melancholy," "forbidding," "odd," "dry," and "sulky." The last two words, "sulky" and "dry," Blagden used most often, and although he applied them to others on occasion, he used them regularly only in describing Cavendish. If Cavendish's usual behavior in company were to be compressed into single words, Blagden's words were well-chosen. "Sulky," according to a dictionary source from 1744, means "silently and obstinately ill-humoured; showing a tendency to keep aloof from others and repel their advances by refusing to speak or act." The first dictionary figurative meaning of "dry" is "feeling or showing no emotion." "Odd," another of Blagden's words for Cavendish,

[1] 1 March 1810, Charles Blagden Diary, Royal Society 5: 428 (back).

R. McCormmach, *The Personality of Henry Cavendish - A Great Scientist with Extraordinary Peculiarities*, Archimedes 36, DOI 10.1007/978-3-319-02438-7_5,

transferred from numbers, odd-even, means that which "exists or stands alone; single, sole, solitary, singular," and from a sixteenth-century usage, "peculiar; eccentric."[2] "Sulky" and "dry" describe moods, with considerable overlap; here we take "sulky" to refer to silence, an alternative way of saying, as Blagden did, "Cav. said little,"[3] and "dry" to refer to emotional coldness. In characterizing Cavendish, Blagden occasionally used words meaning the converse of "sulky" and "dry," indicating approachability: "civil," "civiller," "very gracious."[4] Cavendish did not or could not conceal his moods, a skill most people master early in life. He was not a man without moods, which are feelings; it would seem he was at their mercy.

Sulky or conversational. Blagden wrote in his diary that Cavendish was "sulky," "rather sulky," and "sulky enough."[5] He also wrote, "friendly with Cav," "talked easily with Cav," "talked much with Mr C.," "Mr Cavendish freer than usual," "Cav. talked more to me than usual with greater interest."[6] When Cavendish spoke, it was not to deliver a monologue, or not always. Blagden mentioned the "questioning way of Mr Cavendish."[7] He "asked questions civilly," Blagden noted after a meeting.[8]

Dry or friendly. Just as he was usually sulky, Cavendish was "dry as usual."[9] Blagden also recorded an occasional show of warmth and animation. On a political topic: "Cav. with inflexion: So much for talent without judgment!"[10] On expectations of war in Europe: Cavendish "said he could scarcely refrain from bursting out."[11] At a meeting of the Royal Society Council: "Cav. looked very much vexed."[12]After returning from an extended period abroad, Blagden greeted Cavendish, who "shook kindly my hand; asked if here for any time. I said three months at which he looked disappointed but said nothing."[13] Four days later, at the Royal Society, Cavendish "talked rather friendly."[14] At a meeting of the Monday Club, Blagden spoke to Cavendish of "spirit and independence, & true friends."[15] Blagden did not say if Cavendish responded, but it is clear that on Blagden's side there was friendship.

On a given day, Cavendish's moods could vary. At the Royal Society, Blagden found Cavendish "at first sulky, then came around."[16] At the Royal

---

[2] *Oxford Universal Dictionary*, 568, 1358, 2074.

[3] At the Royal Society Club. 11 December 1794, Charles Blagden Diary, Royal Society 3: 37.

[4] Various dates, Charles Blagden Diary, Royal Society.

[5] 30 November 1794, 30 April and 14 December 1795, ibid. 3: 35, 56 (back), 81 (back).

[6] 16 March 1795, 25 February 1796, 30 March 1803, and 9 May 1805, Charles Blagden Diary, Royal Society 3: 50 (back), 92; 4: 187; 5: 333.

[7] 4 September 1794, ibid. 3: 15.

[8] 26 June 1806, ibid. 4: 455.

[9] 11 February 1796, ibid. 3: 88 (back).

[10] 19 April 1807, ibid. 5: 61 (back).

[11] 20 December 1795, ibid. 3: 82 (back).

[12] 28 May 1807, ibid. 5: 73 (back).

[13] 8 May 1803, ibid. 4: 151.

[14] 12 May 1803, ibid. 4: 152.

[15] 25 August 1794, ibid. 3: 13.

[16] 19 May 1803, ibid., 4: 153.

Society Club, Cavendish was "dry & worse at R[oyal] S[ociety]."[17] At the Royal Society Club, Cavendish "changed tone"; at his house, he had a "different tone."[18] Occasionally Cavendish was said to be "better," with no further explanation.

Cavendish's expressions showed complexity. "Cavs eyes full of resentment, whilst he talked civilly."[19] "Mr Cavendish civil but unmeaning." He was "civil, but rather affectedly." At the Royal Society Club, "Mr C. equivocal."[20] "Cav. tried to cut into conversation, yet cold & sulky."[21] After a meeting of a club at which Blagden found no interesting conversation, he wrote in his diary, "Made nothing of C, cannot understand him."[22]

Blagden's references to Cavendish's feelings could be uncertain. In a conversation about the Monday Club, "Mr Cavendish knew not what to do. Said some men without certain feelings."[23] It is unclear whether the second sentence was said by Cavendish or Blagden, since Blagden frequently eliminated personal pronouns in his hurried diary entries. When Blagden showed and explained drawings of Venus from a paper on astronomy, Cavendish "seemed more than usually jealous."[24] "Jealous" could refer to envy or rivalry, though this seems out of character. More likely it refers to strength of feeling.

"Sulky" and "dry" are words of criticism, not of moral judgment. Blagden used stronger words for failings of other persons, never of Cavendish: "horrid behavior," "horrible character," "character more odious," "moral sentiment the base," "vulgarity," and "cunning & falseness."

Around the turn of the nineteenth century, Blagden's diary entries deal less with science and more with the general agitation over war and politics, even when they record conversation at Banks's house and at the Royal Society. They reveal that Cavendish's interests were less narrowly scientific than Wilson's sources reported. At the Monday Club, Cavendish agreed with his family's political position: he said that ministers and measures needed to be changed, and that they "should have confidence in Fox."[25] At the Royal Society, in Cavendish's hearing, Blagden spoke of the late prime minister William Pitt, "at wch Cav. lift[ed] his hands."[26] Another evening at the Royal Society, Cavendish "said, that if Pitt came in against K[ing]'s inclinations, the K[ing] if quite well, w^{d} soon find the means of getting him out again." Blagden said that North Germany was quiet at present, but Cavendish "still thought Holstein w^{d} be attacked at some moment."[27] Cavendish agreed with Blagden

[17] 19 March 1795, ibid. 3: 51.

[18] 18 May 1795, 8 June 1795, ibid. 3: 59 (back) and 61.

[19] 17 June 1809, ibid. 5: 330.

[20] 13 November 1794, 6 July 1795, ibid. 3: 29 (back) and 64.

[21] 19 February 1807, ibid. 5: 39.

[22] 27 August 1795, ibid. 3: 67.

[23] 12 November 1795, ibid. 3: 76 (back).

[24] 4 September 1794, ibid. 3: 14 (back).

[25] 16 March 1795, ibid. 3: 50 (back).

[26] 1 December 1806, ibid. 5: 15.

[27] 26 March 1804, ibid. 4: 214.

that the conduct of Austria depended on Russia, but disagreed with him about Austria and Prussia, observing "that the only chance of a combined resistance to the French was by a fair intelligence between Austria & Prussia."[28] Concerning a report of war with America, "Cav. said these people now more moderate than their predecessors." Blagden answered that moderation would be "contrary to their principles & sentiments," to which Cavendish "assented & looked in agitation." Blagden then said, "we had but war with all the world, & become a nest of pirates, that perhaps might support ourselves by so… & We might soon be at war with Russia." "To all this [Cavendish] sadly assented."[29] On the condition of the world: to Blagden's observation that "all mankind gone mad together," Cavendish "said, he thought there was a great diminution of common sense of the world."[30] Cavendish may not often have talked about politics, but when he did he revealed his interest, knowledge, and party.

Blagden brought up Cavendish in his diary and letters, but given his access to Cavendish, we might have expected more from him. He provided most of the content of the family obituary of Cavendish, but there he largely limited himself to describing his work, which was public. Blagden's reluctance to leave a personal account of Cavendish may be explained by a comment he made about Boswell's account of his and Johnson's tour of the Hebrides: "Most people would be sorry to have a bosom friend, who kept a journal of their conversations, to publish as soon as they should be dead." The same point is made in a passage he copied out: "the Historiographer of France, being frequently admonished to write history according to the duties of his place, replied 'I love my Masters to well to write their lives.'"[31] Even so, in fragments, Blagden's references to Cavendish confirm his biographers' account of his shyness and related behaviors. They also allow us to form a fuller image of him, showing more complexity of personality and wider interests in the world.

5.2 Dealing with People

Another window onto Cavendish's personality, which Wilson also may not have known of, is his business correspondence. As part of his marriage settlement, Lord Charles received farms and tithes in Nottinghamshire and Derbyshire. The year before he died, he turned over the management of these properties to his oldest son. Henry's letters to his stewards and to a lesser extent to his tenants give us insight into how he dealt with people.

[28] 30 November 1804, ibid. 4: 286 (back).

[29] 15 May 1806, ibid. 4: 442.

[30] 3 April 1804, ibid. 4: 217.

[31] Charles Blagden to Joseph Banks, 9 October 1785, Banks Correspondence, Royal Botanic Gardens, Kew 1: 210. The passage Blagden copied out is in notes in Blagden Papers, Yale, box 4, folder 45.

In the summer of 1782, Cavendish selected as his steward William Gould, who had been recommended by the duke of Devonshire's agent as an able man. To familiarize Gould with the estate, Cavendish sent him a tabulation of the most recent rents and tithes. It listed 23 separate payments, with a total return that year of £1,571.[32] The return next year was £1,649. The income from the estate increased over time: in 1804, it was £2,215; in 1809, it was £3,277. These figures were not all profit; there were ongoing expenses such as repairs of fences, taxes, tithes, payments to the poor, and salary of the steward. Total expenses usually came to under £200 in normal years, but frequently there were extra expenses, which could be considerable.[33] Because Cavendish was a rich man, whatever these sums, they were trifling, but he did not see them that way. He took the management of his estate with the utmost seriousness.[34] Owing to negligence of his father's steward and to large payouts for enclosures, the value of three of the properties had decreased by £1,500 or £2,000 by the time Cavendish took charge.[35] His object was not to maximize his profit but to avoid further "injury to my estate." The estate had come to him through the family, and he would leave it to the family. He was its temporary responsible caretaker. His oversight of the estate tells us several things about him.

Reason, information. Cavendish dismissed his father's steward Thomas Revill. One explanation Cavendish gave him was that he had not given his father enough information. The new steward Gould's first task was to make a complete examination "into the condition of the whole estate."[36] To be "master of the business," Cavendish constantly pressed Gould for more information.

Information, for Cavendish, included reasons for acting. His letters read: "I have good reason," "I have great reason," I have "insufficient reason." He regularly pressed Gould to give him his reasons. Concerning an enclosure, Cavendish told Gould, "If you had given your reasons for recommending to me to do so I might very likely have agreed to them. For this reason I could wish that in future when you recommend any thing to me that you would always give your reasons." It took time to give reasons, but it saved time in the end.[37]

The vicar of a parish complained to Cavendish about having to repair the cancel of his church, claiming it was the custom for the proprietor to bear the expense.[38] Cavendish thought the vicar was wrong, and unless he saw "much more

[32] Copy of rental enclosed. Henry Cavendish to William Gould, 8 and 28 August 1782, Devonshire Collections, Chatsworth, L/34/7.

[33] Itemization of rental income and expenses, 1783–1804. "Account of J^{n}: Renshaw with the Hon: Henry Cavendish." Year-by-year lists of receipts, salary, and expenses: ibid., L/34/10.

[34] His receipt of £3,277 in 1809 went to a "trust account." "The Account of J^{n}: Renshaw with the Hon: Henry Cavendish for the Receipts and Payments on Account of His Estates in the Counties of Nottingham & Derby for One Year Ending Lady Day 1809," ibid., L/34/16.

[35] William Gould to Henry Cavendish, 28 September 1782, Devonshire Collections, Chatsworth, L/34/7.

[36] Henry Cavendish to William Gould, 6 September 1782, draft, ibid., L/34/7.

[37] Henry Cavendish to William Gould, n.d. [after 21 February 1789], draft, ibid., L/34/12?

[38] Stephen Moore to Henry Cavendish, 31 July 1783, ibid., L/34/7. Moore was vicar of Hayton.

reason," he would not consent to repairing it.[39] Whether or not it was the custom for the vicar to pay for the repair was a factual question, and Cavendish looked to information and reason to resolve the disagreement. In his reply to the vicar, Cavendish used the word "reason" six times in a letter of five sentences. In Cavendish's scientific writings, we see a parallel. The brief "Conclusion" to his paper on the theory of heat repeats "strict reasoning" three times.[40] In his dealings with people, in the course of managing his estate, and in his study of the physical world, reasons were ever his guide.

Justice. Cavendish's management of his estate provides examples of his sense of justice. Justice is about rules and laws, and though it arouses feelings, it is not about feelings. A dictionary meaning of justice fits Cavendish perfectly: action in conformity to reason, fact, and truth.[41] In his father's correspondence with his steward, two words appear with notable regularity, "justice" and "reasonableness." Henry's correspondence reads like his father's. With the prospect of enclosure in a parish containing a property of his, Cavendish faced having to exchange tithes, from which he received rent, for land. The proprietors of the parish offered Cavendish a specified allotment of land. His steward advised him not to accept it and proposed an alternative allotment. Cavendish thought that the alternative would be asking for too much, with the result that the enclosure commissioners would be "less disposed to do me justice." Cavendish proposed what he thought was a fair exchange, and it was rejected. He pressed his steward to get more information about acreages, rents, and tithes in the parish so that what he proposed would be seen as "reasonable" and be accepted. Justice in this matter was to be settled by numbers and reason. Justice all around would be served if his estate "should be improved in the same proportion as that of the landowners."[42] The affair dragged on for years. The quantities of money in dispute were of no practical consequence for Cavendish. At issue was a matter of principle: justice all around.

Considerateness. In dealing with his tenants, Cavendish wanted to avoid all "disagreeable dispute," to proceed in the "least vexatious manner." He strove to act in ways that did not hurt people, as the following examples show.

When Cavendish told his father's steward Revill that he was replacing him, he instructed him not to collect the next rents. Revill protested that if he did not collect the rents, it would "raise suspicions to the prejudice of [his] character." Cavendish did not see the "least reason" for his concern, for "it is so natural when a person takes possession of an estate that he should trust to care of it to a

[39] Henry Cavendish to Stephen Moore, 20 November 1783, draft, ibid.

[40] McCormmach, *Speculative Truth*, 191–93.

[41] *Merriam Webster's Collegiate Dictionary*, 10th ed. (Springfield: Merriam Webster, 1993), 636.

[42] Henry Cavendish to William Gould, 20 February 1785, draft, Devonshire Collections, Chatsworth, L/34/7. He told a spokesman for the landholders, "I am far from desiring that my estate should be more benefited by the [enclosure] act than that of the landholders." Henry Cavendish to William Sherbrooke, 16 February 1785, draft, ibid. What was fair could be decided by measurement and mathematics: "The best way of determining the present value of all the lands at Arnold will [be] to inform one of the quantity of inclosure & open field & of how much an acre each lets p.a. at a medium." Henry Cavendish to William Gould, 23 February 1785, draft, ibid.

person whose judgment he can rely on." No one would doubt Revill's "integrity" if he did not collect the rents, but in case the new steward should come across any suspicions of that sort, he would be instructed by Cavendish to put the matter right. When Revill continued to protest, Cavendish wrote back firmly, giving his complete reasons for dismissing him. He did not doubt Revill's "fidelity & good intention," but his laxness in providing his father with the information he needed to manage his estate effectively and the "infirmity" of his temper causing him to "quarrel or behave with petulance" with the tenants, made life hard for the tenants and difficult for the owner.[43]

Cavendish had to concern himself with the financial state of his tenants. Usually he dealt with problems through correspondence with his steward, though occasionally tenants wrote to him and even called on him in person. In Gould's proposals for managing the estate, Cavendish thought him hard. At the beginning, Gould wanted to take much of the tithe land into his own hands by terminating the tenants almost at once, giving them only 6 months' notice. Cavendish was "unwilling to turn out tenants who have not behaved ill upon such short notice." Although 6 months was sufficient legal notice, it would not give tenants enough time to acquire new farms.[44] Concerning a tenant who was "in danger of failing," Cavendish wanted Gould to look into the matter and tell him "what you think of him & whether his distress is owing to extravagance or mismanagement or to his having engaged in an undertaking too expensive for his fortune."[45] If the latter were the case, Cavendish, like his father, would make allowances.

Cavendish inherited a particularly difficult tenant named Rose. His father's steward Revill had made an agreement with Rose for a new enclosed farm. On the face of it, the agreement was to Cavendish's advantage; the rent was higher than it should have been for the land, but Gould thought that Rose would be unable to maintain the property. Cavendish said that he would make allowances for Rose, who had mistaken the value of the land, "if he takes good care of the farm." The buildings on it were in bad shape, and the cellar was 4 ft deep in water. Rose called on Cavendish to make a special case. After the meeting Rose told Gould that Cavendish had ordered him to build a drain in his cellar immediately. Cavendish said he had not given him such an order. Because of his behavior, and because he paid far more for the farm than its true value, Cavendish expected that "he will make it very bad tenant," and he would like to be rid him "if it can be done without inconvenience."[46] Cavendish was right about Rose.

The only tenant who was always behind in his rent, Rose appeared untrustworthy and dishonest to Gould, who wanted to "make a Distress on his Effects," evidently taking over his cattle.[47] Cavendish cautioned his steward that he wanted to remove Rose "in a manner as a little distressing to him as I can." He did not like Gould's

[43] Henry Cavendish to Thomas Revill, 5 September 1782, draft, ibid.

[44] Henry Cavendish to William Gould, 7 September 1782, draft, ibid.

[45] Henry Cavendish to William Gould, 6 September 1782, draft, ibid.

[46] Henry Cavendish to William Gould, 12 December 1782, 15 January 1783, drafts; William Gould to Henry Cavendish, 11 January 1783, ibid.

[47] William Gould to Henry Cavendish, 23 January 1785, ibid.

proposal of seizing his effects, which would "oblige him to part with them to great loss." He wanted Gould to avoid that course if he could, even "though at the expense of ¼ or even more of what is due to me."[48]

Eight years after he had taken charge of the estate, Cavendish told his steward that it was vain to hope for any improvement from Rose and that he should take the farm away from him as soon as he could, but he should give Cavendish his thoughts on the matter first. Cavendish was willing to forgo £80 from the arrears. Gould was at liberty to try to recover the rest, "but if gentle methods will not do I wish you to send me word before you have recourse to others." Cavendish wanted to be free of Rose but not to hurt him. Rose refused to quit at Lady Day, as requested, and instead he called on Cavendish again. Not knowing that he had refused to quit, Cavendish did not ask him what his plans were. Cavendish told Gould, "I did not say much to him."[49] Rose quit his farm in 1794, 12 years after Cavendish was first informed that he was a problem.[50]

The difficult tenant Rose tried Cavendish's patience, but Cavendish gave him every opportunity to make a success of his farm. Rose continued to perform poorly, and in the end Cavendish had little choice but to let him go. As a landlord Cavendish tried to be fair. When problems came up, he considered them from the tenants' point of view as well as his own. Wilson's choice of words – Cavendish was "to appearance a misanthrope" – can be questioned. Cavendish discouraged intimacy, but he was not a misanthrope and his tenants would not have thought he was.

Judgment. In making decisions about his estate, Cavendish used sound judgment. There are many examples of this. To give one: as tenants, he preferred farmers to gentlemen. Another is: under enclosure, he preferred to be allotted a certain proportion of the land rather than land of a certain monetary, or absolute, value. Judging land by absolute value, was difficult, and "different persons may differ much in their opinions so that the commissioners may much injure my property without intending to act dishonestly."[51] The absolute value was partly subjective and variable, and the enclosure commissioners usually overrated it.[52] Another example is: when Cavendish's next steward John Renshaw became very ill, he recommended his son as his replacement. Cavendish was open to the idea but he wanted letters of recommendation for his son. Gould asked several businessmen for recommendations. Gould had misunderstood him, Cavendish said. He did not want Gould to "procure letters of recommendation of [his son] for vague letters of this kind written to a father cannot be considered as of much weight." Cavendish asked the son to ask persons he had done business with to write letters, or to provide names of persons

[48] Henry Cavendish to William Gould, 6 February 1785, draft, ibid.

[49] Henry Cavendish to William Gould, September 1790, 7 March 1791, drafts, ibid.

[50] Henry Cavendish to William Gould, February 1794, draft, ibid.

[51] Henry Cavendish to William Gould, n.d. [After 28 February 1785], draft, ibid.

[52] Henry Cavendish to William Gould, 24 December 1784, draft, ibid.

for Cavendish to write to.[53] The son told Cavendish that what he required was "very reasonable and proper."[54] Cavendish had peculiarities, but so far as we can tell they did not affect his judgment.

[53] John Renshaw to Henry Cavendish, 22 May 1802; John Renshaw, Jr., to Henry Cavendish, 5 June 1802; Henry Cavendish to John Renshaw, 13 July [1802], draft; Henry Cavendish, n.d. [August 1802], draft, Devonshire Collections, Chatsworth, L/34/10.

[54] John Renshaw, Jr., to Henry Cavendish, 6 September 1802, ibid.

Chapter 6
Views of Cavendish

6.1 Interpretations

People who knew Cavendish and people who have studied him have wondered about his strangeness. An indication is their habit of likening him to persons whose lives lie outside the normal range. Bickley, the Cavendish family chronicler, said that Cavendish "lived the life almost of a hermit."[1] Pearson, a later family chronicler, wrote that Cavendish "appears a sort of Merlin figure, discovering the secrets of the world around him for his private satisfaction."[2] With reference to Cavendish's silences, Brougham compared him to Trappist monks. Wilson, who liked the comparison, applied it to Cavendish's preference for solitude, and added prisoners to monks. He "took the self-imposed vows of a Scientific Anchorite, and, like the Monks of old, shut himself up within his cell"; "out of the monk's cell, and the prisoner's dungeon, there have been very few men who have lived for nearly fourscore years, and held so little communication with their fellows, or made so few friendships as Cavendish." Likening Cavendish's cell to a "kingdom," Wilson compared him to a king, who from his "throne … dispensed royal gifts to his brethren." Because of Cavendish's "sense of isolation," Wilson said, "he was like a deaf mute sitting apart from a circle, whose looks and gestures show that they are uttering and listening to music and eloquence, in producing or welcoming which he can be no sharer."[3]

Blagden recorded in his diary a conversation with friends: "Talk about Mr Cavendish, & explanation of character."[4] It is a shame that he did not say what the explanation was, since it would have been the best informed of any we have from the time. Early writers laid stress on the presumed meager allowance his father gave him to the middle of his life. Because of it, Thomson said, he acquired the

[1] Bickley, *Cavendish Family*, 201.

[2] Pearson, *Serpent and Stag*, 121.

[3] Brougham, *Lives* 1: 259. Wilson, *Cavendish*, 178, 186.

[4] 14 July 1795, Charles Blagden Diary, Royal Society 3: 65 (back).

R. McCormmach, *The Personality of Henry Cavendish - A Great Scientist with Extraordinary Peculiarities*, Archimedes 36, DOI 10.1007/978-3-319-02438-7_6,

"habits of economy and those singular oddities of character, which he exhibited ever after in some striking a manner."[5] He might have acquired habits of economy this way, but it is unlikely he also acquired his solitary ways from his allowance, which in any case may have been ample.[6] Another explanation of his behavior was the loss of his mother at an early age, depriving him of warmth that might have tempered his character. The loss may well have left an enduring mark, a possibility supported by a body of literature on the importance of family upbringing, especially the mother-child relationship, on the emotional and mental development of children.[7] Cavendish's brother Frederick forever regretted the absence of maternal affection in his life. Pearson believes that the mother's death condemned Henry and Frederick "to a secluded childhood which undoubtedly contributed to later shyness and problems of communication, even between themselves."[8] Wilson saw merit in the loss of Cavendish's mother as an "outward circumstance," but he discounted it as the main explanation.[9] The strongest influence on Cavendish's character, he said, was an inward circumstance, "the strongly marked original elements of his nature."[10] That is, Cavendish was born that way.

Cavendish's biographers usually described and evaluated his major publications at some length; they discussed his character only briefly and separately. Brougham believed that scientific and literary men gave the age "greater glory" than did statesmen and warriors, and that to understand the age, it was necessary to look at the lives of these men, which are "truly the history of their times." Thinking that Cavendish's life was "of most interesting contemplation," he made conscientious inquires, going so far as to borrow Cavendish's scientific manuscripts from the family. What he found interesting to contemplate in Cavendish's life was the science, his "personal history" being of no great account.[11] In Brougham's sketch, Cavendish's personality appears as the sum of its oddities, with no evident connection with his science. In their biographical dictionary, published 4 years after Cavendish's death, John Aiken and William Johnston recognized a connection: Cavendish's odd "man-

[5] Thomson, *History of Chemistry* 1: 336.

[6] According to Thomson, Cavendish received an annuity of £500, which Thomson called "rather narrow circumstances." Ibid. It should be pointed out that when his father Charles Cavendish was a young man, he received an annuity of £300, which was standard for younger sons of the aristocracy. Charles's father, the second duke of Devonshire, intended to increase the annuity to £500 at his death but moved it ahead when Charles married. Wilson heard from a second source, a "gentleman" who knew Cavendish from the British Museum, who put his annuity at £120 until he was 40, which would have been narrow, though Cavendish was living at home. *Cavendish*, 160.

[7] In a study of children from age 2 to 7, those who were shy, timid, and unsociable at age 2 were the same at 7, indicating that shyness is long-lasting. Shyness at age 2 arises either from temperament or from some stress in the child's environment, such as the death of a parent; Cavendish lost his mother at age 2. Jerome Kagan, J. Steven Reznick, and Nancy Snidman, "Biological Bases of Childhood Shyness," *Science* 240 (1988): 167–71, on 167, 171.

[8] Pearson, *Serpent and Stag*, 117.

[9] Wilson, *Cavendish*, 171–72.

[10] Ibid., 171–72.

[11] Brougham, *Lives*, London edition, 1: ix, xi, 444.

ners were not the result of irritability or peevishness, but of a total want of common feelings," a temper which in the philosopher had "some admirable effects," rendering him "patient, cautious, and exact."[12] Two centuries later, Oliver Sacks made a similar observation. Some of the characteristics that distinguished Cavendish's autism were the "very traits he used so brilliantly in his pioneering scientific research."[13]

Cuvier made a connection, though only by way of an analogy: Cavendish's "way of life was as regularized and precise as his experiments." Cuvier admired Cavendish for his single-minded dedication to science; he imagined that Cavendish was too busy with scientific projects to conduct an active social life, evidently an explanation of his preference for solitude. Passing over Cavendish's peculiarities, which he apparently relegated to the "trivia of life," he kept to what he saw as important, the scientific work and the scientist. He said that Cavendish "demonstrated to the world a model of what a scholar should be, as well as an example of the happiness he should enjoy."[14] His elevation of Cavendish to a happy model scholar was made easier by the fact that he did not know Cavendish. Like Cuvier, Larmor thought that Cavendish's work affected his personality. The time and concentration Cavendish needed to prosecute all his studies "were amply sufficient to account for the strengthening of Cavendish's habits of reserve and isolation"; "he was a natural philosopher so profound and universal as to have no time to be anything else."[15] "The cultivation of the highest domains of physical science is rarely consistent with dispersal of interest in other directions."[16]

Wilson took Cavendish's "peculiarities" seriously, calling his character "very remarkable and interesting,"[17] and offering the only unified account of Cavendish. For this purpose, he invoked the ancient Greek concept of macrocosm and microcosm, according to which the same order of things can be observed at every scale of existence, from the universe to the smallest particles, with man situated in the middle, a reflection of the whole. He said that Cavendish "in the Microcosm of his own nature … tried to reflect and repeat the subjection to inflexible rule, and the necessitated harmony, which are the appointed conditions of the Macrocosm of God's Universe."[18] Cavendish insisted that the trivial routines of his life follow "a law as inflexible and imperative as that which rules the motions of the stars." Cavendish's trivial routines signified to Wilson something that was neither trivial nor routine, "a strongly developed will, which gave a singular consistency and unity to all the proceedings of its possessor."[19] It has been observed that the personality of

[12] John Aiken and William Johnston, *General Biography; or, Lives, Critical and Historical*, vol. 9 (London, 1814), 284.

[13] Sacks, "Henry Cavendish."

[14] Cuvier, "Cavendish," 236, 238.

[15] Joseph Larmor in Cavendish, *Scientific Papers* 2: 399.

[16] Larmor, "Preface to Volume 1," ix.

[17] Wilson, *Cavendish*, 165.

[18] Ibid., 186–87.

[19] Ibid., 187–88.

a scientist can affect his work in certain ways, and conversely a scientist's work can affect his behavior or make more pronounced certain personality traits.[20] Wilson seemed to suggest that by acts of will, Cavendish modeled his life after his study of nature, in conformity to the rhythms of the universe. He ordered his daily rounds to reflect the unvarying cycles of the universe, the daily rotation the earth on its axis, the yearly rotation of the earth about the sun.

Wilson's full characterization of Cavendish's "vocation," quoted in part above, is: "His Theory of the Universe seems to have been, that it consisted *solely* of a multitude of objects which could be weighed, numbered, and measured; and the vocation to which he considered himself called was, to weigh, number, and measure as many of those objects as his allotted three score years and ten would permit."[21] From the testimonies he collected, Wilson distilled the essence of his subject: Cavendish's brain "seems to have been but a calculating engine."[22] Ever since Wilson's book, writers have called Cavendish's brain a calculating engine.

The metaphor captures Cavendish's propensity to calculate. It can also be misleading, for it leaves out much. Cavendish's travel journal of 1785, which is reproduced at the end of this book, contains copious geological observations, without calculations. An example of work of a different sort makes the same point. Cavendish published a brief paper on a slowly moving luminous arch, believed to be the only kind of aurora whose height can be measured. He calculated the height of this aurora, but instead of ending his paper there, he incorporated the height in a theoretical "conjecture." The conjecture was an analogy with the familiar explanation of another kind of aurora, the "common aurora borealis," as "parallel streams of light shooting upwards." If the analogy held, Cavendish said, the luminous arches in question should form a long thin band of light, and his calculated height was qualified in a way that he showed. "I do not offer this as a theory of which I am convinced; but only as an hypothesis which has some probability in it, in hopes that by encouraging people to attend to these arches, it may in time appear whether it is true or not." Cavendish offered a hypothesis about the nature of the phenomenon together with encouragement to observers to test the hypothesis and possibly elevate it to a true theory.[23] His paper contained a calculation, and it also had a qualitative objective, to decide the nature of a puzzling aurora. Wilson's likening of Cavendish's brain to a calculating engine is furthest from the mark when it is applied to his theoretical work. Cavendish amply demonstrated that he was capable of constructing subtle theories in natural philosophy, and although he expressed them mathematically, calculation had little to do with it. Wilson's statement that Cavendish

[20] Feist, *Psychology of Science,* 123.

[21] Wilson, *Cavendish*, 186. Quantity being the distinguishing mark of Cavendish's work, Wilson may have looked to the bible for a passage to give it proper emphasis: "Thou hast ordered all things in measure, and number, and weight," though he could have found it elsewhere too. (Wisdom 11.21).

[22] Wilson, *Cavendish*, 185.

[23] Henry Cavendish, "On the Height of the Luminous Arch Which Was Seen on Feb. 23, 1784," *PT* 80 (1790): 101; in Cavendish, *Scientific Papers* 2: 233–35.

was unable to investigate any question other than quantitatively is too strong. Quantity was a preference not a constraint.

In a figurative language similar to Wilson's, Michael Fitzgerald, a psychiatrist who has written extensively on autism, describes Ludwig Wittgenstein as an autistic philosopher whose mind was "mechanical." Wittgenstein was trained as an engineer and was interested in machinery, but the main evidence for his thinking is the organization of his treatise the *Tractatus*, a system of propositions similar to Euclidean geometry, which has been likened to a "calculating machine." Fitzgerald hypothesizes that the thinking of an autistic person is "in the style of a calculating machine."[24] The comparison may fit Wittgenstein better than it does Cavendish. In his laboratory, Cavendish worked with measurements, numbers, and calculations, but he also took account of things that lie outside mechanical calculation: the selection of the phenomena to study, the handling of instruments, the registering of sense impressions and their interpretations. When working in his study, he called upon his powers of invention and insight to formulate concepts for describing the natural world, operations beyond mechanical calculation. At some stage he might organize his thoughts in the form of Euclidean-like definitions, hypotheses, and propositions and then call on his mental calculating machinery to match or predict quantitative phenomena, but that is only one part of his thinking. The expression "calculating machine" suggests the mind of a savant more than it does Cavendish's. Alternative terms such as "mathematical" mind or Temple Grandin's "music and math" mind mean much the same thing, but being more open they describe Cavendish better. Better suited yet, as suggested earlier, is a curious mind in search of truths of nature arrived at with the aid of strict reasoning.

Wilson was a fine writer, with a gift for metaphor, especially visual metaphor. Trained as a physician, he carried out researches on vision, publishing a valuable book *Researches on Colour-Blindness* and a popular "ornate account of the senses" *Five Gateways to Knowledge*.[25] He used light as a metaphor to characterize both science and Cavendish, establishing a poetic identity of the two.

> I would liken science and poetry in their natural independence to those binary stars, often different in colour, which Herschel's telescope discovered to revolve round each other. 'There is one light of the sun,' says St. Paul, 'and another of the moon, and another of the stars: star differeth from star in glory.' It is so here. That star or sun, for it is both, with its cold, clear, white light, is SCIENCE: that other, with its gorgeous and ever-shifting hues and magnificent blaze, is POETRY. They revolve lovingly round each other in orbits of their own, pouring forth and drinking in the rays which they exchange; and they both also move round and shine towards that centre from which they came, even the throne of Him who is the Source of all truth and the Cause of all beauty.[26]

---

[24] Michael Fitzgerald, *Autism and Creativity: Is There a Link between Autism in Men and Exceptional Ability?* (Hove and New York: Brunner-Routledge, 2004), 122–24.

[25] P. J. Hartog, rev. R. G. Anderson, "Wilson, George (1818–1859)," *DNB* 59: 536–39, on 539.

[26] George Wilson quoted in Natural History Society of Montreal, Reviews and Notices of Books, *The Canadian Naturalist and Geologist* 6 (1861): 393.

His metaphor for science, the "cold, clear, white light" of a star, which he derived from Scripture, he applied to Cavendish: "He was not a Poet, a Priest, or a Prophet, but only a cold, clear Intelligence, raying down pure white light, which brightened everything on which it fell, but warmed nothing – a Star of at least the second, if not of the first magnitude, in the Intellectual Firmament."[27] At the level of metaphor, Wilson's cold, clear white light does better justice to Cavendish's mind than a calculating machine. His setting was the eighteenth-century Enlightenment, the era of enlightening, with its stress on reason. Cavendish was, as Wilson says, one of its luminaries.

The qualities Cavendish needed to pursue his scientific interests were all that Wilson could discover in his man. He summed up his character. It

> can be described only by a series of negations. He did not love; he did not hate; he did not hope; he did not fear; he did not worship as others do. He separated himself from his fellow men, and apparently from God. There was nothing earnest, enthusiastic, heroic, or chivalrous in his nature, and as little was there anything mean, groveling, or ignoble. He was almost passionless.... An intellectual head thinking, a pair of wonderfully acute eyes observing, and a pair of very skilful hands experimenting or recording, are all that I realize in reading his memorials.... He is more to be wondered than blamed.[28]

Wilson's vivid portrait has been the standard interpretation of Cavendish's character. The chemist A. J. Berry, author of a book-length biography of Cavendish, wrote of his "striking deficiencies as a human being," and he quoted from Wilson's passage above. Thorpe, editor-in-chief of Cavendish's scientific papers, wrote to one of the contributing editors that he had enormous admiration for Cavendish's "intellectual powers," but not for his character: Cavendish "was not a man other men are, but simply the personification and embodiment of a cold, unimpassioned intellectuality."[29] Despite the "atrophy or non-development" of faculties possessed by most humans, Wilson wrote, Cavendish was a genius no less than the artists on whom the world fawned.[30] On this point, Cavendish's interpreters were agreed.

Who Was Henry Cavendish? There are two answers to this question. One is Cavendish as others saw him, his social identity, about which we have considerable testimony. The other is Cavendish's own answer, his sense of self, about which we can only surmise; the evidence suggests that it was not much different from how others saw him. Most people form their ideas of who they are largely from how they affect other people. Cavendish was raised in a society that deferred to rank. He was not in the direct line of descent of dukes of Devonshire, but he was close to it; he did not need to pick up subtle clues to know its meaning. His sense of himself in this respect was outwardly expressed in familiar ways: nice houses, lots of servants, carriages, and free time. The esteem he earned from his scientific accomplishments was equally evident. His course of life linked the two main reference points of his identity, his rank and his work: in the organizations where he performed his

[27] Wilson, *Cavendish*, 186.

[28] Ibid., 185.

[29] Edward Thorpe to Joseph Larmor, 7 February 1920, Larmor Papers, Royal Society, 1972.

[30] Wilson, *Cavendish*, 185.

aristocratic duty of service, he was welcomed as a natural philosopher bringing useful knowledge, skill, and intelligence.

His reference points were secured to firm foundations. We have noted that the English aristocracy was in ascendancy, and during his lifetime its power was not seriously threatened. In the century after the Scientific Revolution, which had exhibited the power of experiment, observation, and mathematics to build solid structures of knowledge, natural philosophy was in ascendancy in the world of learning. In his time, Cavendish was enviably placed in English society.

6.2 Feelings

Views of Cavendish are colored by feelings he arouses, which vary by individual, era, and occupation. Following are examples of the range. Upon hearing of Cavendish's death, the president of the Royal Society Banks lamented the loss to science, but otherwise "felt nothing." Blagden, by contrast, "felt much affected." He said he "continued all day to feel the effect of this event on my spirits."[31] From his window 2 weeks later he watched the "funeral procession of my late friend … with much emotion."[32] Young was confident that Cavendish knew his worth, but he regretted that he was denied the gratifications of the normal run of humanity. He missed the sympathy that comes from friends and "looks that we love," which enhance the beauty of the world, and his wealth deprived him of any belief that scientific triumphs would gain him "some great object in life that he earnestly desired; a gratification generally indeed illusory, but which does not cease to beguile us till we become callous as well to the pleasures as to the sorrows of existence."[33]

Wilson had strong feelings about the man. The year after his biography of Cavendish appeared, he published a biography of the physician and physiologist John Reid, whom he described as a person of "Courage, Hope, and Faith."[34] He had looked hard for signs of those same qualities in Cavendish, but he could make no emotional connection. In a letter he wrote while working on his biography of Cavendish, he described the great scientist as "a man without a heart." But even a man without a heart, he said, "I think about, and read about, and dream about, and picture to myself in all possible ways, till he grows into a living being beside me, and I put my feet into his shoes, and become for the time Cavendish, and think as he thought, and do as he did." In the end, he regretted giving "so great an amount of time and thought to so small a matter." If he had known the burden it would become, he would not have accepted the invitation to write it. Like all his past efforts, he

---

[31] 24 February 1810, Charles Blagden Diary, Royal Society 5: 426.

[32] 8 March 1810, ibid. 5: 431 (back), 432.

[33] Young, "Cavendish," 445.

[34] This description is taken from a poem Wilson wrote about him, "The Late Dr John Reid," included in his sister Jessie Aitkin Wilson's *Memoir of George Wilson*, 343.

experienced this one as "bleak and dark," and his image of Cavendish corresponds.[35] That feeling was picked up in a lecture in 1855 on Cavendish's discovery of the composition of water. The lecturer Lewis Steiner apologized to his students for bringing before them a dropout from society. Having read Wilson's new biography, and referring to Cavendish as a "calculating engine," he asked: "Are we not created for grander and nobler purposes than to be mere negations? … Is it enough *not* to be a villain, a debauchee, a murderer? Or, rather, is it not our duty to be *something* that shall create an influence for *positive good* on our fellow-men? To this the answer must be made, that the character of Cavendish is not introduced as a subject of admiration, or for imitation, but rather as a warning to all men who cultivate the intellect, that they do not neglect the social portion of their nature."[36] Bickley in his Cavendish family history in 1911 based his account of Henry Cavendish on Wilson's book, which left him with a feeling of pity and loss: "There is something pathetic about such an existence as Henry Cavendish's, so fruitful and yet so utterly barren. He was an eminent scientist, a society of learned men was called after him, and his name is still held venerable. Small profit to set against his losses. He understood the structure of the universe. But the stars had no song for him, neither had earth any laughter."[37] Pearson in his history of the family in 1983 took issue, finding Cavendish inspiring not pathetic: "the more one learns about him, the more enviable and admirable he seems; and the more to be pitied those around him. For Henry Cavendish really was the most fortunate of men, possessed of a great mind, unceasing curiosity, and mental powers that lasted till his death. He was fortunate in being able to stay free from the cares and passions that enslave most lesser men. He had material freedom, and was lucky to have lived in a period when a dedicated aristocratic amateur like him could make discoveries on such a scale and over such a range of subjects."[38] Oliver Sacks said that after rereading Wilson's biography of Cavendish, he was "deeply moved."[39]

[35] Quotation from a letter Wilson wrote at the time, ibid., 340–41. Wilson who suffered from bad health died at age 41. His limited scientific research did not show "marked originality." His biography of Cavendish and his book on color blindness were his most important publications. His most important public work was his directorship for 4 years of the Industrial Museum of Scotland, later renamed the Royal Museum of Scotland. Hartog, rev. Anderson, "Wilson," 538.

[36] Lewis H. Steiner, *Henry Cavendish and the Discovery of the Chemical Composition of Water* (New York, 1855), 6.

[37] Bickley, *Cavendish Family*, 207–8.

[38] Pearson, *Serpent and Stag,* 121.

[39] Sacks, *Uncle Tungsten*, 120.

Chapter 7
Eccentricity

In a study of several British scientists, Victoria Carroll shows how they came to be regarded as eccentric. The setting is the early nineteenth century when the sciences were becoming more clearly demarcated, their practitioners more self-consciously professional, and the boundaries between orthodox and unorthodox science more carefully observed. Crossings of boundaries were perceived as eccentric. Carroll's study succeeds in showing that eccentricity can be a "revealing lens" for understanding social developments of science.[1]

Cavendish belonged to an earlier stage of science, and his eccentricity arose from a different source, one having to do with character rather than scientific boundaries. Young supposed that Cavendish's severe habits of study alone spared him from "absolute eccentricity."[2] I take him to mean that outside of science, Cavendish was eccentric. In this chapter, I speak of Cavendish as "eccentric." Alternatively, I could speak of his "difficult character," as Wilson does,[3] but "eccentric" is acceptable, and it has the advantage of precedent.

If we think of eccentricity as lying on a scale, at one end we find ticks, fanciful beliefs, and odd likes and dislikes. A person who will touch nothing yellow has a minor eccentricity. At the other end of the scale, we find behavior so strange as to cause general wonderment.[4] Clifford Pickover, in his book on scientific eccentrics, *Strange Brains and Genius,* says that Cavendish was the "shyest of all scientists," whose shyness went "beyond shyness, beyond ordinary social phobia"; this "shabby, eccentric man" was "not entirely human," "more an alien robot than a member of

[1] Victoria Caroll, *Science and Eccentricity: Collecting, Writing and Performing Science for Early Nineteenth-Century Audiences* (London: Pickering and Chatto, 2008), 164.

[2] Young, "Cavendish," 444.

[3] Wilson, *Cavendish*, 170.

[4] We can imagine another scale, with traits associated with absolute conformity at one end and those associated with absolute eccentricity at the other. Such a scale is considered in Weeks and James, *Eccentrics*, 11.

R. McCormmach, *The Personality of Henry Cavendish - A Great Scientist with Extraordinary Peculiarities*, Archimedes 36, DOI 10.1007/978-3-319-02438-7_7,

the human race."[5] If we believe half of what Pickover says, Cavendish was notably eccentric with respect to shyness.

We use the word "eccentricity" loosely to refer to persons who are notably odd in some manner. Before it was used in that sense, it had technical meanings. In geometry it meant a circle that is not concentric with another circle. In astronomy it meant paths of planets not centered on the earth or sun or not circular. It was applied to comets because their paths are not circular, and it is from comets that the figurative use of "eccentric" derived, entering English usage in the late seventeenth century. To the middle of the eighteenth century, it was applied to persons who were bold and innovative. The *London Times* in 1787 gave it the wider meaning we accept today, "a departure from the general conduct of society." It was not until the early nineteenth century that "eccentric" was used as a noun applying to a person exhibiting eccentric behaviors.[6]

Although the word "eccentric" originated in science, in its figurative meaning it has had scant scientific study. To make a start, in 1988 David Weeks and Kate Ward, and again in 1995 Weeks with another co-author Jamie James, published book-length studies of eccentrics as psychological subjects, based on a sample of about 1,000 self-professed British eccentrics. They also studied around 150 eccentrics from the past, persons who were called eccentric or were given a comparable description in their time. Cavendish qualified as one of their historical eccentrics by reason of his social behavior: he was shy and introverted "to a highly eccentric degree," and his "selective avoidance of people probably amounted to a social phobia."[7]

The authors found eccentricity in all ranks of society and in both sexes. Historically, aristocracy and landed gentry dominated the ranks of eccentrics, but that is because the upper class was written about more than other people and because they had more money, required for some kinds of eccentricity. The reason there seemed to be more male than female eccentrics is probably prejudice. The authors recognized that eccentricity lies on a continuum, that it varies over time, place, and social stratum, and that the perception of eccentricity is subjective. Their work was descriptive, not analytic.[8]

---

[5] Pickover thinks that obsessive behavior like Cavendish's is probably a "disorder" as well as an eccentricity, having its origin in abnormal physical characteristics of the brain. He also thinks that eccentricity probably plays a positive role in some great scientists. Clifford A. Pickover, *Strange Brains and Genius: The Secret Lives of Eccentric Scientists and Madmen* (New York: Quill, 1999), xi, 106 112, 252.

[6] Sophie Aymes-Stokes and Laurent Mellet, "Introduction," *In and Out: Eccentricity in Britain*, eds. S. Aymes-Stokes and L. Mellet (Newcastle upon Tyne: Cambridge Scholars Publishing, 2012), 5. Caroll, *Science and Eccentricity*, 12–13. Anon., "Eccentricity (Behavior)," http://en.wikipedia.org/wiki/Eccentricity_(behavior).

[7] Ibid., 107–8. An atypical eccentric in their sample, Cavendish was an introvert who held normal ideas, whereas most of their eccentrics were extroverts who held eccentric ideas.

[8] David Weeks and Jamie James, *Eccentrics: A Study of Sanity and Strangeness* (New York: Villard, 1995), 10–12, 42, 49–50.

By examining eccentrics from the perspective of more or less non-eccentric persons, the authors constructed a profile of 15 most common characteristics, 5 of which stand out. Of the full 15 characteristics, 8 apply to Cavendish. Of the short list of five, four apply to him: nonconformity, creativity, strongly motivating curiosity, and obsession with one or more hobbyhorses. The one characteristic that does not obviously apply is idealism, a desire to change the world. As a Whig, Cavendish would have supported liberal reforms, and he welcomed changes in the material world through applied science and technology, but there is no sign he wanted to change the social world, in which he had a comfortable place. Of the four traits that do fit, the two that best describe him are curiosity and obsession. Of the range of possible motivations, curiosity is the only one that is primarily intellectual, and when it is combined with obsession we have our man: Cavendish the eccentric investigator, who goes to obsessive labor to satisfy his curiosity, the end result of which is truth, its own reward.[9]

The common idea of the eccentric Englishman in Cavendish's time was a highly unusual person "whose faults were on the right side." Then as now, the harmless misfit was the type, though misanthropy was often present, and eccentrics could be disturbing.[10] Beyond that, it is easier to say what the idea was not than what it was, and any generalization has exceptions. Eccentricity was not conscious exhibitionism. It was not libertine; libertines were occasionally included in collections, but for other reasons.[11] It was not criminal or sinister, though robbers, swindlers, and murderers were occasionally included.[12] It was not revolutionary, though it could be political. It was not religious, though individual clergy could behave eccentrically.[13] Of particular relevance to Cavendish, eccentricity was not maladaptive. Richard Kirwan, a colleague of his, is an example. We read in Blagden's diary in 1795, "Kirwan as eccentric as ever." Kirwan was a hypochondriac given to marvelous routines to maintain body heat. He had his servant sleep in his room with instructions to wake him every few hours to pour hot tea down his throat.[14] Kirwan's

[9] Ibid., 27–28, 32–33, 181–82. The empirical findings of Weeks's studies tell us about categories of eccentricity and about the personality traits that accompany them, but his method of selection of eccentric persons fails to identify some kinds of eccentrics. If Cavendish had been alive at the time of his studies, he would not have been included, for he would not have volunteered as a self-defined eccentric to undergo an interview with the researchers.

[10] Paul Langford, *Englishness Identified: Manners and Character 1650–1850* (Oxford, New York: Oxford University Press, 2000), 301–6. Miranda Gill, "Rethinking Eccentricity," http://www.cam.ac.uk/research/features/rethinking-eccentricity.

[11] Edward Wortley Montague, author, politician, and world traveler, had a wife at home and also "wives of almost every nation." Anon., *Biographical Sketches of Eccentric Characters* (n.p., 1832), 121.

[12] An example is Jonathan Wild, who headed a gang of thieves, and was hanged at Tyburn. Ibid., 338–44.

[13] The Reverend Mr. Trueman of Daventry is included in a collection of eccentrics. He stole from his parishioners when he called on them to offer spiritual comfort. He earned only £400 a year but left £50,000 when he died. Edith Sitwell, *English Eccentrics* (1933; Harmondsworth: Penguin Books, 1976), 250–51.

[14] 14 July 1795, Charles Blagden Diary 3: 65 (back). Pickover, *Strange Brains and Genius*, 96.

eccentricity did not affect his work: he was competent and productive in several scientific fields. His eccentricity was also benign: at worst, his hot-tea routine interrupted his servant's sleep, and the servant was paid for the discomfort.

Eccentricities were often comical or absurd, not least because their behavior was unnecessary. Kirwan could have slept through the night with no ill effects. Lest we write off eccentricity as quirkiness and nothing else, we should be aware that it has its champions. During a time of industrial leveling and Victorian conformity, the philosopher John Stuart Mill welcomed eccentricity as an antidote to oppressive popular opinion, regarding it as a measure of "strength of character." He quantified its benefits: "the amount of eccentricity in a society has generally been proportional to the amount of genius, mental vigour, and moral courage it contained."[15] In the next century, Edith Sitwell, author of *English Eccentrics*, and herself an eccentric, wrote an appreciation of eccentricity that could apply to Cavendish: "the man of genius and the aristocrat are frequently regarded as eccentrics because genius and aristocrat are entirely unafraid of and uninfluenced by the opinions and vagaries of the crowd."[16] More recently, in the face of conformity of another kind, Weeks and Ward wrote that the existence of eccentricity points to a valued quality of human nature, its allowance for individuality. "In an era when human beings seem typecast by their culture or genes, eccentrics are a refreshing reminder of everyone's intrinsic uniqueness. By heedlessly flouting norms of behavior that most of us never question, they remind us how much of our liberty we forfeit without thought, and how great our ability is, in fact, to forge our own identities and shape our own lives."[17] John Timbs, in his *English Eccentrics and Eccentricities*, wrote, "how often do we find eccentricity in the mind of persons of good understanding." However "outlandish, odd, queer" the eccentric appears, he "may possess claims to our notice which the man who is ever studying the fitness of things would not so readily present." We do well to keep Timbs's observation in mind where we consider Cavendish's eccentricities.[18]

The English spoke of "character" where we speak of "personality." Both words imply wholeness, though modern psychology does not identify personality with character alone. Personality is said to express temperament as well as character, the former shaped more by inheritance, the latter by social experience.[19]

---

[15] John Stuart Mill, *On Liberty* (London, 1859). Quoted in Carroll, *Science and Eccentricity*, 11.

[16] Edith Sitwell, *Taken Care of: The Autobiography of Edith Sitwell* (New York: Atheneum, 1965), 145.

[17] David Weeks and Kate Ward, *Eccentrics, the Scientific Investigation* (East Kilbridge, Scotland: Stirling University, 1988); quoted in Pickover, *Strange Brains and Genius*, 279.

[18] Timbs, *English Eccentrics* 1: iii–iv.

[19] Character and temperament have long been distinguished by psychologists: character is what people become intentionally; temperament is their inborn emotional predisposition. For the purposes of psychobiological research, the distinction is put differently, though not incompatibly. Character and temperament each have distinct brain systems, and each has several independent dimensions. Temperament is the "dynamic organization of the psychobiological systems that regulate automatic responses to emotional stimuli," and it is "moderately heritable and stable throughout life." Character, by contrast, is "moderately influenced by family environment and only weakly

Paul Langford has written about how certain traits came to be seen as distinctive of the English character in the eighteenth century. Foreigners largely agreed with the English on what these traits were. The word Langford uses for them is "Englishness," which entered the dictionaries only at the end of Cavendish's life. During his lifetime, the expression "national character" was preferred. Today the idea of national character is unpopular, suggesting behaviors rooted in race and ethnicity, but in the eighteenth century it did not have this forbidding association.[20] It is, of course, a stereotype, but useful nonetheless here.

National character implied a degree of conformity, which appeared at odds with English individuality. Priestley said that the English had the reputation of having "least of an uniform national character, on account of their liberty and independence, which enables every man to follow his own humour."[21] The English resolved the seeming contradiction by adopting the idea of "eccentricity," which by assuming a norm of behavior allowed for individual diversity without threatening social harmony. Eccentricity was "a kind of safety valve for pursuits that might be plausibly tolerated." In practice, a departure from the norm was tolerated so long as it could be "classified as harmless."[22] Cavendish's departure from the norm met the test.

Langford organizes his study of national character around six "supposed traits of Englishness": eccentricity, taciturnity, reserve, candor, decency, and energy. He acknowledges an arbitrariness in his selection; for example, for candor he could as well have used other traits such as honesty, simplicity, sincerity, and directness. Cavendish scores high on all of these traits, if higher on some than others. By making eccentricity an English trait, Cavendish is seen both as a model of an eighteenth-century Englishman and as someone who differs in some striking ways from the average Englishman.

Some traits of the eighteenth-century English national character look eccentric today, and because Cavendish shared them he looks more eccentric to us than he did to his contemporaries. In what follows, we place Cavendish in his social setting. We look to the national character for help in clarifying what was seen as eccentric in his behaviors. For our purposes, we regard the national character as behaviors that were commonly observed in England and which foreigners and the English alike saw as characteristic. In addition, we compare Cavendish's behaviors with those of Englishmen who lived in or around Cavendish's time who were seen as eccentric, giving us an idea of the range of eccentric behaviors. We view Cavendish's eccentric

---

heritable," and it develops into adulthood. To temperament belong the "automatic associative responses to emotional stimuli that determine habits and moods"; to character belong "self-aware concepts that influence our voluntary intentions and attitudes." C. Robert Cloninger, "Temperament and Personality," *Current Biology* 4 (1994): 266–73, on 266–67.

[20] Langford, *Englishness*, 1–2, 7–8, 26.

[21] Ibid., 291. Before Priestley, the philosopher David Hume used almost the same words: because of the "great liberty and independency which every man enjoys," the English "of any people in the universe, have the least of a national character." Ibid., 22.

[22] Langford, *Englishness*, 292, 300, 303.

behaviors as departures from an assumed norm, leaving aside the possibility of their arising from a neuropsychological condition.

We know the names of many English eccentrics of the eighteenth century thanks to a genre of popular biography that began about 1800 and is still going strong: collections of biographical sketches, each a group of anecdotes, more or less jumbled together, demonstrating eccentricity of various sorts. The subjects were a catchall of persons who showed odd ideas, habits, or physical traits. They included persons who today are considered eccentric such as misers and recluses, but also giants, hermaphrodites, strong men, and freaks of all kinds. They included both male and female eccentrics, the latter often for masculine traits and irrationality. They often included Irish, Scots, and Welsh as well as English.[23] Their criteria for eccentricity were vague, as seen in the advertisement in the revised second edition of *Eccentric Biography, Or, Memoirs of Remarkable Characters*. Evidently in response to criticism of the first edition, the editor removed characters whose "peculiarities are not sufficiently striking," replacing them with characters whose "eccentricities render them truly worthy of insertion." His decisions about removing and adding characters were based on what readers wanted, striking peculiarities. The purpose of his collection of eccentric characters was to entertain and instruct, with care taken to exclude any account that has the "least umbrage to delicacy." "Singularities of human nature" is as close as he came to defining eccentricities.[24] Timbs, in *English Eccentrics and Eccentricities*, a mid-nineteenth century standard of the genre, began his sketch of Cavendish, "Eccentricity in men of science is not rare," following which he mentioned Cavendish's demonstration of the composition of water. Discovery could possibly count as an eccentricity, seeing things that others do not, but discovery is not why Timbs included Cavendish. To portray Cavendish's eccentricity, he strung together entertaining anecdotes from what he called Wilson's "clever memoir."[25]

Appearance. Character, in contrast to temperament, was flexible, thought to be revealed in individual choices in the moral sphere but also in the relatively trivial things of life such as dress. The most common kind of eccentricity of character was conservatism, and probably the most common conservatism was dress of an earlier age.[26] Cavendish's dress, as we have seen, was described as "that of the gentleman of the preceding half-century."[27] A few years older than Cavendish, Henry Lee Warner, a politician and landowner, wore the dress "precisely that of the English gentleman of the last age": a coat and waistcoat with gold lace, complete with curved-toed shoes. Warner's fashion was earlier than Cavendish's and more conspicuous.[28] In

---

[23] Caroll, *Science and Eccentricity*, 14, 36, 40.

[24] Advertisement in Anon., *Eccentric Biography, or, Memoirs of Remarkable Characters ...*, 2d rev. London ed., 1st American ed. (Boston, 1804).

[25] Timbs, *English Eccentrics* 1: 142–46.

[26] Langford, *Englishness*, 305.

[27] John Davy, quoted in Wilson, *Cavendish*, 168.

[28] Anon., *Kirby's Wonderful and Eccentric Museum; or, Magazine of Remarkable Characters ...*, vol. 2 (London, 1820), 471–72.

retirement, the naturalist Charles Waterton "wore an old-fashioned swallow-tailed coat, refusing to adopt the formal fashion of his day."[29] A related eccentricity was carelessness of dress. Cavendish's dress was said to be "a little neglected."[30] The eminent Cambridge classical scholar Richard Porson was "careless" and "negligent" about his appearance throughout his life. Known equally for his excellent emendations of classical texts, his wit, and his capacity for drink, "his clothes were shabby, his linen dirty," and the "skirts of his rusty black coat hung with cobwebs."[31] Cavendish's habit was nothing so disreputable.

Speech and Silence. Cavendish was one of several notable aristocrats who were known for their silences. James Hamilton, eighth earl of Abercorn, 1712–89, was called "his taciturnity, the Earl" by Horace Walpole, though "succinctness rather than mere taciturnity" might better describe him, and his letters to his estate agents reveal him to be a man of "good sense."[32] The same can be said of the taciturn Cavendish's letters to his estate agent. Another Abercorn, John James Hamilton, first marquess of Abercorn, 1756–1818, was taciturn on principle. Extravagant and rank-conscious, he welcomed celebrated people to his home at Stanmore, where they were free to ride, hunt, or do anything they liked, but if they encountered him they were under no circumstance to speak to him; he spoke only at meals.[33] A "supreme example" of late eighteenth-century taciturnity was Francis Rawdon Hastings, known as earl of Moira, 1754–1826, an Irish-British politician, military officer, Governor-General of India, and fellow of the Royal Society. He enjoyed company, he was always courteous, and he spoke eloquently in public, but at his own table he said nothing beyond inviting a guest to drink a glass of wine; he would sit for hours in silence.[34]

Langford says that even foreigners who were sensitive to English inconsistencies "made an exception for taciturnity, one constant characteristic of an Englishman." Here is a sample of observations of this English characteristic. The English preferred silence to wasted words, which they regarded as wasting time. If they had servants, they expected them to be mostly silent. Regardless of their rank in society, persons who talked too much were considered bores. Conversation at dinners descended to whispering to neighbors. At parties, instead of holding conversation, cards were played. Foreigners regarded an English conversation as a long silence. The proper conduct of the English was "in large measure about silence, suspicion, and separateness." The English associated silence with wisdom, good manners, and modesty; nothing, it seemed, could draw them out.[35] If Brougham's comparison of Cavendish with the

---

[29] Yolanda Foote, "Waterton, Charles (1782–1865)," *DNB* 57: 573–575, on 575.

[30] Young, "Cavendish," 444.

[31] Timbs, *English Eccentrics* 2: 150–61, on 160. Geoffrey V. Morson, "Porson, Richard (1759–1808)," *DNB* 44: 919–24, on 920.

[32] David Huddleston, "Hamilton, James, eighth earl of Abercorn (1712–1789)," *DNB* 24: 854.

[33] Timbs, *English Eccentrics* 2: 286. Donaldson, *Brewer's Rogues*, 3.

[34] Langford, *Englishness*, 304. Roland Thorne, "Hastings, Francis Rawdon, first marquess of Hastings and second earl of Moira (1754–1826)," *DNB* 25: 744–49.

[35] Langford, *Englishness*, 175–77, 179–80, 185–88, 216.

monks of La Trappe was apt,[36] then his silence was conspicuous even in a society that looked favorably on silence. Observations of English taciturnity were often perceptive, though their generalization to all English was, of course, in the imagination.

Feeling at ease at his club, "where he would not encounter strangers in any number," an Englishman became communicative, accessible, "almost loquacious." Removed from that comfort, he reverted to form, the sulky English male.[37] To a visitor, even English clubs could seem strangely quiet: a man belongs "not so much because he likes to speak, as that he possesses the art of holding his tongue," members being expected to respect one another's silences. At a dinner with colleagues, Cavendish suddenly broke his silence, addressing his neighbor, "I am told that you see the stars round, Dr. Herschel." "Round as a button," Herschel replied. Cavendish lapsed into silence again until nearly the end of dinner. "Round as a button?" he asked in a doubtful voice. "Exactly, round as a button," Herschel replied.[38] The exchange is an example of Cavendish's silence and his occasional departure from it, but since Herschel said no more than Cavendish, it is also an example of English dinner conversation. Notoriously silent, Cavendish had the manners of a proper English gentleman carried to lengths.

This description of club life and Cavendish's conversation with Herschel need to be set beside recollections of Cavendish talking freely to colleagues at his clubs. He was normally reserved, but when he was familiar with a person, his "conversation was lively, and full of varied information."[39] This was the other side of a silent English gentleman, who could be "almost loquacious" in a reassuring setting.

Emotions. A foreign visitor observed, "England is not the country of emotions."[40] By the end of Cavendish's life, English stolidity, impassivity, and self-control were identified with the national character, which was an object of pride as evidence of rationality and disproof of superficiality. Described as the "coldest and most indifferent of mortals,"[41] Cavendish had an overabundance of this English virtue.

Weeks and his associates found that eccentrics by and large are optimistic, confident, happy, and buoyant. This is unsurprising since their sample of self-identified eccentrics consisted mostly of extroverts, who it is well known feel comfortable in their own skin and get more enjoyment from living than other people. Cavendish, Weeks's historical eccentric, was "confident" in science, but none of the other words describing extroverts apply to him. Among the reports of Cavendish's behaviors, one mentions a "shrill laughter," but it was not caused by anything we would find amusing.

Shyness. Timbs observed that the "eccentric character" is "a man who always casts his eye towards the ground, as if he were ashamed of looking any one in the face;

[36] Brougham, *Lives* 1: 259.

[37] Ibid., 179, 215–16.

[38] Constance Lubbock, *The Herschel Chronicle: The Life Story of William Herschel and His Sister Caroline Herschel* (Cambridge: Cambridge University Press, 1933), 102.

[39] John Davy, *Memoirs* 1: 222.

[40] Langford, *Englishness*, 250.

[41] Quote from one of his informants, Wilson, *Cavendish*, 173.

and who pretends, when accosted, to be near-sighted, so that he does not know even the friend that had served him. This short-sightedness is very common."[42] Cavendish might have been taken for this eccentric character at the beginning, but his eccentricity was more profound. Thomson described Cavendish as "shy and bashful to a degree bordering on disease" and seeming to "feel himself unfit for society."[43]

Strangers. Edmund Burke said that the English were reputed to be "a sullen, unsocial, cold, unpleasant race of man." The philosopher Immanuel Kant explained why the English merited this description. "The Englishman behaves insolently toward everyone else because he thinks that he is self-sufficient, that he does not need anyone else and so can dispense with being pleasant to other people." He "easily becomes an eccentric not out of vanity but because he concerns himself little about others, and does not easily do violence to his taste out of complaisance or imitation."[44] Kant captured an aspect of Cavendish's eccentricity. Vanity played no part in it, but unconcern did; Wilson called it Cavendish's "*passive* selfishness."[45]

Strangers were advised not to become "familiar," an un-English relationship implying the right to intrude. A foreigner observed that if a stranger "should venture to address them [the English], they receive it with the air of an insult." Strangers were told that unless Cavendish spoke to them first, they should not speak to him "as he would be offended."[46] On a visit to London, having obtained permission to use Cavendish's library Alexander von Humboldt was cautioned that if he should encounter the owner "he was on no account to presume so far as to speak, or even greet" him.[47] Cavendish's reaction to strangers went beyond manners: he had a "perfect horror" of a strange face.[48]

Solitude. Cavendish's physical privacy was protected by his household rules. Research, writing, and reading, his principal activities, were private activities, temporary withdrawals from the public world. In his laboratory, he worked alone or with an assistant; alone in his study, he read and wrote. In either place, he was master of his own time. Eighteenth-century writers were interested in another kind of privacy, psychological, which being privacy of the mind could be summoned in a social setting. This kind of privacy was normal for Cavendish too. In the company of colleagues, he conveyed a desire to be apart but also a desire to communicate, a shifting balance.[49]

---

[42] Timbs, *English Eccentrics*, 40–41.

[43] Thomson, *History of Chemistry* 1: 337. Barrow, *Royal Society*, 144.

[44] Langford, *Englishness*, 226, 291.

[45] Wilson, *Cavendish*, 173.

[46] Pepys, quoted in Wilson, *Cavendish* 168. Langford, *Englishness*, 238, 249, 255.

[47] From K. Bruhn's *Life of Alexander von Humboldt*, in James Thorne, *Environs of London* (London, 1876) 1: 111.

[48] Children, quoted in Wilson, *Cavendish*, 169.

[49] Patricia Meyer Spacks, *Privacy: Concealing the Eighteenth-Century Self* (Chicago, London: University of Chicago Press, 2003), 7–8, 21, 27–28.

Jealous of his privacy, the Englishman "could not tolerate ease of access to his home." Standing for his freedom and independence, his home was his castle; in dread of intrusion and interruption, he shut himself up. For an acquaintance to drop in at his house at dinnertime and expect to be fed, or for a banker to call unannounced and expect to do business, was in the worst of taste. In dining houses in the city, spaces for guests were separated off as stalls. London squares separated their occupants from the city. The single-family separate house was the English ideal. England was divided up by hedges, brick walls, and iron fences.[50] Cavendish's solitude was firmly anchored in English social customs, distinguished only by its insistence.

Jeremy Bentham, 1748–1832, social reformer, founder of modern utilitarianism, and eccentric, worked alone and saw little company, and in these ways he resembled Cavendish. He was described by the essayist William Hazlitt as "an anchoret in his cell," the same image Wilson used to describe Cavendish. Bentham referred to himself as a "hermit," another word used for Cavendish.[51] In his desire for separateness, Cavendish was probably even more determined than Bentham. The impression he gave was as "a solitary being in the world," whose sole "desire seemed to be to be alone and to be left alone."[52]

Women. When Cavendish was small, he probably had a governess, but all trace of encounters with women in his formative years is lost. The death of his mother might bear on his later reaction to females. There is no doubt that English traditions contributed to his difficulty. His schooling was carried out in an all-male setting, as were all his later scientific activities. Males were brought up to behave "with extreme caution where women were concerned," and some never learned how to relate to them. In high society, mixed company was the exception rather than the rule, men congregating in one place and women in another. Men tended to place little value on women's conversation. Samuel Richardson's said that "there are very few topics that arise in conversation among men, upon which women ought to open their lips. Silence becomes them." The relations between men and women entered a running debate on the unsociability of the English.[53] Sightings of Cavendish avoiding women are perfectly credible.

---

[50] Langford, *Englishness*, 106–8.

[51] Philip Lucas and Anne Sheeran, "Asperger's Syndrome and the Eccentricity and Genius of Jeremy Bentham," *Journal of Bentham Studies* 8 (2006): 1–37, on 7, 17. The authors make the case that Bentham had Asperger's syndrome. John Stuart Mill, who knew Bentham well, said, "In many of the most natural and strongest feelings of human nature he had no sympathy; from many of its graver experiences he was altogether cut off; and the faculty by which one mind understands a mind different from itself, and throws itself into the feelings of that other mind, was denied him by his deficiency of Imagination." Ibid., 7. This is a textbook description of a "theory of mind" deficiency, commonly present in persons with Asperger syndrome. Other than for their hermit-like behaviors, Bentham and Cavendish were not very much alike. Bentham treated human emotions and motivations in his work. Cavendish studied the laws of the physical world. The authors point to a number of Bentham's traits that, so far as we know, do not apply to Cavendish: vanity, overconfidence, egocentricity, and unclear expression in writing.

[52] Dr. Sylvester, quoted in Wilson, *Cavendish*, 170. Barrow, *Royal Society*, 144.

[53] Langford, *Englishness*, 200, 248, 304.

The reason he was singled out was the extent of his avoidance: from the agreement of several of several sources, Wilson concluded that Cavendish showed “extreme repugnance to encounter females.”[54] In his *English Eccentrics*, Timbs headed his entry, “The Woman-Hating Cavendish.”[55]

Regularity and Routine. Typically a person was eccentric not for any one remarkable deed but for a habit. A person was eccentric who always walked through doors backwards or who was always miserly. Rigidity at the dining table is another example: an unvarying diet, if maintained, counted as eccentric. The British raised cattle and sheep in their pastures, and they liked to have beef and mutton on their tables. British statesman thought that their nation’s success in arms owed in part to its diet of red meat.[56] The following two cases illustrate eccentricities centered on meat. The reclusive Francis Edgerton, eighth earl of Bridgewater, 1756–1829, was a manuscript collector, fellow of the Royal Society, and a relative of Cavendish’s. He had his cook prepare nothing but boiled beef and potatoes, this whether or not he had guests. On one occasion while living in France, he set out from Paris with 16 carriages filled with luggage and 30 servants on horseback, and when after a few hours he could find no dining house that served boiled beef, he ordered the whole procession to return to Paris. His dining eccentricity went beyond diet. Preferring dogs to people, he gave dinner parties for them, dressing them in high fashion including miniature fine shoes, and assigning a servant to each dog, which was given a napkin.[57] The second example of a dietary eccentric is the Scottish physician George Fordyce, 1736–1802, fellow of the Royal Society and of the Royal College of Physicians, and well known to Cavendish for his important experiments on heat and in chemistry. His scientific work convinced him that humans eat more often than they need to. Following the example of the lion, which gorged and fasted, he ate only once a day, at the same hour and at the same place. His main dish was always a pound and a half of rump steak. While the steak was being prepared, he ate a trifle, which might be half a broiled chicken or a plate of fish. To drink he had a tankard of strong ale and a bottle of port wine, and he bought a quarter pint of brandy to take with him. When he finished eating, he stopped at three coffee houses, where he sipped his brandy with water. Then he returned home to give a lecture in his private course on chemistry.[58] Cavendish’s reputed inflexible menu for dinner, leg of mutton, was eccentric, but as we see here not extreme. Moreover, his inhospitality and the uninviting arrangement of his houses can be questioned. In 1785, on

---

[54] Wilson, *Cavendish*, 170.

[55] Timbs, *English Eccentrics* 1: 142.

[56] Aymes-Stokes and Mellet, “Introduction,” 12, 48–49.

[57] Donaldson, *Brewer’s Rogues*, 99. Timbs, *English Eccentrics* 1: 111–13. Jonathan R. Topham, “Egerton, Francis Henry, eighth earl of Bridgewater (1756–1829),” *DNB* 17: 994–95. Victorian Web, “Francis Egerton, 8th Earl of Bridgewater,” http://www.victorianweb.org/science/science_texts/bridgewater/b1.htm.

[58] Sitwell, *English Eccentrics*, 184–86. Timbs, *English Eccentrics* 2: 4–7. N. G. Coley, “Fordyce, George (1736–1802),” *DNB* 20: 358–60, on 359.

a visit to London, Watt said he dined "at Mr. Cavendish's, who lives very elegantly, and gave us a good English dinner."[59]

House and Servants. To avoid direct communication with female servants, Cavendish left a note at a certain hour on the hall table with instructions for his dinner.[60] Since we are told he always ate leg of lamb, the instructions must have been more or less the same, perhaps with some adjustment in time to agree with his researches and meetings. Charles Seymour, sixth duke of Somerset, 1662–1748, politician and courtier, avoided intimacy by communicating with servants through hand gestures rather than speech. The reason behind his method was the inordinate pride he took in his birth and rank and a corresponding disdain for common people.[61] Sir Henry Harpur Crewe communicated his instructions to his servants by letter because he was afraid of them.[62] Somerset's and Crewe's methods may have been atypical, and their motivations different, but they both acknowledged a distance between master and servant required by English manners. Relations between masters and servants in English homes were characterized by an absence of human warmth.[63]

According to a neighbor at Clapham, "Cavendish would never see a female servant, and if an unfortunate maid ever showed herself she was immediately dismissed." This behavior might properly belong under the heading of misogyny, which is where Wilson places it, but if so it also belongs here under servant and master behavior. If Lord Crewe saw a maid after 10 in the morning, he immediately dismissed her.[64]

When Cavendish encountered one of his maids with cleaning tools on the stairs, he immediately had a back stairs built.[65] This too has been taken as evidence of his misogyny, but it can be seen another way. The addition of back stairs in British houses was common, the object being to remove servants as far as possible from their masters except when they were called to present themselves. For this reason, it has been termed a "revolutionary invention." By the time Cavendish built his back stairs, it was no longer revolutionary, having been around for 100 years.[66]

Wealth. Persons who amassed fortunes or were profligate were not considered eccentric,[67] but wealthy persons who were miserly were. A well-known example is the eccentric John Elwes, 1714–89, a land owner, politician, and builder of several London squares, who inherited over £250,000. A man of integrity and considerable accomplishment, he gave generously to friends and gambled and speculated

---

[59] Crowther, *Scientists of the Industrial Revolution*, 311.

[60] Brougham, *Lives* 1: 258–59.

[61] Donaldson, *Brewer's Rogues*, 584–85, on 584. R. O. Bucholtz, "Seymour, Charles, sixth duke of Somerset (1662–1748)," *DNB* 49: 857–60, on 859.

[62] Langford, *Englishness*, 303.

[63] Ibid., 241–44.

[64] Wilson, *Cavendish*, 169–70. John Timpson, *Great English Eccentrics* (Norwich: Jarrold Publishing, 1991), 84.

[65] Wilson, *Cavendish*, 170.

[66] Spacks, *Privacy*, 6.

[67] Wilson, *Eccentric Mirror,* 1.

prodigiously, but on his own needs he spent less than £100 a year. When he traveled, he put two or three hard-boiled eggs in his coat pocket, avoided toll gates as much as possible, rode his old brood mare on the grass to spare the horseshoes, and stopped at streams to spare him the cost of inns.[68] Another example is James Lowther, first earl of Lonsdale, 1736–1802, an in-law of Cavendish's and the occasion of a law suit over money with his father. Possessed of immense wealth, he spent money to buy power, but otherwise his goal in life was to hoard money, not spend it. Confident of his social superiority, he felt no need to exhibit it. To show his contempt for common opinion, he rode in a dilapidated carriage, pulled by un-groomed horses.[69] Owing to his frugality, the earl of Lonsdale's great benefactor Sir James Lowther of Whitehaven, known as the richest commoner in the kingdom, was also known as "Farthing Jemmy."[70]

Cavendish spent what he needed for the life he chose, no more and no less. Around the time his father died, his income was about £15,000 a year. At the time of his own death, it was around £40,000.[71] Apart from inheritance, the origin of his wealth, the reason he grew ever richer was that he spent less than he received from his Bank of England annuities. His wealth drew attention because of its size, not because of what he did with it. His unaffected way of living was in keeping with the English character. In the way he treated his wealth, Cavendish was not interesting enough to be eccentric.

Religion. Cavendish gave the impression that religion meant nothing to him.[72] In his one reference to religion in his youth, in his poem on the crown prince, he subordinated religion to his special interest, nature and natural philosophy. Religion, as mentioned, occasionally entered eccentric biographies as wayward clerics,[73] who were usually Anglicans of the latitudinarian variety, their eccentricity consisting of an over-fondness for drinking, racing, and hunting. Religious enthusiasts were disquieting and for that reason they were rarely considered eccentrics.[74] Persons like Cavendish who took no outward interest in religion were in a minority but not eccentric for that reason.

Physical World. Cavendish was focused. Throughout history, we find individuals with obsessive interests and work habits like his. His colleague the astronomer William Herschel, 1738–1822, was as focused, and he worked at least as hard, probably harder. John Wilkinson, 1728–1808, an ironmaster, whom Cavendish visited

---

[68] Donaldson, *Brewer's Rogues*, 241–42. Anon., *Biographical Sketches*, 169–76. Alexander Gordon, rev. Anita McConnell, "Elwes, John (1714–1789)," *DNB* 18: 370–71, on 370. Life of John Elwes, 36-page section, in G. H. Wilson, *The Eccentric Mirror, Reflecting a Faithful and Interesting Delineation of Male and Female Characters …*, vol. 3 (London, 1807).

[69] Donaldson, *Brewer's Rogues*, 409.

[70] Anon., "Sir James Lowther, 4th Baronet," http://en.wikipedia.org/wiki/Sir_James_Lowther,_4th-Baronet.

[71] 2 March 1810, Charles Blagden Diary, Royal Society 5: 429.

[72] Wilson, *Cavendish*, 181–82.

[73] Several clerics are included in Timbs, *English Eccentrics*, vol. 1.

[74] Langford, *Englishness*, 306.

on his journey in 1785, was said to be possessed of "iron madness." A great advocate of new uses of iron, he surrounded himself with objects made of the metal. Among his "more eccentric creations" were several coffins and an obelisk memorial built for him out of iron, insuring that his physical surroundings would not change in the next life.[75] Cavendish's manner of work in science was not itself eccentric. It contributed to his eccentricity because it excluded almost every other interest in life.

Knowledge. Library owners who showed a rage for collecting books were eccentric. An example is Richard Heber, 1773–1833, scholar and heir to a fortune, who filled several houses in London and elsewhere with his books. A visitor to one of them described what he saw: "Up to the very ceiling the piles of volumes extended, while the floor was strewn with them, in loose and numerous heaps." Heber believed that no man should be without three copies of a book, one for show, one for use, and one for borrowers.[76] Cavendish was satisfied with one copy. He had the money to be an eccentric collector, but not the desire. He bought books to learn about the physical world, not to collect them. He was known for his remarkable breadth of knowledge in natural philosophy acquired in part by a lifetime of reading books from his library shelves.[77] A man could be eccentric for holding fanciful beliefs but not for having a mastery of the sciences.

Honesty and Honor. An Englishman's word was his bond, a valued trait of the English character,[78] though it was not specifically English. A contemporary of Cavendish's wrote of the "hereditary probity of the Cavendish family." The reference was to a less gifted cousin of Cavendish's, Lord George Augustus Cavendish, a Member of Parliament, in whom "probity … supplied the place of ability."[79] Blagden spoke of Cavendish's "strict principles of sincerity," a synonym of honesty.[80] When Cavendish's cousin Lord Frederick repaid him for a loan of £3,500, he added £37 to the amount as interest. Cavendish wrote to him that there had been no mention of interest in the promissory note and so he could not "with honour accept it."[81] His probity was a trait of the English character.

Openness. The English expected openness from the press, politicians, and fellow countrymen, though they might allow a limited case for secrecy in government and commerce. Their case against secrecy in general was the suspicion that it concealed discreditable behavior.[82] The English preferences for openness and privacy were

---

[75] J. R. Harris, "Wilkinson, John (1728–1808)," *DNB* 58: 1010–13, on 1012. Anon., "John Wilkinson (industrialist)," http://en.wikipedia.org/wiki/John_Wilkinson.

[76] Timpson, *Great English Eccentrics*, 126–28.

[77] Brougham, *Lives* 1: 250.

[78] Langford, *Englishness*, 122.

[79] Nathaniel Wraxall, quoted in the entry on Lord George Augustus Cavendish, in *The History of Parliament: The House of Commons 1790–1820*, ed. R. Thorne (London: Boydell and Brewer, 1986), http://www.historyofparliamentonline.org/volume/1790-1820/cavendish-george-augustus.

[80] Blagden to Cavendish, August 1789.

[81] Henry Cavendish to Lord Frederick Cavendish, 3 December 1798, Devonshire Collections, Chatsworth.

[82] Langford, *Englishness*, 96, 99.

compatible, the former being a matter of candor, the latter of sociability. Numerous examples of Cavendish's practice of openness are given earlier in the book. In this, Cavendish again exemplified the English character.

Scientific Society. English liberty had two foundations: one was individuality, the freedom of the Englishman to be himself, the rationale of his eccentricity, and the other was association, the ability of self-governing Englishmen to cooperate to achieve a common good.[83] English liberty was seen in the organization of scientific talent in England. The Royal Society differed from scientific academies in Europe by its independence, its president Banks explained to a foreigner. Scientific academies in Europe were assembled by their monarchs and were kept at their beck and call, under obligation to answer any questions put to them by their governments. By contrast, Banks said, "we are a set of free Englishmen elected by each other & supported at our own expence," under no obligation to "receive orders or directions from any department of government be it ever so high." That was their "chief boast."[84] We see an idiosyncratic version of English liberty in Henry Cavendish and scientific society, the eccentric solitary willingly forming successful associations with colleagues to further the advancement of science.

Simplicity. Cavendish had "a truly philosophical simplicity of manners," Blagden said.[85] A foreign visitor observing the "proud aristocracy of England" found their manners to be "perfectly simple." In a time of the cult of nature, simplicity was valued for its naturalness, as revealed in a person's unaffected bearing. English society being relatively free, a person could speak openly and say what he thought, simply. In language, simplicity or plainness was preferred over figure and metaphor, making the point of openness. Simplicity expressed desirable English traits: absence of artificiality, presence of good sense, and moderation of emotions.[86] Cavendish's preference for simplicity in all things was exemplary not eccentric.

Cavendish Doubles. So far, we have compared Cavendish with eccentric individuals one trait at a time. Here we look at eccentric individuals bearing several resemblances to Cavendish.

The reclusive scholar is a common stereotype in fiction and in life. Like Cavendish, John Barrett, 1753–1821, a distinguished classicist at Trinity College, Dublin, was said to have had a "simplicity of manners." Reclusive and regular in his behavior, he left the walls of his college only three times a year, twice to cross the street to his bank and once to accompany his fellows to an observatory. He was indifferent to dress, his suit threadbare and patched. Unmarried, he died wealthy, though he differed from Cavendish in how he viewed his wealth. His income was £2,000 a year, of which he spent no more than £20 except for books. His outstanding eccentricity was

[83] Mandler, *English National Character*, 53.

[84] Joseph Banks to Count W., 2 June [n.y.], Banks Correspondence, Royal Botanic Gardens, Kew, vol. 3, 1.3.

[85] Charles Blagden to William Cullen, 8 May 1784, Blagden Letters, Royal Society, C.70.

[86] Langford, *Englishness*, 87–92.

miserliness, which did not interfere with his work. Among his accomplishments was the discovery in the library of a rare copy in Greek of a large part of the Gospel of St. Matthew's, which he translated and edited.[87]

If an eccentricity was sufficiently pronounced, its bearer could become famous solely for being an eccentric. This was the case with a future distant relative of Henry Cavendish's, William Cavendish-Scott-Bentinck, fifth duke of Portland, 1800–79. The two Cavendishes had a number of similar behaviors. They did not like people to look at them. They gave their servants instructions in writing and ordered them to keep out of sight. They stayed away from women. They disliked meeting people and rarely invited people to their homes. Both preferred to be alone, though in this desire the fifth duke far outdid Henry. Solitude was his distinguishing eccentricity. To avoid human contact, he excavated 15 miles of underground tunnels and rooms where he could roam at will, unseen and undisturbed. When he surfaced at night, he followed a lantern held by a servant, who was required to keep a distance of 40 yards. He traveled in a closed wagon, which was driven to the train station through his private tunnel. Around the garden of his town house, he built a screen so that he could not be seen by neighbors.[88] He lived in a few rooms of his mansion, scores of other rooms remaining empty of furniture, looking forlorn. The main difference between the two Cavendish eccentrics was that in his solitude Henry did important work.

Our next double is Francis Galton, fellow of the Royal Society, 1822–1911. Wilson said that Cavendish saw the universe as consisting of things that could be quantified. It would be more accurate to say that Cavendish was interested in those things that can be quantified, and even that wording needs qualification, but it is close. Galton saw the universe similarly, though in his zeal to quantify he went further than Cavendish. Being wealthy, he was free to pursue his "obsessive urge to quantify everything he saw around him," based on his belief that "almost anything could be counted, correlated, and made into some sort of pattern." To give an example, he drew up a boredom index based on the number of fidgets per minute of an audience. In the end, he left a mark on several fields: statistics, inheritance, eugenics, experimental psychology, exploration, anthropology, meteorology, and inventions. The president of the Geographical Society summed him up: his mind was mathematical, he was very smart, he had a strong sense of duty, he had no vanity, he had no tact, and he had little or no imagination. He could not connect emotionally with people. In these several respects, he bore a likeness to Cavendish.[89]

The tireless prison reformer John Howard, 1726–90, led a more typical life than Cavendish, marrying twice, serving as county sheriff, and practicing religion, but in

---

[87] Donaldson, *Brewer's Rogues*, 45–46. H. T. Welch, "Barrett, John (1753/4-1821)," *DNB* 4: 45–46.

[88] Donaldson, *Brewer's Rogues*, 507–8. Raymond Lamont-Brown, *A Book of British Eccentrics* (Newton Abbot, Devon: David & Charles, 1984), 52–53. "William Cavendish-Scott-Bentinck, fifth Duke of Portland," http://www.en.wikipedia.org/wiki/William_Cavendish-Scott-Bentinck_5th_Duke_of_Portland. Anon., "A Leader among British Eccentrics," http://www.suite101.com/article/a-leader-among-british-eccentrics.

[89] Pickover, *Strange Brains*, 134, 138, 140. Galton's obsessive thinking and behavior are analyzed in Lennard J. Davis, *Obsession, A History* (Chicago and London: Chicago University Press, 2008), 86–94. Clement Markhams was president of the Geographical Society.

Fig. 7.1 John Howard. By Mather Brown. Prison reformer (Wikimedia Commons)

other ways he was Cavendish's double (Fig. 7.1). He followed his special interest, prisons, with obsessional intensity. His daily activities were exactly regular and bound to routine. He conducted himself in an old-fashioned manner and in plain dress. On journeys, he traveled alone or with one servant. He was highly uncomfortable with his celebrity, refusing to sit for a portrait, his wish being to "remain in obscurity and silence." Because of his privileged standing in society, he was able to live as he wished.[90] In all these ways, he resembled Cavendish, who may have met him, since he was a member of the Royal Society and published papers on the weather in its journal.

We come to Cavendish's closest double. Not in his biography of Cavendish but in another publication, Wilson compared Cavendish with the English chemist William Hyde Wollaston, 1766–1828: the similarity in the behaviors of the two men was sufficiently precise and detailed they might have been "twin brothers." Wollaston's caution, like Cavendish's, was proverbial: according to Wilson, his "idea of a truth was

[90] Uta Frith makes the case that Howard had Asperger's syndrome. *Autism: Explaining the Enigma*, 2d ed. (Malden, MA, Oxford, Victoria: Blackwell, 2003), 55–57. Ioan James, *Asperger's Syndrome and High Achievement: Some Very Remarkable People* (London and Philadelphia: Jessica Kingsley, 2006), 53–61. Rod Morgan, "Howard, John (1726–1790)," *DNB* 28: 390–94, on 392–93.

not so much something proved true, than something which could not be proved not to be true." He protected his solitude even more rigorously than Cavendish did his, refusing to let friends into his laboratory. Wilson listed the similarities of the two men: a strong love of truth, taciturnity and reserve, precision in writing, catholicity of scientific interests, relish for scientific study, disinterest in subjects lying outside of science, frugal habits, perfect integrity, and candor and justice in dealing with men of science.[91] To another observer, the mathematician Charles Babbage, Wollaston's defining traits were "extreme caution," "precision," and avoidance of "errors," traits equally pronounced in Cavendish. On the distinction between what he knew and did not know Wollaston was never in doubt, and he was in no hurry to bring results before his fellows, having no fear of being anticipated by them. His air of infallibility led his colleagues to refer to him as "the pope." His knowledge of mathematics was more limited than Cavendish's, but he used it with such effect that he was considered by some as one of the greatest mathematicians of the age. He started out practicing medicine, but soon abandoned it. With a partner he developed a way to make the new element platinum malleable and practical, with the result that he was well-off, free to follow his scientific interests (Fig. 7.2).[92]

Cavendish knew Wollaston and thought well of him, proposing him for membership in the Royal Society.[93] An incident shows his appreciation of Wollaston's work. Wollaston and the Irish chemist Richard Chenevix were in disagreement over the nature of palladium: Wollaston claimed that he had isolated a new element; Chenevix said it was an alloy not an element. In 1803 Chenevix sent a paper on palladium as an alloy to the Royal Society, and in 1805 he sent a second one. In between these papers, Wollaston communicated to the Royal Society his identification of palladium as an element. Cavendish thought that Chenevix's second paper should not have been published because his first paper made his experiments doubtful. He was the only member of the Royal Society Council to vote against its publication, and, as Blagden saw it, he took Wollaston's part in the dispute.[94] The "twin-brother" chemists Wollaston and Cavendish, were right: palladium is an element, as foreign chemists would confirm before long. Another incident points to a similarity, their strong reserve. When Cavendish died, several younger chemists evidently hoped that he had remembered them in his will. The news that he had not remembered them affected Wollaston and Davy differently, according to Blagden. "Davy said, Mr. C. has at least remembered one man of science [Blagden], in a tone of voice which expressed much," but "Wollaston's countenance was unchanged."[95] Parallels

[91] George Wilson, *Religio chemici. Essays*, ed. J. A. Wilson (London, 1862), 286–87, 291, 293–94.

[92] Charles Babbage, *Reflections on the Decline of Science in England, and on Some of Its Causes* (London, 1830), 203–6. I thank Geoffrey Cantor for calling my attention to this passage. Trevor I. Williams, "Wollaston, William Hyde (1766–1828)," *DNB* 59: 991–93.

[93] Certificates, Royal Society 5 (9 May 1793).

[94] 1–3, 6–8, 17, 23 February; 9, 17 March 1810, Charles Blagden Diary, Royal Society 5: 307–9, 312–13, 318, 320.

[95] 6, 8 March 1810, Charles Blagden Diary, Royal Society 5: 430 (back)-432.

Fig. 7.2 William Hyde Wollaston. By John Jackson. Chemist and natural philosopher (Wikimedia Commons)

between Wollaston and Cavendish are striking but not surprising. Cavendish's solitary ways would place him in a population today of shy persons with obsessive interests, many of them scientists. His solitary ways would have done the same in his time. People who knew Cavendish were likely to have come across other individuals who shared his kind of strangeness.

His Brother. Henry Cavendish's younger brother had a connection with royalty. As a young man, Charles Cavendish was a gentleman of the bedchamber to the Prince of Wales Frederick, after whom he named his second son, the prince serving as his sponsor. Frederick had the same education as Henry: first Hackney Academy, then Peterhouse, Cambridge. While at the University, he had a bizarre accident. He fell from an upper window onto a court, sustaining a deep indentation on his head, which left him unable to return to the university or enter public life. For his last 40 years he lived with two domestics in a village in Hertfordshire, Market-Street.

The brothers had a fraternal regard, but they seldom saw one another. Frederick understood that the way to engage his brother was through family duty and science. The scant surviving letters between them are mostly about Frederick's finances and health (Chap. 13), with the exception of a letter by Frederick, which begins, "As I know you observe the Aurora Borealis with much attention, I send you an account of one which appeared last Night."[96] Frederick's head injury affected his competence, but it is unknown if it affected his personality too. In any case, he shared a number of traits with Henry. He took up two abiding interests, literature and drawing. Although his interests differed from Henry's, his pursuit of them was similar. Gifted with a powerful memory, he was called a "living edition" of modern English poets, and he built a substantial library of classics and modern literature. He developed considerable proficiency in drawing, taking pride in his accuracy in rendering natural subjects, plants and leaves. He preferred to "live plainly." His dress was of "the old-school; – deep ruffles, a bag wig, and cocked hat." He was a bachelor. He was proud of his family name and of his family's accomplishments. On the point of his high descent, he showed "some eccentricities," which became more marked in later life. At parties, he liked for the guests to toast to "the noble house of Cavendish, root and branch." He liked his friends to refer to him as "the honorable," and in his neighborhood most persons referred to him as "my lord," a mistake he did not correct. The parallels between the brothers are evident: active interest in nature, love of reading, excellent memory, book collecting, plain living, old-fashioned dress, bachelor life, family pride, and reputation for eccentricity. In two respects, the personalities of the brothers were different. One was in habits of charity. Although Frederick was not rich, he gave liberally, but even here there was a similarity: Henry depended on other people to tell him when and how much to give, and Frederick depended on the judgment of friends in giving to charity. The main difference in their personalities was Frederick's "very social disposition," though again there was a similarity: Henry too was sociable, though in a restricted sense, voluntarily joining the company of scientific men.[97] Whereas on the surface, the two brothers looked to be opposites, complementing rather than reflecting one another – Frederick artistic, convivial, generous, and obsessed with title; Henry scientific, aloof, shy, without any display of generosity, and outwardly indifferent to title – in personality they were broadly similar.

[96] Frederick Cavendish to Henry Cavendish, 1 March 1780; in Jungnickel and McCormmach, *Cavendish* (1999), 493–94.

[97] Anon., "Memoirs of the Late Frederick Cavendish, Esquire," *Gentleman's Magazine* 82 (1812): 289–91. The anonymous obituary was written by a friend of Frederick's.

Chapter 8
Autism

Fitzgerald writes that "many people with [Asperger's syndrome] are described as being eccentric," though in his experience most people "who have the word 'eccentric' attached to them do not meet diagnostic criteria."[1] By consensus, Cavendish was eccentric. If he met diagnostic criteria for autism is a question we consider here.

Sacks's finding that Cavendish was probably autistic was written up in *The New York Times*, "A Disorder Far beyond Eccentricity."[2] The title serves to mark the transition to our new subject. Our attention shifts from colorful personalities to persons whose behavior qualifies as a medical condition. In this chapter, we tentatively agree with Sacks and look at Cavendish from the perspective of autism.

Most of us have an idea of autism. We think of people who are remote, unable to bond with others, unresponsive to others' feelings, have trouble speaking, and are given to repetitive behaviors. No two autistic persons have identical traits, but a generalization can be made. They have difficulty understanding thoughts and feelings of other persons; they have a strong desire for truth, knowledge, and perfection, a preference for understanding the natural world over interpersonal relations, and useful skills and social vulnerabilities. In general, they have a different way of experiencing the world, a different way of thinking.[3] In the following paragraphs, we look more closely into what it means to be autistic.

Definitions and diagnostic criteria are given in the fourth edition of the *Diagnostic and Statistical Manual of Mental Disorders* (*DSM*-IV), published by the American Psychiatric Association, and in the tenth edition of the *International Classification of Diseases* (*ICD*-X), published by the World Health Organization. The manuals define "classic" autism as a disorder with three areas of difficulties. The first is social interaction: this includes unresponsiveness to others, lack of friends, disinterest in sharing, and atypical gestures, eye contact, facial expressions, and responses to the emotions of other

---

[1] Fitzgerald, *Autism*, 35.

[2] Erica Goode, "Cases: A Disorder Far beyond Eccentricity," Health Section, *The New York Times*, 9 October 2001.

[3] Tony Attwood, *The Complete Guide to Asperger's Syndrome* (London and Philadelphia: Jessica Kingsley Publishers, 2007), 333.

R. McCormmach, *The Personality of Henry Cavendish - A Great Scientist with Extraordinary Peculiarities*, Archimedes 36, DOI 10.1007/978-3-319-02438-7_8,

people. The second is verbal communication: this includes difficulty with language and conversation, and atypical aspects of speech in intonation, pitch, and emphasis. The third is repetitive behaviors: this includes preoccupation with narrow interests, mannerisms such as hand flapping, and insistence on fixed routines. There are other difficulties commonly found in autistic persons: intellectual disability, heightened or diminished sensitivity to sensory stimuli such as sight and sound, and perceptual problems in making sense of sensory stimuli.[4]

At one extreme of the "autism spectrum disorder" are persons who are unable to speak and are otherwise severely handicapped. The autism we are interested in is at the other extreme, the high-IQ end, which in *DSM*-IV and *ICD*-X enters as two separate and closely related categories, high-functioning autism and Asperger's syndrome.[5] The three areas of difficulties of the previous paragraph apply to Asperger's syndrome, with qualifications. Speech may or may not be normal, and certain behaviors are less evident; for example, hand flapping, preoccupation with objects, and impairment in intellectual development. The diagnostic criteria for autism and Asperger's syndrome are not identical, but the two disorders cannot always be distinguished,[6] and experts disagree whether to combine them or to keep them separate. The *DSM*-IV gives preference to the diagnosis of autism if the characteristics of a person fit both autism and Asperger's syndrome, though most clinicians reverse the order.[7] In this book, in acknowledgment of the revised *DSM*-V (see Chap. 16), wherever the word "autism" appears, it is to be understood as belonging to the part of the "autism spectrum disorder" that *DSM*-IV calls "Asperger syndrome" and "high-functioning autism."

Classified as a developmental disorder, autism is usually diagnosed early, the average age being between 3 and 4, Asperger's syndrome later, often at age 6 or older.[8] The one reference we have to Cavendish's early years comes from the family obituary, where Blagden wrote that Cavendish's "habits had, *from early life*, been secluded."[9] Blagden, if anyone, was in a position to know. It tells us that autism is not ruled out, but nothing more than that; some people who are not autistic live

---

[4] Ilona Roth, *The Autism Spectrum in the 21st Century: Exploring Psychology, Biology and Practice* (London and Philadelphia: Jessica Kingsley, 2010), 3–4, 38–41.

[5] "High-functioning autism" refers to autism with an above-average IQ and with language delay; Asperger's syndrome is without delay. The distinction between high-functioning autism and Asperger's syndrome may depend on the circumstances of the individual, and in practice the terms are interchangeable. For completeness, I mention the other sub-types of autism. "Pervasive developmental disorder – not otherwise specified" (P.D.D.N.O.S) – is the term used when autistic features are insufficiently pronounced for a definitive diagnosis of autism or Asperger's syndrome. "Atypical autism" is used when autistic features are only partly seen. The "autism spectrum" includes all these types. Simon Baron-Cohen, *Autism and Asperger Syndrome* (Oxford: Oxford University Press, 2008), 14, 21–26. Roth, *Autism Spectrum*, 42.

[6] Roth, *Autism Spectrum*, 41–42.

[7] Attwood, *Asperger's Syndrome*, 44.

[8] Temple Grandin, *The Way I See It*, 2d rev. ed. (Arlington, Texas: Future Horizons, 2011), 8. Baron-Cohen, *Autism and Asperger Syndrome*, 37.

[9] Blagen's contribution to the family obituary. Italics added.

secluded lives from early age. More specific information about his early years is lacking, and to that extent his autism is undeterminable. There are diagnostic criteria specifically for adults with autism, but their departures from the standard criteria depend on evidence we do not have for Cavendish.[10] The case for his autism depends on the degree of match between his adult behaviors and those of persons known to be autistic. If we say that Cavendish was autistic, this implies that if we knew about his childhood, we would find autistic traits there. In what follows, we revisit certain of Cavendish's traits, including those that made him appear eccentric, and compare them with traits commonly found in persons diagnosed with autism.

Before we begin, a word of caution is in order. To make a diagnosis of autism, it is not enough to check off the symptoms in the diagnostic criteria as laid out in the *DSM* or *ICD* or the Gillberg list. Agreement between the symptoms and the criteria is important in making a reliable diagnosis, but so are the severity of the symptoms and their effect on the disorder. The determination requires clinical training. We will find that a good many of Cavendish's personality traits are similar to ones commonly found in persons who are diagnosed with autism. The match is suggestive and perhaps significant, but on this basis alone we cannot conclude that Cavendish was or was not autistic. In the next chapter, the Conclusion (Chap. 9), we see what we can make of our findings.

Who was Henry Cavendish? Contemporaries knew him as a rich aristocrat, a talented natural philosopher, and an individual with peculiar behaviors. In this chapter, the tentative answer to the question adds another description, autistic.

Because autistic people have difficulty interpreting other people's reactions to them, Fitzgerald says that they have "a very distorted sense of self and a very unusual identity," "a deficient sense of self." They are "unable to form a coherent picture of themselves."[11] He gives as a historical example the philosopher Ludwig Wittgenstein, whom he associates with "identity diffusion and lack of sense of self." Throughout his life, Wittgenstein went off in various directions, taking on a wide range of occupations, struggling with the problem of identity, and frequently reinventing himself. His identity and sense of self were "fragmented in the same way that his view of the world was fragmented."[12] Cavendish showed none of Wittgenstein's restlessness, and his view of the world would seem to have been no more fragmented than natural philosophy, on which it was largely based. There is no sign that he wrestled with the question of who he was. If when he was young he had difficulty forming a coherent picture of himself, the difficulty was resolved when he elected a life of natural philosophy. Cavendish probably would have answered our question with the obvious: he was a member of a distinguished aristocratic family, and he was a natural philosopher. He was conscious of his peculiar speech, according to observers, but what he made of it we have no idea.

---

[10] Attwood, *Asperger's Syndrome*, 48–50.

[11] Fitzgerald, *Autism*, 46–47.

[12] Ibid., 95–97.

Appearance. Cavendish's dress was always the same, regardless of fashion. He walked with a slouch, with one hand behind his back and possibly the other inside his coat.[13] Autistic people "tend to be unconcerned about fashion or whether what they wear is contemporary."[14] They tend to wear the same clothes all the time, a habit explained in part by their adherence to routine and also in some cases by their sensitivity to touch. Temple Grandin, a scientist diagnosed with Asperger's syndrome, dressed like a slob for years without realizing it.[15] Autistic persons can differ from others in their way of moving, a matter of poor balance and coordination.[16] Clumsiness is one of Christopher Gillberg's criteria, an alternative to the *DSM*'s criteria for Asperger's syndrome.[17] Walking with scarcely any arm motion is an Asperger trait.[18]

Speech, Silence, and Eyes. Cavendish's speech was shrill and hesitant. He was averse to conversation, entering into it usually only when it was on a scientific topic, and then diffidently. Eye contact could bring an immediate end to conversation. In company, Cavendish uttered cries as he moved about, and when approached by a stranger, he might abruptly turn away with a cry. He repeated parts of speech.[19]

Withdrawal upon eye contact, involuntary vocalizations and repetitive patterns of speech are common autistic behaviors. *DSM* criteria for Asperger's syndrome refer to "abnormalities in inflection," "talking too much" or "too little." Speech can be "unusually high-pitched" and have unusual "stress and rhythm." Gillberg's criteria single out "odd prosody, peculiar voice characteristics." The most obvious difficulty autistic persons have in speaking is to adjust what they say to the social situation, failing to recognize that other people have other interests. Conversation for them is a chance to exchange information, to learn something, and if there is no exchange there is no point to it, and they prefer not to waste their time. "Idle chit chat is boring," Grandin finds. Autistic persons speak in facts, and without wishing to, they are often tactless. They frequently fall silent for no clear reason.[20] They do not look at the eyes as much as ordinary people do, and when they do look, they are not good at reading them; because the eyes give important clues about emotional messages, they are at a social disadvantage.[21]

---

[13] A chemist, quoted in Wilson, *Cavendish*, 168. In the sketch of him, his other hand is inside his coat. It is possible that the drawer invented the hand inside the coat, as it was a common pose for formal portraits.

[14] Asperger Management, http://www.aspergermanagement.com/personal-appearance.

[15] Temple Grandin and Sean Barron, *The Unwritten Rules of Social Relationships*, ed. V. Zysk (Arlington, Texas: Future Horizons, 2005), 377.

[16] Attwood, *Asperger's Syndrome*, 259.

[17] For many clinicians, Christopher Gillberg's diagnostic criteria are preferable to the *DSM*'s, being closer to Hans Asperger's original descriptions. Attwood, *Asperger's Syndrome*, 53.

[18] Ledgin, *Asperger's and Self-Esteem*, 46.

[19] Wilson, *Cavendish,* 167–68, 175. Barrow, *Royal Society*, 144.

[20] Frith, *Autism*, 128–29. Attwood, *Asperger's Syndrome*, 37, 206, 224, 266–67. Grandin, quoted in Ledgin, *Asperger's and Self-Esteem*, xiii.

[21] Frith, *Autism*, 104–5.

Emotions and Moods. Cavendish came across as cold, without normal emotions.[22] Hans Asperger wrote of his subjects, "They are strangely impenetrable and difficult to fathom. Their emotional life remains a closed book."[23] Autistic persons have difficulty in understanding emotions and how to act in response to them.[24] An emotion that is especially hard for them is "intentional empathy," which depends on the situation and on the minds of others. They respond to sufferings of others, but they may not know how to show sympathy in a given case, and accordingly they may appear callous. They do show empathy of a direct sort, an intuitive emotional response that does not depend on knowing what is inside another person.[25]

The most difficult lessons for autistic persons to learn are "social and emotional."[26] Marc Fleisher, a mathematician with Asperger's syndrome, writes, "Socializing is harder than any Maths equation for me," and social confidence is "most daunting."[27] Autistic persons learn "social skills" by conscious observation and study, whereas other persons know them instinctively, having the benefit of "emotional relatedness." Emotional relatedness is about the ways "people interact emotionally with each other," about their "sense of connection to others," and typically involves "expressions of affection."[28] The distinction is illustrated by Cavendish's colleagues' observation that he showed "no affections, but always meant well." We could say that Cavendish lacked the emotional relatedness we call "affections," and that he learned compensating social skills, which translated as "always meant well."[29]

Grandin's emotions are simple and often strong.[30] She can become very emotional over situations that are "directly tied" to her "whole sense of being," to her professional identity and to the work she does. Her actions, however, are directed by intellect, not by emotions.[31] In Chap. 3, we met examples of Cavendish's emotional responses to a threat to his work, chemistry, and to an affront to his honor, matters in which his sense of being as a natural philosopher and a Cavendish was strong.

Autism is associated with mood disorders, especially anxiety and depression, which often go together.[32] It is estimated that more than half of the persons with Asperger's syndrome experience depression, partly because of "their greater awareness of being different and greater awareness of their difficulties."[33] Grandin was in a constant state of anxiety until she discovered antidepressants, which turned her life around

[22] Wilson, *Cavendish*, 173.

[23] Asperger, quoted in Attwood, *Asperger's Syndrome*, 316.

[24] Roth, *Autistic Spectrum*, 80, 82.

[25] Frith, *Autism*, 111–12.

[26] Attwood, *Asperger's Syndrome*, 242.

[27] Marc Fleisher, *Making Sense of the Unfeasible: My Life Journey with Asperger Syndrome* (London and New York: Jessica Kingsley, 2003), 110.

[28] Grandin and Barron, *Unwritten Rules*, 38–39.

[29] 15 September 1794, Blagden Diary.

[30] Grandin, *Thinking in Pictures*, 89.

[31] Grandin and Barron, *Unwritten Rules*, 35.

[32] Attwood, *Asperger's Syndrome*, 129.

[33] Baron-Cohen, *Autism and Asperger Syndrome*, 98.

and eliminated "debilitating panic attacks." Her co-author Sean Barron, who also has Asperger's syndrome, ultimately overcame fear as a driving force in his life, replacing it with "interests, goals, and curiosity."[34] Cavendish exhibited considerable anxiety in certain social settings,[35] and he may have shown signs of mild depression. He evidently dealt with them the way Barron did, with interests, goals, and curiosity.

Solitude. Wilson said that Cavendish's most striking peculiarity was "a singular love for solitariness."[36] A craving for solitude can be a sign of autism. In solitude, autistic persons experience no significant impairment. The clinical psychologist Tony Attwood tells of a way of removing autistic traits from a person, and it requires no therapy and is very simple. You place the person in his room and leave him there alone, closing the door behind you as you leave. While he is there, he does not have the core criterion of autism, "qualitative impairment in social interaction," though presumably he still has nonsocial behaviors such as repetitive routines. Solitude is a powerful "emotional restorative," above all if the person is occupied with an absorbing interest. Difficulties come up only when he leaves the room and encounters people again.[37]

Women. With the rare exception, Cavendish did not keep company with women. He was thought to be a woman hater.[38] Autistic persons have difficulty with intimacy, emotional and sexual. Lack of sexual desire is more common than excess.[39]

Regularity. In "trivial" as well as scientific activities, Cavendish's life passed "with the utmost regularity." He was not, however, always good at putting order in his affairs. Left to his own devises, his library was in disarray.[40]

Autistic persons prefer regularity in all aspects of their lives, and they adopt inflexible routines to insure it. The need for sameness goes "to the heart of autism," the developmental psychologist Simon Baron-Cohen says.[41] Autistic persons often find organizing and planning difficult, however.[42]

House and Servants. Cavendish's instructions to his servants eliminated unwanted social interaction.[43] Except for the servants' wing, Cavendish's houses were spaces of solitude. The corresponding autistic trait is mentioned above.

Wealth. Cavendish's great wealth came by inheritance and sound investment, not by the pursuit of wealth.[44] His riches afforded him leisure and means to cultivate his

---

[34] Grandin and Barron, *Unwritten Rules*, 378, 380.

[35] Wilson, *Cavendish*, 169.

[36] Barrow, *Royal Society*, 144. Wilson, *Cavendish*, 165.

[37] Attwood, *Asperger's Syndrome*, 55–56.

[38] Wilson, *Cavendish*, 169–70, 178.

[39] Attwood, *Asperger's Syndrome*, 304–10.

[40] Thomson, *History of Chemistry* 1: 338. Wilson, *Cavendish*, 174.

[41] Attwood, *Asperger's Syndrome*, 173. Baron-Cohen, *Autism and Asperger Syndrome*, 40.

[42] Attwood, *Asperger's Syndrome*, 296.

[43] Wilson, *Cavendish*, 169–70.

[44] Family obituary.

special interest; otherwise he did not allow it to affect his life. Anecdotes tell of Cavendish's way of dealing with charity.

A special interest such as science can dominate the time and attention of an autistic person to the exclusion of most other matters, wealth included.[45] In the case of charity, Cavendish revealed the familiar failure of autistic persons to understand the thoughts of other persons.[46] He was not ungenerous, only incapable of spontaneous generosity; to give, he needed prompting.

Religion. Cavendish gave the impression that religion meant nothing to him.[47] Religious belief or its absence is not an autistic trait.

Death. On his deathbed Cavendish had no parting words for anyone. Consistent to the end, he banished his servant so that he could experience his last moments in perfect solitude.[48] An autistic person's preference for solitude is mentioned above.

Physical World. The only subjects Cavendish took much interest in were scientific,[49] or were related to science, and the sciences that interested him were physical. His interest was lifelong and steady. His relation to the physical world was his most important relation, the bedrock of his personality, on which his active life was built.

Many autistic children have learning difficulties. Although we know next to nothing of Cavendish's childhood, the outcome tells that any difficulties he had then did not stand in the way of the adult we do know. It is likely that his teachers recognized in him an uncommon interest in mathematics and the physical sciences and a distinct talent for them. The most common interest in the children Hans Asperger studied was nature, and some of them showed remarkable abilities and specialized knowledge in natural science, chemistry, and mathematics. He found that some were more interested in the physical world and its sciences, others more interested in animals and the life sciences. Either way, their interest could not be called a mere hobby; on the contrary, it took over their lives.[50] In Cavendish's day, it would have been natural for an autistic boy who acquired a special interest in the physical world to take up the study of natural philosophy. This was particularly likely to happen if the boy had a mentor and the mentor was a father who was a natural philosopher.

Parents can become impatient with their autistic children's endless talk about their special interest and their constant demands for items related to it. Cavendish's father might have done what parents today often do, reduce his son's access to his special interest and encourage him to engage in a variety of activities.[51] As it happened, his father had the same interest, and he owned instruments and books to support it. He probably did not find his son's talk and needs unfamiliar and burdensome.

[45] Attwood, *Asperger's Syndrome*, 172–73.

[46] Frith, *Autism*, 79–80.

[47] Wilson, *Cavendish*, 181–82.

[48] Young, "Cavendish," 445–46. Wilson, *Cavendish*, 182–84.

[49] Wilson, *Cavendish*, 182.

[50] Attwood, *Asperger's Syndrome*, 172.

[51] Ibid., 189, 192.

The botanist Robert Brown said that Cavendish's father appreciated his talent and never treated him unkindly.[52] Such a mentor, Grandin says, makes "the difference between a successful and satisfying life and one in which a person becomes a loner, remaining on the fringes of society and feeling miserable."[53] Attwood writes that in the outcome of a life, "it is not the severity of expression [of autism] that is important but the circumstances, expectations and coping and support mechanisms."[54] With his father's encouragement and financial support and later the support of colleagues, Cavendish realized his goal, a life of science.

The special interest of autistic persons is usually self-taught at the beginning, and it is pursued with feeling. The attention they devote to it can be a disruption of everyday life if it is without a larger meaning; this would be the case, for example, if the interest were match covers. Or it can be the source of great intellectual enjoyment and constructive work, as when the interest is, say, the physical sciences. Over time, it becomes more complex, sometimes evolving into two or more simultaneous interests.[55] In Part II of this book, we see Cavendish take an interest in industry and mining, especially in aspects that relate to his scientific studies. The *DSM* advises diagnosticians to determine that the special interest of a person is abnormal in intensity compared with interests of most other people.[56] All witnesses agreed that Cavendish pursued his scientific interest with uncommon intensity. He was acknowledged by all as an authority on the physical sciences, exhibiting a characteristic of autistic persons of being "natural experts."[57]

The special interest of an autistic person offers an alternative to social engagement, a haven. Its benefits are many: it overcomes anxiety, provides relaxation, and gives pleasure; it provides a sense of identity and builds self-esteem; it facilitates conversation, signaling intelligence; it gives coherence, order, and predictability to life; and it can lead to the study of the natural sciences. Many autistic adults recall the chaos they experienced early in their lives and the relief they found in the pattern and stability of the physical world, so unlike the social world in these key aspects. It was "natural" for them to have taken an interest in physical objects at an age when typical children are learning about interpersonal relations. They had an "innate interest" in the sciences of physical forces that affect their world such as meteorology, and in the sciences of patterns such as mathematics. They maintained their interest through their lives as an essential element in their "lifelong search for the pattern or meaning of life." The gratifications that come with their special interest are greater than the other pleasures of life.[58] We have good reason to think that Cavendish's special interest afforded him a rewarding and acceptably complete life.

---

[52] Wilson, *Cavendish*, 161.

[53] Grandin quoted in Ledgin, *Asperger's and Self-Esteem,* xiii.

[54] Attwood, *Asperger's Syndrome*, 54.

[55] Ibid., 60, 174, 180.

[56] Ledgin, *Asperger's and Self-Esteem*, 22.

[57] Attwood, *Asperger's Syndrome*, 178–80.

[58] Ibid., 182–88.

Thinking and Mind. Cavendish had a keen analytical mind, evident both in his mathematical-theoretical work and in his experimental work.[59] He brought a mathematical way of thinking to his interest in the physical world, with obvious success. His example was a distinctive contribution, recognized as such by his contemporaries.

There are several theories of autism, all of which address the thinking of autistic persons. The mind-blindness theory holds that children with autism are slow to form a "theory of mind," or the ability to imagine the feelings and thoughts of others. The executive dysfunction theory holds that autistic people are unable to plan actions well. The weak central coherence theory holds that they find it difficult to integrate information and form a coherent global picture, and instead focus on small details. The empathizing-systematizing theory holds that autistic people are deficient in empathy and skilled at systematizing. The latter two theories have marked similarities. Both see strengths as well as weaknesses in autism. Both recognize attention to details as a common autistic trait; both acknowledge that attention to details is a style of thinking, a preference that is not for all autistic persons. They initially differed over global processing of information, the empathizing-systematizing theory allowing for it, and the weak coherence theory considering it a deficiency in autistic persons; the latter theory has been revised, replacing a deficiency with a style of thinking, which allows for global processing.[60] The case for Cavendish's autism would fail without the recognition that there are autistic people who do not always direct their attention to detail.

Because Cavendish's thinking is known primarily through his scientific writings, the emphasizing-systemizing theory has the advantage that it makes explicit a parallel between autistic thinking and scientific thinking. It is a continuum theory with two variables: empathy, a social variable; and systematizing, a non-social variable. Empathy has its everyday cognitive and emotional meaning. Systematizing is a drive to make and understand systems, and a skill in recognizing patterns, regularities, and structures. In its most general meaning, a system is something that follows rules governing input, operation, and output. There are many kinds of systems; for example, mechanical systems such as windup clocks and wagons, natural systems such as weather and tides, and abstract systems such as physics and mathematics.[61] Systematizing is inductive, "specific-to-general thinking." The systematizer after obtaining a series of specific results formulates a general rule; and if a result breaks the rule, the rule is reexamined. Once the systematizer knows the rules of his system, he can predict how it works. Autistic persons are inclined to "analyze or construct systems," ordering details through rules; for example, a physical system through natural laws.[62] The practice of physical

---

[59] Humphry Davy, quoted in John Davy, *Memoirs* 1: 221.

[60] Frith, *Autism*, 157–68. Baron-Cohen, *Autism and Asperger Syndrome*, 53–57, 62–69. Attwood, *Asperger's Syndrome*, 241–43. Roth, *Autism Spectrum*, 122–35.

[61] Baron-Cohen, *Autism and Asperger Syndrome*, 62–63.

[62] Roth, *Autism Spectrum*, 128. Temple Grandin, "Comments," in Norm Ledgin, *Diagnosing Jefferson: Evidence of a Condition That Guided His Beliefs, Behavior, and Personal Associations* (Arlington, Texas: Future Horizons, 2000), 199.

science requires “high systematizing abilities.” The goal of autistic persons is to understand the system, and their reward is to discover the “*causes* of things.” The causes are not an end in themselves; rather they give autistic persons control over their world.[63] Natural philosophy and the sciences it contained were the origin of the systems Cavendish analyzed.

On questionnaires that measure empathy and systematizing, autistic persons score low on empathy and high on systematizing compared with the general population. This accounts for their uncommon attention to detail, which is needed to understand a system. It likewise explains their difficulty in generalizing; they understand each system as a unique system, and generalize only within that system.[64] On the empathizing part of a questionnaire Cavendish no doubt would score low, and on the systemizing part he would score high. In the laboratory, he paid uncommon attention to detail, as we would expect of an experimenter, and he generalized from experiments to all phenomena of the same kind, a form of reasoning widely practiced by natural philosophers. His scientific work was compatible with a diagnosis of autism. Outside of his work, in situations where the integration of information often depends on knowing other people’s thoughts and feelings, he may have had more difficulties than we know about.

Grandin observes that autistic minds are specialized, predisposed to think in one of three ways. A person may think in pictures; he is often good at drawing and poor in algebra. A second person may think verbally; he writes well, is skilled in verbal logic, memorizes facts easily, translates languages fluently, and is poor at drawing. A third person may think in patterns; he is good, for example, at mathematics, physics, and engineering.[65] It is not obvious where Cavendish would fit in Grandin’s scheme. He probably did not think in pictures in Grandin’s sense. He did not reveal any particular skill in freehand drawing, but what he drew was usually apparatus, which did not call for much skill, and with drawing tools he made precise scientific drawings as an aid to mathematical reasoning. His manuscripts contain a good many studies in pure mathematics, a preponderance of which deal with plane and spherical geometry, a pictorial form of mathematics with wide applications in physical science. He probably did not think verbally in Grandin’s sense either, but he had an excellent memory for facts, and he wrote with flawless grammar. His interest in learning and translating languages was probably not strong, typical of his countrymen. He learned to read and write Latin in school, and he read French and German, as we know from several extracts of scientific papers in his handwriting, but when he wrote a letter in French to the editor of a journal, the writing was in another hand, and when he had to read a hand-written letter in German he needed help from a German. If we were to limit ourselves to one of Grandin’s three preferred ways of thinking, we would choose thinking in patterns, which is done by the kind of mind she calls the “music and math mind.”

---

[63] Simon Baron-Cohen, *The Essential Difference: Male and Female Brains and the Truth About Autism* (New York: Basic Books, 2003), 61–67.

[64] Baron-Cohen, *Autism and Asperger Syndrome*, 69–71.

[65] Grandin and Barron, *Unwritten Rules*, 102.

Mathematics and Music. We know that Cavendish was recognized for the "depth and extent of [his] mathematical knowledge.[66] Autistic persons have a strong desire for certainty. They avoid situations in which there is no one right way to act. Early in life they are often drawn to mathematics with its certain truths, and they may develop a talent in it.[67] If they enter a field of science, in their search for truths of nature they settle for Cavendish's expression "tolerable certainty."

David Tammet, an autistic young man with remarkable mental abilities, has strong feelings about numbers. The number pi, for example, he finds "extremely beautiful and utterly unique," like a Mozart symphony. The number is "its own reason for loving it."[68] This does not sound like Cavendish, but whereas a diagnosis of autism would tell us much about Cavendish, much of his inner life would still remain a closed book.

We turn to the other half of Grandin's "music and math" mind. Interest in music and interest in mathematics are often found together in autistic persons. Both subjects are characterized by order, pattern, and an imagined world. Grandin is interested in music, particularly in musical form, and she is especially drawn to the music of Bach. Her interest is intellectual, and the pleasure she gets from music is intellectual; music does not move her emotionally.[69] Cavendish's probable interest in music is discussed earlier. If he was musical, he may not have had an emotional response to music either, but we cannot know that for certain.

Senses. In his scientific work, Cavendish made extensive use of the standard five senses and of other senses as well. Autistic persons often experience abnormal responses to sensory stimuli, which can have a large impact on their lives. Almost all children with Asperger syndrome are extremely sensitive to certain sounds, and the majority of them are sensitive to certain touches, tastes, and smells, and about one in five is sensitive to light.[70]

Grandin finds that at one end of the autism spectrum, sensory and cognitive problems are worse but emotional relatedness is more normal, and at the other end intelligence is high and sensory problems are lesser but emotional relatedness is weak or absent.[71] We would expect Cavendish with his high intelligence and weak emotional relatedness to show minimal sensual abnormalities, and that is what we find. If he did have any sensory abnormalities they could have helped him in his work. There is evidence that autistic people "are faster and more accurate on visual search tasks"; likewise, they are good at detecting variations in vision, sound, touch, and even smell.[72]

---

[66] Humphry Davy, quoted in John Davy, *Memoirs* 1: 221.

[67] Atwood, *Asperger's Syndrome*, 241.

[68] Daniel Tammet, *Born on a Blue Day: Inside the Extraordinary Mind of an Autistic Savant, a Memoir* (New York, London, Toronto: Free Press, 2007), 185.

[69] Oliver Sacks, *Musicophilia: Tales of Music and the Brain*, rev. ed. (New York: Vintage Books, 2008), 318.

[70] Attwood, *Asperger's Syndrome*, 271, 275, 279, 282, 284.

[71] Grandin and Barron, *Unwritten Rules*, 43.

[72] Baron-Cohen, *Autism and Asperger Syndrome*, 56.

Objectivity. This is a requirement in scientific research. Even so, Cavendish's insistence on it stands out, as examples in this book show. The desire for objectivity is a well-known trait of autistic persons. Grandin says that in her field, she has "a reputation for being totally objective"; when forming a judgment, she concerns herself with the details of a case, and her emotions play no part.[73]

Accuracy and Perfection. These are well-established goals in scientific research. Each of Cavendish's works was said to be "perfect at the moment of its production."[74] The accuracy of his observations and reasoning was singled out by his contemporaries. Autistic persons frequently are "perfectionists with high self-imposed standards of achievement." Perfectionism and accuracy, Attwood says, are two qualities that recommend autistic persons for employment. They prefer accuracy to speed in accomplishing any task.[75]

Knowledge and Truth. Cavendish was knowledgeable in all branches of physical science.[76] Autistic persons can acquire encyclopedic knowledge in their fields of special interest.[77] According to Cavendish's contemporaries, his sole motive for carrying out scientific researches was a disinterested "love of truth and of knowledge." Love of truth is a common autistic trait, joining other traits such as honesty and desire for certainty. The quest for truth comes naturally, as does persistence in detecting and avoiding errors.[78]

Writing. So far as we know, other than for business, science and subjects related to science were all Cavendish *wrote* about. When he wrote, he was by himself, at home in his study, undistracted, doing what he enjoyed. Anyone who knew him only through his writing would have no clue that he might be autistic. Autistic persons can write fluently, and they can become fine writers on their subject of interest.[79] They are drawn to the language of science for its freedom from emotional associations.

Scientific Society. Almost daily, Cavendish met with scientific colleagues in the city. Autistic persons talk at length about their special interest, unaware of the effect they have on their listeners. If not turned off, their listeners are likely to be impressed.[80] Cavendish listened far more than he talked, but under the right circumstances, he "conversed a great deal."[81] He probably did not tire his listeners because he talked almost exclusively to people with scientific interests who wanted to hear what he had to say. This was the social side of Cavendish.

---

[73] Grandin, "Comments," in Ledgin, *Diagnosing Jefferson*, 202.

[74] Humphry Davy, quoted in John Davy, *Memoirs* 1: 221.

[75] Attwood, *Asperger's Syndrome*, 141, 254, 295.

[76] Playfair, *Works* 1: lxxxiv.

[77] Attwood, *Asperger's Syndrome*, 179–80.

[78] Ibid., 117–18, 230. Humphry Davy, quoted in John Davy, *Memoirs* 1: 221.

[79] Frith, *Autism*, 125.

[80] Attwood, *Asperger's Syndrome*, 116, 188.

[81] Thomson, *Chemistry* 1: 337.

Autistic persons who develop a high level of skill in a field are valued and sought out. They enter a small society of persons who accept them on their own terms. Their single-mindedness is an asset, and their strange behaviors are tolerated as eccentricities. Their special interest is a way for them to "achieve some social life and friends." Grandin's social life revolves around people who share her special interests.[82] We could view Cavendish as an eighteenth-century instance of this positive outcome.

Talent. Cavendish was amply endowed. Near the end of his life, Hans Asperger wrote, "It seems that for success in science or art, a dash of autism is essential. For success, the necessary ingredient may be an ability to turn away from the everyday world, from the simply practical, an ability to re-think a subject with originality so as to create in new untrodden ways, with all abilities canalized into the one specialty."[83] Norman Ledgin, the author of a book on historical persons with autistic traits, says that in researching their lives he found that certain traits probably contributed to their success.[84] That finding comes as no surprise to Grandin, who observes that geniuses are an abnormality. Without genes responsible for autism, she thinks that the civilized world probably would be reduced to "a place full of highly social yakkity-yaks who would never do anything new or creative."[85] Baron-Cohen writes: "People with autism, whose minds differ from what we consider typical, frequently display both disability and exceptional aptitude. Genes that contribute to autism may overlap with genes for the uniquely human ability to understand how the world works in extraordinary detail – to see beauty in patterns inherent in nature, technology, music and math." He suggests that genes associated with autism persist over generations because they are co-inherited with genes responsible for mathematical and technical talent, which society welcomes.[86]

Way of Life. So far as we know, all that Cavendish ever wanted to do was to study and to advance natural philosophy. If he was autistic, he would have had inborn preferences, among them the common preference for the natural world over the social. Nothing, however, in his autism would have turned him into a natural philosopher. Special interest, which appears as a psychological trait in this chapter, does not by itself explain the turn his life took. Cavendish had to will it.

Interpretations. Wilson's comparison of Cavendish to a deaf mute is insightful. The deaf mute is a stand-in for a person with autism at a loss in a world of emotional strangers and unable to speak about it.[87] What Wilson missed is a possibility he could not have known: as an autistic person, Cavendish's insistence on regularity in his life would have originated in a peculiar structure of his brain, not in a strong will

[82] Grandin, quoted in Ledgin, *Asperger's Syndrome and Self-Esteem*, xiii, 26.

[83] Asperger, quoted in Attwood, *Asperger's Syndrome*, 199.

[84] Ledgin, *Asperger's and Self-Esteem*, 18, 32.

[85] Grandin, "Comments," in Ledgin, *Diagnosing Jefferson*, 200.

[86] Simon Baron-Cohen, "Autism and the Technical Mind," *Scientific American* 307 (2012): 72–75, on 74–75.

[87] Wilson, *Cavendish*, 186.

to repeat nature; he would have followed fixed routines not because he wished to emulate the earth's revolution about the sun but because he had a disorder, the same disorder that inclined him to study the regularities of nature.

Feelings. If we believe that Cavendish was autistic and that this is why he came across as cold, our feelings about him will reflect that belief. If we think of him as a haughty aristocrat who kept his own counsel, we will have different feelings about his silence and tactlessness, and different interpretations. We might regard his habit of rarely speaking not as a quirk of personality but as an expression of social superiority; or his writing a check for £10,000 to give to a former librarian as a gesture of contempt; or his need never to be wrong as intellectual arrogance. I do not think these interpretations are supported, but they are not out of thought.

Chapter 9
Conclusion

9.1 Personality and Science

In Chap. 2, we looked at personality traits of scientists in today's world. We recall them briefly here. Under social traits: scientists, when compared with non-scientists, are found to be more dominant, assertive, hostile, self-confident, autonomous, introverted, and deviant. Under cognitive traits: they are more open, flexible, intelligent, curious, and imaginative. Under motivational traits: they are more driven and ambitious. Persons are more likely to choose a career in science if they are firstborns, if they are Protestants, Jews, or irreligious, and if they have models for a life of science. Among scientists, those who study the physical world tend to be less social, oriented more to objects than to people.

Cavendish was a firstborn, and in making science his special interest he had a model. His father made scientific experiments, socialized with men of science, and served as an administrator in the national scientific society. If Cavendish had a religious persuasion, it was Protestant, and he had no interest in organized religion. He was less social than non-scientists, preferring solitude to company, and when he did interact it was with small groups of colleagues. In keeping with his shyness, he had a preference for the physical sciences. He was open to new scientific problems and findings. In his work, he was cautious, curious, perseverant, and self-confident, and, of course, he was intelligent. To this point and to the extent that psychological traits of scientists today apply to their counterparts in the eighteenth century, Cavendish fits the personality profile, departures from it having to do with how strongly he expressed the traits. Traits in the profile he did not have are interpersonal. He was not dominant, arrogant, hostile, or deviant, and if he was ambitious at all, it was for his work and not for himself. We may conclude that Cavendish was fairly typical of scientists as we know them today, though he was missing those traits we associate with competitiveness, a drive familiar in science today. His aristocratic rank more than his psychology could help explain this difference.

R. McCormmach, *The Personality of Henry Cavendish - A Great Scientist with Extraordinary Peculiarities*, Archimedes 36, DOI 10.1007/978-3-319-02438-7_9,

The classical virtues he was taught at school he incorporated into his life with one exception, balance. He was immoderate in his pursuit of scientific knowledge to the exclusion of every other course of life. In a period sometimes called "the age of reason," this imbalance was thought odd but worthy and even admirable. Certain traits of his were subordinated to this pursuit, and supported it; for example, self-confidence, autonomy, perseverance, and introversion. Other traits were largely independent; for example, coldness, shyness, taciturnity, and solitariness, though by discouraging intimacy they assured him ample privacy to pursue his work.

We see a continuity of traits in his work and in his life outside it. Blagden recognized this when he spoke of his friend's "philosophical simplicity of manners." His participation in the politics of the Royal Society was characterized by objectivity and emotional detachment, the perspective of the researcher. His reliance on reason, evidence, and measure is seen in his management of his estate no less than in his scientific work. The scientific curiosity he brought to the physical world in his laboratory he brought to the world of industry and manufactures on his journeys. The simplicity, caution, and circumspection that characterized all his scientific activities we see in the everyday conduct of his life. The same holds for the standards he held for his scientific results: accuracy, precision, and perfection. He had a strong will; he was able to combine elements of his personality to make a life of natural philosophy. The moral code of his way of life was universal: truthfulness, sincerity, probity, and openness. His characteristic behaviors and values were personality traits, and not just a repertoire specific to his line of work, though his way of life would have made some of them more pronounced than others. We next consider how he came by his traits, some of which obviously served him well in his chosen way of life.

9.2 Nervous Disorders and Eccentricity

Nervous Disorders. Brougham observed in Cavendish's facial expression a "nervous irritation" he seemed to feel.[1] In describing his manner, Davy used the same word, "nervous."[2] Eighteenth-century meanings of the word "nervous" were: characterized by an agitation or disordered state of the nerves; suffering from a disorder of the nerves; excitable, easily agitated, timid.[3] Cavendish's speech was described as "excited," and his manner as having "an air of timidity" and "a quickness and sensibility almost morbid."[4] "Morbid" meant unwholesome, sickly. Cavendish was "shy and bashful to a degree bordering on disease," Thomson said.[5]

Some of the above words relate to what were called "nervous disorders," which included hypochondria, hysteria, and dyspepsia, minor illnesses leading to frequent

[1] Brougham, *Lives* 1: 259.

[2] Humphry Davy, *The Collected Works of Sir Humphry Davy, Bart.*, ed. J. Davy, vol. 7 (London, 1839), 139; quoted in Wilson, *Cavendish,* 167.

[3] "Nervous," *Oxford Universal Dictionary,* 1321.

[4] Pepys, quoted in Wilson, *Cavendish,* 168. Young, "Cavendish," 444.

[5] Thomson, *History of Chemistry* 1: 337.

visits with physicians. It was thought that the seat of these illnesses was the nervous system, hence the name. Through the century, increasing interest was shown in temperament, and the predisposition to nervous disorders was called a "nervous temperament," understood to be inherited or caused by external circumstances such as trauma to the head. The difference between a nervous disorder and a nervous temperament was subtle, raising the question of the reality of nervous disorders as illnesses requiring a physician. The physician William Heberden, a friend who had recommended Cavendish for membership in the Royal Society, was decided on this point: he wrote in 1802 that nervous disorders were as real as any illness the physician confronted, and that the difficulty in recognizing and treating them arose from general ignorance of the connection between mind and body.[6] Cavendish was evidently seen by his colleagues as having behaviors in common with persons with nervous disorders, but they stopped short of labeling him with a disorder. Young may have had this in mind when he associated Cavendish's speech mannerisms with the "constitution of his mind" rather than a "deficiency of his organic powers."[7] He considered Cavendish to be an eccentric person, not a person normally considered ill.

For completeness, we consider another category of mental abnormality, lunacy. A foreign visitor in 1789–90 observed "numerous English eccentricities, which in some other place would be called madness, but here are called only capriciousness or whim."[8] The distinction between "lunacy and mere eccentricity is not easy, particularly in a period celebrated for odd behavior," according to a historian of British nobility in the eighteenth century, who names several peers who seemed indisputably insane.[9] Brougham said that Cavendish's neighbors thought him "out of his mind."[10] From what authority he got this information he did not say. In any event no one who knew Cavendish suggested he was out of his mind. There was an implicit understanding that his outward peculiarities were a feature of a native constitution grounded on the bedrock of reason, the laws of logic.

Eccentricity. However typical Cavendish might appear as a scientific investigator, he was not a typical person. Brougham said that Cavendish's servants thought him "strange."[11] So did others. We know a good deal about his behaviors from a variety of sources, though without Wilson's biography he might not be remembered as an eccentric today; although Wilson did not use the word "eccentric," the eccentric Cavendish is largely his creation, which is not to say it is wrong. Citing Wilson's biography published 15 years earlier, Timbs included Cavendish in his *English Eccentrics*.

[6] There was a problem of deciding on the symptoms of nervous disorders. Low spirits and despondency, symptoms of hypochondria in the eighteenth century, belong to what we call "affective disorders" such as depression. The symptom of anxiety did not feature largely in eighteenth-century discussions of nervous disorders. Edward Hare, "The History of 'Nervous Disorders' from 1600 to 1840, and a Comparison with Modern Views," *British Journal of Psychiatry* 159 (1991): 37–45, on 37–38, 42–43.

[7] Young, "Cavendish," 444.

[8] Langford, *Englishness*, 289.

[9] Cannon, *Aristocratic Century*, 11–12.

[10] Brougham, *Lives* 1: 259.

[11] Ibid.

Ever since Timbs, if not before, collections of eccentric biographies have commonly included an entry on Cavendish drawn from Wilson. Most have followed Timbs's example, which is to relate briefly Cavendish's peculiarities, as Wilson describes them, without singling out any one of them as defining his eccentricity. With the exception of John Timpson, who in *Great English Eccentrics* says that Cavendish's only eccentricity was his "inordinate shyness,"[12] it would seem that Cavendish's behaviors taken as a whole are what people find eccentric about him, an eccentric *life*.

In Chaps. 3 and 4, we looked at a number of Cavendish's behaviors, and in Chap. 7 we compared him with persons who showed similar behaviors and were considered eccentric. We found that his eccentricity largely took the form, first, of exaggerations of certain accepted and generally admired traits of the English national character and, second, of certain behaviors that lay outside the national character. The behaviors that would have appeared mildly eccentric at the time included his old-fashioned dress, some of his routines, his relations with servants, and his aversion to women. Only three or four of his behaviors would have appeared very eccentric: his special interest to the near exclusion of other interests; his shyness and penchant for solitude; his coldness; and possibly some of his regularities. Just how extreme these traits were is open to question.

If Cavendish's eccentricity was not due to a disorder, its origins may be complicated and elusive. We are told that he had secluded habits from an early age. This would suggest that the first place to look is at home. Charles Cavendish was not eccentric. He was competent in natural philosophy without making the social error of seeming to be a technical man. Described in his obituary as amiable, he hosted dinners for his friends at home, and he was comfortable taking charge of meetings. He and his son Henry had quite different personalities. Henry's brother Frederick was sociable like his father. He had eccentricities, but they differed from Henry's and were less pronounced and may have been affected by an early injury. The possible effect of the loss of Henry's mother is discussed in an earlier chapter and below, but given what little is known about Henry's home life, his eccentricity cannot be fully explained that way.

Eccentricity is compatible with high intelligence and achievement. If we agree that Cavendish was eccentric, we could stop here and assume that his personality was one of a myriad of possible personalities consistent with a normal brain. Alternatively we could look for an explanation of his eccentricity; that is what we do below.

9.3 Shyness and Introversion

In this section, we consider two personality traits, shyness and introversion. They do not rise to the level of disorders, but they can be mental handicaps, which can be severe. Cavendish's reputation for eccentricity stems in large measure from behaviors that fall under shyness and introversion.

[12] Timpson, *Great English Eccentrics*, 83.

Shyness. "Shyness" is a word that appears repeatedly in accounts of Cavendish, and his shyness among strangers was by all accounts excessive. He might show embarrassment, painful self-consciousness, and tension, avoid eye contact, fall silent, and on occasion flee. If a stranger had interesting information, Cavendish reacted to him in a more complicated way, showing the mix of avoidance and attraction typical of very shy people.[13]

Shyness, a feeling of apprehension, discomfort, and awkwardness in company, may be a lifelong personality trait or it may begin in childhood in reaction to physical anxiety. The only event in Cavendish's childhood we know of is the loss of his mother. If children are isolated, or if they are bullied by caretakers, or if they are ridiculed for lack of social skills, they can acquire a fear of strangers that never leaves them.[14] Since the home environment, especially parental support and sibling relationships, is decisive for the outcome of shyness, the absence of the mother is potentially significant in Cavendish's case.[15] Heredity was probably a factor in Cavendish's shyness: of all personality traits shyness has the strongest genetic component,[16] and we know that taciturnity, one of the forms shyness takes, was a Cavendish family trait.

Shy persons feel strong anxiety in the presence of strangers, especially strangers of the opposite sex, and in situations that either call for assertive action or invite evaluative judgments;[17] likewise, persons who are otherwise shy can feel comfortable in a circle of people they know.[18] These observations agree with Cavendish's social behavior. Being stared at, being conspicuous, and breach of privacy are all occasions for shy behavior;[19] we know they were for Cavendish's. In the company of scientific colleagues he knew, he was fine.

In other respects, Cavendish's shyness does not fit the common pattern. Shyness is correlated with low self-esteem; Cavendish was shy, but there is no indication that he was uneasy with who he was. Shy persons are inclined to spend much time thinking about their feelings and actions and how they appear to other persons; Cavendish

[13] Caroll E. Izard and Marion C. Hyson, "Shyness as a Discrete Emotion," in *Shyness: Perspectives on Research and Treatment*, eds. W. H. Jones, J. M. Cheek, and S. R. Briggs (New York and London: Plenum Press, 1986), 147–60, on 151, 153. Distinctions are made between shyness, embarrassment, and social anxiety, though shyness and embarrassment are similar and are often considered together. W. Ray Crozier, "Social Psychological Perspective on Shyness, Embarrassment and Shame," in *Shyness and Embarrassment: Perspectives from Social Psychology*, ed. W. R. Crozier (Cambridge: Cambridge University Press, 1990), 19–50, on 48, 54. Hereafter the short title *Shyness and Embarrassment* is used for specific contributions to this volume.

[14] Arnold H. Buss, "A Theory of Shyness," in *Shyness: Perspectives on Research and Treatment*, 39–46, on 44–45.

[15] Jonathan M. Cheek and Stephen R. Briggs, "Shyness as a Personality Trait," in *Shyness and Embarrassment*, 315–37, on 326.

[16] Ibid., 329. Kagan, Reznick, and Snidman, "Biological Bases of Childhood Shyness," 167–71.

[17] Cheek and Briggs, "Shyness as a Personality Trait," 316.

[18] Crozier, "Social Psychological Perspectives on Shyness," 32.

[19] Buss, "Theory of Shyness," 44.

gave no indication of this. They tend to be conformist, accepting the opinion of the majority, adopting attitudes of persons in authority, descriptions which conflict with Cavendish's work in natural philosophy. They would rather get along than get ahead, and avoid criticism than earn praise. Being unambitious, they are often unemployed and when they enter a career they do not do well, the worst consequence of chronic shyness.[20] Thorpe suggested that Cavendish's "shy retiring nature" was a handicap in his work: his "taciturnity and reluctance to enter into conversation must have largely prevented him from obtaining information orally even if it were available at the period."[21] Thorpe was undoubtedly right about the handicap, though there is considerable evidence that Cavendish dealt with his handicap rather successfully. He may not have spoken often, but he was always present and listening. By working around his shyness, he escaped the common fate of shy persons; he had a satisfactory life of science, though if he had not been shy he might have been more productive.

There can be side benefits. Shyness protects privacy, affording shy persons the luxury of solitude. By not offending people, they less often run into conflict, giving fewer opportunities for others to criticize them. Shyness allows them to be selective about engaging with people, saving them time and sparing them aggravation. Having discovered the benefits, some shy persons like being shy,[22] and it is possible that Cavendish was one of them.

Introversion. Shyness is a blanket term for a group of related afflictions. What is taken for shyness by others may be an expression of reserve or introversion, behaviors that enable their owners voluntarily to reduce social contacts. For their own reasons, introverts gain no reward from social interaction, and so they avoid it. The shy person is not unsociable but fearful; the introvert is unsociable by choice.[23]

Usually regarded as one extreme of a personality spectrum, the other extreme of which is extroversion, introversion like shyness is largely inborn. The introvert is a familiar type, the loner. He is, according to one study, retiring, introspective, cautious, insistent on ethical standards, reliable, has few close friends, and keeps feelings to himself. Cattell extends the list to include silent, serious, self-sufficient, individualistic, critical, cold, precise, objective, skeptical, rigid, and prone to sulk. The introvert, he says, stands by his ideas, makes his own decisions, is rule-bound, has restricted interests, is retiring with the opposite sex, and has strong work habits. Because he has low expectations of human relations, he is drawn to the orderly, impersonal sphere

[20] Buss, "Conception of Shyness," 47. Cheek and Briggs, "Shyness as a Personality Trait," 328.

[21] Thorpe in Cavendish, *Scientific Papers* 2: 326.

[22] In a survey of about 5,000 people, it was found that 10–20 % of shy persons prefer being shy. Philip G. Zimbardo, *Shyness: What It Is, What to Do About It* (Reading, Mass.: Addison Wesley, 1977), 20, 141–42.

[23] Telltale feelings of shyness – anxiety, unease, and awkwardness – are absent in introversion. Anon, "Extroversion and Introversion," http://en.wikipedia.org/wiki/Extroversion_and_Introversion. Anon. "Shyness," http://en.wikipedia.org/wiki/Shyness.

of science, logic, and music.[24] The words above – "cautious," "silent," "cold," "precise," "objective," "prone to sulk," "insistent on ethical standards," "rule-bound," "restricted interests," and "retiring with the opposite sex" – are ones we have come across in accounts of Cavendish.

Another interpretation of the introvert gives a different picture of his inner life. Carl Gustav Jung, who developed the concept of introverted personality, said that the introvert seeks an "internal world of subjective experience." According to the Myers-Briggs extroversion-introversion preference indicator, which is based on Jung's work, the introvert is questioning, taciturn, shy, and impenetrable, a description which applies to Cavendish, but he is also "inwardly directed, focused on the inner world of ideas and understanding," and his conduct is "governed by subjective values." He is "intense and passionate," a "bottle of emotions," with a "tendency towards impracticality."[25] What we can infer of Cavendish's mind from contemporaries' accounts conflicts with the subjective interpretation.

As in the case of shyness, there is a positive side to introversion. Solitude is conducive to concentration, originality, and accuracy, desirable traits. If the introvert has a strong desire, he can overcome his handicap to achieve worthy goals. As far as it goes, introversion in its non-subjective interpretation describes Cavendish fairly well. Its limitation is what it does not include, telltale feelings of shyness, which we see in Cavendish: anxiety, unease, and awkwardness. Perhaps his contemporaries confused introversion for shyness, though that seems unlikely. Neither trait, shyness nor introversion, is a complete description of Cavendish's strangeness.

9.4 Autism and Other Disorders

Autism. Because eccentricity is a cultural category and autism and other disorders are a medical category, they are hard to compare. Some persons who were called eccentric in the eighteenth century would today be diagnosed with a mental disorder; even today we cannot always distinguish clearly between the two.[26] In this book, we accept that just as eccentrics can be shy and introverted they can have disorders, and that persons with disorders can be shy and introverted and eccentric.

Take the autistic trait of special interests. Attwood makes the point that special interests are regarded differently in different cultures, and that in some, notably in British culture, they are more likely to be treated as benign eccentricity than as signs of a psychological disorder in need of treatment. "As a culture," he writes, "the British have always accepted and admired eccentric people." An example of persons

[24] Anthony E. Kemp, *Musical Temperament: Psychology and Personality of Musicians* (Oxford: Oxford University Press, 1996), 36–39, 49. Pervin, *Personality, Theory*, 283.

[25] Kemp, *Musical Temperament*, 34. Pervin, *Personality, Theory*, 148–49. Anon., "Extroversion and Introversion."

[26] Weeks and James, *Eccentrics*, 14–15.

with special interests who are seen as eccentric is the avid collectors on the British television series *Antiques Roadshow,* some of whom seem to Attwood to exhibit signs of Asperger's syndrome.[27]

Historical diagnoses of autism are undertaken for two main reasons: they give encouragement to persons who have the disorder by calling attention to successful autistic persons in the past; and by recognizing that strange behaviors of biographical subjects fit a clinical pattern, they enlarge our knowledge of the past. Despite their worthy ends, historical diagnoses are a hard sell. The psychological biographer of a possibly autistic person is usually at a disadvantage, lacking both the training and experience of the clinician and a living subject. Ledgin thinks that the disadvantage is not a disqualification, that the clinicians' criteria allow laypersons to decide if someone belonged to the "autism/Asperger's continuum." The key, he says, lies in accounts of the subject's early years.[28] The cognitive neuroscientist Uta Frith and Fitzgerald agree that some historical persons can be diagnosed with hindsight, drawing on current understanding and criteria, though they may not always agree on which ones.[29]

The gold standard of historical diagnoses of autism is *Autism in History: The Case of Hugh Blair of Borgue*, written by the early modern historian Rab Houston and Frith, published in 2000. An English contemporary of Henry Cavendish's, Hugh Blair was brought to court in connection with a family feud. The legal issue was whether or not he had been mentally competent to enter into a marriage contract; if not, his marriage was to be annulled. An abundance of witnesses, 29 all told, gave testimony, on the basis of which the court decided that Blair was a "natural fool," "void of common sense," and the marriage was annulled. Although there were no direct sources from his youth, people remembered that he had always been that way, a fool. The authors conclude that Blair would be given an unequivocal diagnosis of autism today. To summarize the evidence: Blair insisted on sameness; he was deficient in social understanding and communication, though he was not socially withdrawn; he had peculiarities of speech and language; he showed repetitive behavior; and he had narrow interests. His behavior was not consistent with alternative explanations. Although he could read and write and had good rote memory, he was severely autistic. In speaking, he was barely able to string words together. The detailed evidence in this case, which compares with that of clinical cases today, may be unique for a historical person; if not unique, it is rare.[30] By comparison, the evidence in Cavendish's case is rather thin. The mathematician Ioan James, who has written about historical persons with signs of Asperger's syndrome, puts his finger on a major difficulty of the project where he says that "the right kind of biographical information is not easy to find."[31]

---

[27] Attwood, *Asperger's Syndrome*, 199.

[28] Ledgin, *Asperger's and Self-Esteem*, 28.

[29] Rab Houston and Uta Frith, *Autism in History: The Case of Hugh Blair of Borgue* (Oxford and Malden, MA: Blackwell, 2000). Fitzgerald, *Autism and Creativity*, 13–14.

[30] Houston and Frith, *Autism in History*, 1–5, 146–50.

[31] James, *Asperger's Syndrome*, 9.

Let us see how the manuals address the less severe form of autism Asperger's Syndrome. According to the *DSM*-IV, for a person to be diagnosed with the syndrome, certain social and behavioral features must be present. To begin with, the person has to show considerable impairment in his social, work-related, or other functioning. Under social impairment, at least two of the following four criteria must be met. (1) Impairment of nonverbal behaviors such as eye-to-eye contact, facial expression, postures, and gestures. (2) Lack of relationships. (3) Lack of spontaneity in seeking out and responding to persons with shared interests. (4) Lack of social or emotional reciprocity. Cavendish meets all four criteria. The evidence is the behaviors described in Chap. 3. On behavioral traits that do not depend on social interaction, a diagnosis requires the presence of at least one of the following four traits. (1) Interests that are limited in subject and abnormal in intensity or direction. (2) Adherence to nonfunctional routines. (3) Stereotypical, repetitive physical mannerisms. (4) Preoccupation with parts of objects. Cavendish meets the first of these criteria and possibly the third.[32] The evidence is given in Chap. 4. Of Gillberg's 20 criteria, distributed between 6 categories, for Asperger's syndrome, probably 9 of them apply to Cavendish, and these are covered by the *DSM* criteria above. However, according to one category "speech and language peculiarities," Cavendish would seem to be eliminated. This category has five parts, at least three of which must be present for a diagnosis of Asperger's syndrome: delay in the development of speech, superficial perfection in expressive language, pedantic language, impaired comprehension of language including misinterpretations of literal meanings, and "odd prosody, peculiar voice characteristics."[33] Cavendish meets only one criterion for certain, the last. It is unclear how much we can read into the matches for Cavendish, whether good or poor. *DSM* and Gillberg diagnostics apply to living subjects, and diagnoses are made by persons who are trained in it.

With Sacks's diagnosis of Cavendish, the reservations are not the same, and we can read more into the matches. He uses different terms and gives more weight to Cavendish's way of thinking than to his social behavior, and as a result the agreement between his observations of Cavendish and the *DSM* criteria is not immediately obvious. Sacks bases his diagnosis of Cavendish's autism on the following seven characteristics. (1) "A striking literalness and directness of mind." (2) "Extreme single-mindedness." (3) "A passion for calculation and quantitative exactitude." (4) "Unconventional." (5) "Stubbornly held ideas." (6) "Rigorously exact (rather than figurative) language." (7) "A virtual incomprehension of social behaviors and human relationships."[34] Sacks uses different words than the *DSM*, and he pays more attention to Cavendish's way of thinking than to his social behavior, and so his agreement with the manual is not immediately obvious. The evidence for the working of Cavendish's mind comes mainly from his writings on science, technology, and business, where we would not normally expect figurative language or non-literalness. For the same reason, characteristics 1 and 6 have a large overlap. It is not clear what "stubbornly held ideas" refers to. Cavendish was evidently the first British chemist

[32] Attwood, *Asperger's Syndrome,* 41.

[33] Ibid., 37.

[34] Sacks, "Henry Cavendish."

to abandon the phlogiston theory, while Priestley and other colleagues held firmly to it. "Virtual incomprehension of social behaviors" is too sweeping. He demonstrated a good understanding of human motivations during the dissensions of the Royal Society. Studies show that autistic people understand basic motivations quite well, their difficulty coming with more complex emotions and points of view.[35] Sacks does not say what he means by "unconventional." As we saw in the section above on eccentricity, Cavendish could well be considered conventional, if somewhat out of date, his eccentricities being expressions of conventional behavior carried to excess. One exception, which may be Sacks's meaning, is that given his birthright, Cavendish chose a life of science rather than a career in politics; this was highly unconventional. The *DSM* requires a match with at least two social criteria, and Sacks has only one, the seventh characteristic, but because of his wording, it could cover all of the *DSM* social criteria. With these comments in mind, Sacks's list matches the *DSM* criteria. Cavendish's work was quantitative; his language was literal and rigorously exact; he pursued science single-mindedly; and he had difficulty with social behaviors and human relationships. The same caution about matches in the previous paragraph applies in this case. Diagnostic criteria in the *DSM* being brief and general, they by themselves are an incomplete basis for an accurate diagnosis. For that, experience is required.[36] The person making the diagnosis here is Sacks, who supplies the experience.

Let us consider the possibility that the biographical information and other evidence do not support Sacks's diagnosis. There are, I think, five major arguments that call into question a diagnosis of autism. It should be noted that the arguments have their own weaknesses as well. We look at the arguments and the counter-arguments next.

(1) The testimony about Cavendish came from people who knew him late in his life, and the central developmental feature of autism goes unaddressed for lack of evidence. He may have shown autistic behavior as a child, but we cannot infer it from his adult behavior, and we have only Blagden's word that his habits were secluded from an early age. (2) Outstanding scientists are often obsessive, follow routines, exhibit social anxieties, and in general evoke autistic-like traits. They behave this way to do their work, about which they are passionate; they are rarely autistic. Cavendish may have been autistic, but if we cannot tell the difference between him and persons who were not, we find it hard to make a case for it. (3) He formed a habit of meeting frequently with scientific colleagues in the city, a level of social activity unusual for an autistic person. A counter-argument is that his interaction was highly selective, consistent with his private ways and narrow interests, his social world being a direct extension of his special interest, the physical world. When he was with colleagues, he did what autistic people do, he entered into conversation to exchange information about his special interest. This most unsocial of men, as Wilson described him, could be said to be sociable in scientific company at times. (4) Cavendish's strange ways did not seriously interfere with his course of life, at odds with what we mean by a disorder. He lived a life of natural philosophy,

[35] Anon., "Autism," http://en.Wikipedia.org/wiki/Autism, p. 7.

[36] Attwood, *Asperger's Syndrome*, 40–41.

and all evidence points to it as the life he desired. People diagnosed with autism have a hard time managing their lives, requiring help with their daily affairs, work, and finances. No one who knew Cavendish mentioned his needing help, yet there is an element of uncertainty here. Consider the counter-argument. For some years when Cavendish came to meetings of the Royal Society Club, he had only the exact change for the dinner in his pocket, five shillings. It has been said that his father gave him the five shillings, and that this is an example of his father's parsimony.[37] Alternatively, the five shillings might be seen as an example of his father's practice of assisting his son in managing his daily living. By this time, since Cavendish had lived away from home at school and university, his father's assistance probably would have been a holdover from an earlier pattern rather than a present need. If his father played such a role in his life, we would expect to see changes when his father died. There were indeed changes, which brings us to the fifth argument.

(5) An autistic person tends to abhor major changes, and if he does make them, he is unlikely to have initiated them. Until his early 50s, Cavendish lived in an apartment behind his father's house in London, a manner of living consistent with autism. In early 1783 his father died. This was a major change in Cavendish's life, and to judge from its consequences it was the most important change. In 1782 his father turned over to him the management of his estate in Nottinghamshire and Derbyshire. That year he took a house in the suburb Hampstead, which served temporarily as a country house. In 1785, he bought a permanent country house on Clapham Common. Presumably he could have retained his father's large house with its large garden for his townhouse. Instead, in 1784, he bought a house not far from his father's, on Bedford Square. (Cavendish's new houses are the subject of Chap. 14.) Cavendish made two more related major changes in the same few years. In 1782 he took on an associate, Blagden, to help him with his experiments and generally to help him in his scientific activities. Their association, which came to an amicable end after several years, is the only known close contact Cavendish allowed. In three summers, in 1785–87, Cavendish and Blagden journeyed to different parts of Britain to observe industrial sites and geological formations. (Their journeys are the subject of Part II of this book.) Cavendish's moves, his association with Blagden, and his journeys were major changes in his way of living, and all were initiated by him. The counter-argument is that although his moves certainly disrupted some of his routines, they may have been a necessary part of his scientific plans. In the case of his journeys, he may have found them sufficiently interesting to distract him from the break in his routines. As for his associate, it is significant that he took on a scientific assistant not a personal secretary. It is not uncommon for gifted autistic persons to do things that are atypical of autistic persons. The biography of the physicist Paul Dirac by Graham Farmelo gives us an idea of what a modern day, mildly autistic person of high intelligence might be like. Dirac did some atypical things – he married, he went to work in Florida in his old age – but many detailed instances in his life match the diagnosis. A diagnosis for Cavendish is less persuasive, necessarily. Information about his life is sparser, and the world he lived in was very different than Dirac's and ours.

[37] Wilson, *Cavendish*, 160.

Two more arguments against Cavendish's autism have been raised by Fred Volkmar, a psychiatrist at the Yale Child Study Center and an expert on autism. He thinks that autistic diagnoses of historical persons have been overdone, having grown into a cottage industry. "Certainly, Henry Cavendish sounds like a very strange person," Volkmar says, but even in Cavendish's case he remains skeptical. The reasons he gives are sensible, though not conclusive. One is that Cavendish was taciturn, whereas autistic persons talk endlessly about their special interest. An objection to this is that if persons approached Cavendish the right way, they could set him going, and among people he knew he could talk freely. The second of Volkmar's reasons is that Cavendish was successful, whereas autistic persons usually do not accomplish much.[38] An objection is that autistic persons occasionally are very accomplished. Hans Asperger, who foresaw positive outcomes for the children he studied, followed the adult lives of several: one who had shown spontaneous talent in mathematics became an outstanding astronomer, another received a Nobel Prize in literature.[39]

Other Disorders. Cavendish's autism can be questioned on its own ground, medical. Sacks's diagnosis implies that no other disorder can account as well for Cavendish's behavior. We need to consider what has been ruled out. The *DSM* provides us with a number of possibilities. Of personality disorders, the most promising is "schizoid personality disorder," which is characterized by lack of interest in social relationships, desire for solitude, taciturnity, unresponsiveness to social cues, and emotional coldness. This again is a good description of Cavendish as far as it goes. Other characteristics of this disorder are bizarre ideas and lack of motivation and underperformance at work, which effectively rule out Cavendish.[40]

Another promising *DSM* disorder is "social phobia," which arises from the fear of making a fool of oneself. Everyone feels this at times, but the social phobic feels it intensely, distinguishing it from shyness. Characterized by a desire for social relationships, it is also distinguished from schizoid personality disorder. The social phobic feels under constant threat of "the human gaze"; he has a pressing concern of "how to react to the gaze of others, as well as how to control [his]own eyes." He is ever ready to flee or fend off attacks. He is shy, introverted, self-conscious, prone to embarrassment, and fearful of giving away his anxiety; he is reticent to speak when spoken to, rarely speaks in group settings, possibly stammers, and avoids being the center of attention. Social phobia can be mistaken for snobbery, but it is a "terror of social interaction," with debilitating consequences such as isolation and depression.[41] Social phobia and its close relative "social anxiety disorder" describe part of Cavendish's abnormal behavior, but not all, and in some cases they are at odds with it. They say nothing about obsessive interests and clocklike rou-

[38] Goode, "A Disorder Far beyond Eccentricity."

[39] Roth, *Autism Spectrum*, 10.

[40] Fitzgerald, *Autism*, 37–39.

[41] John R. Marshall, *Social Phobia: From Shyness to Stage Fright* (New York: Basic Books, 1995), xviii, 23–24, 56, 110.

tines, prominent characteristics of Cavendish's. Further, they are accompanied by low self-esteem, which we rejected earlier.[42] He had a mastery of natural philosophy, and he had important relatives. He was assured an ample measure of esteem, from within and without.

Obsessive-compulsive disorder looks plausible on first glance. Without question, Cavendish exhibited obsessive behavior in pursuing his special interest. However, there is no evidence that he tried to resist the obsession or was disturbed by it – on the contrary, his life was scientific study, which by its nature is obsessive[43] – and resistance is required for a diagnosis of obsessive-compulsive disorder. Further, the disorder fails to account for unusual eye contact and facial expressions found in autistic persons, and in Cavendish. Fitzgerald gives an extensive table of disorders that share some but not all of the traits of autism, noting where they agree and disagree with those traits. Few of the disorders in the table show one of Cavendish's most conspicuous traits, a preoccupation with a special interest, and those that do show it, for example, obsessive-compulsive disorder, fail in other respects.[44] We conclude that of all potential disorders, autism fits Cavendish best.

9.5 How Do We Talk About Cavendish?

We consider the choices. The first is autism. Granted that Cavendish had autistic-like traits, we may ask, Were these traits the result of a neurodevelopmental disorder? Said another way, Is a posthumous diagnosis of Cavendish's autism based on anything more than a superficial match of his behavior with autistic traits according to current texts on autism? If our answer is yes, yes the evidence does support a diagnosis of a disorder, we can say that Cavendish was probably autistic, acknowledging at the same time that the dividing line between autism and normality is subjective. If our answer to the above question is no, or the evidence is insufficient to decide, we have an alternative to fall back on. We can say with confidence that Cavendish showed eccentric behaviors, which were variations on a generally accepted mode of conduct, his genetic make-up and his choices accounting for the variation.

Autism. Of the arguments favoring a diagnosis of autism for Cavendish, two are compelling. The first is Oliver Sacks's opinion that the evidence for such a diagnosis

[42] The relationship between autism and schizoid personality is given in Sula Wolff, *Loners: The Life Path of Unusual Children* (London: Routledge, 1995). Anon., "Schizoid Personality Disorder," http://www.mayoclinic.com/health/schizoid-personality-disorder. Anon., "Social Phobia," http://www.behavenet.com/social-phobia. Anon., "Social Anxiety Disorder," http://www.Wikipedia.org/wiki/Social_anxiety_disorder. Anon., "Social Anxiety Disorder," http://www.socialanxietyinstitute.org/dsm.html. A third to one half of persons who suffer from this disorder experience depression, and also frequently anxiety, panic, and embarrassment. Jeralyn Ross, "Social Phobia: The Consumers Perspective," *Journal of Clinical Psychiatry* 54 (1993), Supplement, 5–9, on 5–7.

[43] Davis writes that science "is itself an obsessive activity" characterized by repetitive focusing on one subject. He develops this idea in many places in his book, *Obsession, a History.*

[44] Fitzgerald, *Autism*, 36–41. Frith, *Autism in History*, 147.

Fig. 9.1 Oliver Sacks. Photograph taken in 2009. British by birth, he has held professorships in neurology and psychiatry and in clinical neurology. He has published several collections of studies of persons with neurological disorders. He thinks that Henry Cavendish was in all likelihood autistic (Wikimedia Commons)

is almost overwhelming. Based on his work and reputation, and on his observations on Cavendish, his conclusion carries very considerable weight (Fig. 9.1). His interpretation of Cavendish's behaviors is a stand-in for the clinical evaluation required of any reliable diagnosis meeting *DSM* standards. Second, there is a correlation between most of Cavendish's distinctive behaviors and autistic behaviors. Taken one or several at a time, his behaviors invite alternative explanations, but there is only one explanation that agrees with most of them, autism.[45] If the correlation were due solely to Cavendish's choices, this would be a remarkable coincidence. It does not prove there is no other way of looking at Cavendish's behavior, but it is impressive.

For the present, let us accept with Sacks that Cavendish was probably autistic. Let us also accept that there are inescapable uncertainties attaching to any diagnosis of autism for a person living in the eighteenth century. At such remoteness, and in ignorance of Cavendish's early development, we cannot address him with the confidence of a psychiatrist who diagnoses and treats autistic patients today. In light of this qualification, we might reword Sacks's diagnosis. Given the ever unsettled state of medical definitions and diagnostic criteria of autism, and given the current view of autism as a spectrum of traits, it is prudent to speak of Cavendish as having a cluster of traits rather than to make the essentialist claim that he was autistic, in the way we say a person was blind, for example. We would regard the cluster of traits known from Cavendish's adult years as sufficient for us to talk about him as a person who very likely had autistic traits, not just autistic-like traits, at the same time acknowledging that the label "autism" is problematic. We would recognize that whatever wording we adopt and whatever weight we give to the historical evidence,

[45] Lidbetter would seem to have something like this in mind: only by acknowledging that Cavendish was autistic "can we get anywhere near attempting to understand Cavendish 'the complete man.'" "Henry Cavendish," 786.

we cannot alter a basic reservation: any autistic diagnosis of Cavendish has an irreducible speculative element.

This approach, I believe, is compatible with the scientific caution of Baron-Cohen, who writes:

> there are clues that Cavendish may have had some degree of Asperger's syndrome. He shows abnormalities in social relationships, communication, and some routine-bound repetitive behavior. We must assume that his scientific pursuits were strongly obsessional in nature. However, missing from the historical record are any details of his childhood.[46]

There are "degrees of autism," Baron-Cohen writes, and "you could have a little or a lot of it."[47] How much Cavendish may have had is hard to know, since he was highly intelligent, and by the time we get to know him he had had a long while to learn how to conceal any difficulties. If, as we assume here, the hypothesis of autism holds the advantage, Cavendish had a sufficient degree of autism that if he were to undergo a psychological evaluation today, he would probably be diagnosed with autism. This statement is hypothetical in another way too: he would not have sought clinical help, since he was getting along fine with his life.

Cavendish's autism would have been either Asperger's syndrome or high-functioning autism, both located at one end of the "autism spectrum disorder," according to the official terminology of the *DSM*-IV. Baron-Cohen uses an alternative terminology. Because he considers it uncertain if the two high functioning classes should be called "disorders" in the first place, he prefers the expression "autism spectrum condition." Persons who are diagnosed with the condition would, by definition, have social difficulties, but they would often have above-average nonsocial skills. The term "condition" acknowledges that they have a disability arising from neurobiological factors, but it is not a "*global* disability, and may in some individuals result in talent."[48] As, I trust, this book makes clear, given the way of life he chose, Cavendish exercised his talent with little or no sign of disability. Baron-Cohen's "condition" describes Cavendish's behavior better than "disorder."

Decision. How do we decide between the interpretations, or can we? Did Cavendish behave strangely because he was autistic, in which case he had no choice? Or did he have a choice in the matter? It seems to me that the sources on Cavendish's life support both interpretations about equally well, and that the sources are too limited to decide between them, at least with confidence. Coming to psychology as an outsider, I am a part of the limitation. For my part, I think it is doubtful we can ever know the answer.[49]

When I set out to write this book, I did not foresee the last sentence. I had a provisional answer to the question of Cavendish's autism, but the more evidence I brought to bear, the more I wondered about it. Over time, my view of what this

---

[46] Baron-Cohen, quoted in James, *Asperger's Syndrome*, 63. Consistent with the quotation, James says that his profiles "are not to be regarded as case studies," 11. Cavendish*'s* profile is on 63–68.

[47] Baron-Cohen, *Essential Difference,* 157.

[48] Baron-Cohen, *Autism and Asperger Syndrome*, 14.

[49] This is not the same as saying we cannot consider hypotheses.

book is about changed. I have taken pains to keep my original preconceptions from becoming a bias. If I have succeeded in my aim, I would add nothing to the book to talk about these preconceptions here, and I think readers will be just as glad that I do not.

Many readers will surely agree with Sacks that the evidence for Cavendish's autism is almost overwhelming. Those who do not agree may still have a preference: based on what they know about Cavendish and about autism, and trusting to their *intuition,* they may decide that Cavendish was or was not autistic.

By bringing a psychological view to our understanding of Cavendish, aspects of his life and work are brought together through a common explanation rather than through metaphor and analogy if at all. This is clear if our preference is for an autistic diagnosis, but if we find Cavendish to be an English eccentric of his time who lacked the biological basis of autism, or about whom we lack information to make a determination, we can see his peculiarities as stereotypical English manners carried to excess. Because in many respects the manners agree with personality traits of shyness and introversion, it is difficult if not impossible to distinguish what part of Cavendish's behaviors is owed to character and what part to temperament, what part is chosen and what part is given. Yet regardless of the uncertainty, clinical observations of shyness and introversion contribute a measure of understanding of our subject. He appears less weird, and he joins the human race. The latter would be a truism if it were not for a popular characterization of him as robot. Far from disparaging him, a psychological view endows this truth-seeker with considerable dignity. His strangeness is seen as normal behavior for a minority of persons who share his disorder or personality traits. We have a different view of him and a different feeling about him.

With any psychological evaluation of a historical figure, a red flag comes up. The path I take through a familiar minefield of objections to psychologizing the nonliving allows me to introduce scientific literature into the sources on Cavendish's life without giving him a label, which constantly undergoes revision.

If the understanding of Cavendish in this book must be hedged with reservations, the question arises, Why bother with it at all? He carried out certain experiments on gases, and for most purposes in the history of science that is all that interests us. My answer is that Cavendish was a remarkable scientist who had a remarkable personality, and between his science and his personality there were important connections. Independently of the question of Cavendish's autism, readers of this book will learn much about this man, viewed from a new perspective. That, finally, is the justification of this book.

Part II
He Saw What Interested Him

Chapter 10
The Traveler

Part I of this book relies on reports by others of Cavendish's behaviors. There is no reason to doubt the reports and the conclusions drawn from them, but by themselves they are missing what only Cavendish can tell us. His scientific papers do that, but they reveal only part of his world, and his personality is largely concealed. A partial exception is journals he and Blagden kept of travels in the 1780s. Because his travels exposed him to a common world, what he selected from it is revealing of his personality.

"The eye sees what it brings the power to see." This passage is by Thomas Carlyle, quoted in *The Five Senses; or, Gateways to Knowledge*, an elementary account of the main human senses, the "living inlets of learning." The author is George Wilson, Cavendish's biographer, who understands Carlyle to mean that the eye is educated to see what interests its owner. He gives examples: "The sailor on the lookout can see a ship where the landsman sees nothing; the Esquimaux can distinguish a white fox amidst the white snow.... The astronomer can see a star in the sky, where to others the blue expanse is unbroken; the shepherd can distinguish the face of every sheep in his flock; the mosaic worker can detect distinctions of color, where others see none."[1] What the natural philosopher Cavendish's educated eye saw on his journeys, he and Blagden recorded in their journals. In this chapter, we look at what it was.

We should be aware of an ill-defined category of persons known as "eccentric travelers." According to an author of a collection of biographies of such persons, eccentrics and travelers are often one and the same because they want the same thing, to escape from conventions of society. Eccentric travelers commonly approach new scenes with a more open mind than do typical travelers. They tend to be directed and well informed, illustrating a central fact about eccentric behavior, that it "is not the result of ignoring logic but of pursuing it to unusual extremes."[2] Following are several examples of well-known eccentric travelers. Charles Waterton, the English naturalist we met before, wrote a book about his travels, *Waterton's Wanderings in*

[1] Wilson, *The Five Senses*, 6, 32–33.

[2] John Keay, *Eccentric Travellers* (London: John Murray, 1982), 13–15.

R. McCormmach, *The Personality of Henry Cavendish - A Great Scientist with Extraordinary Peculiarities*, Archimedes 36, DOI 10.1007/978-3-319-02438-7_10,

South America, which inspired the schoolboys Charles Darwin and Alfred Russell Wallace. An enemy of pollution and early friend of the environment, he built a 9 ft wall 3 miles in length around his estate, enclosing a nature reserve. There are many stories of his odd behavior. On a visit to Italy, he showed his "increasing eccentricity" by balancing on one leg on the head of the Guardian angel mounted on top of the castle Saint Angelo. He was said to disappear under the dinner table and bite the legs of guests, imitating a dog, and he invented a new method of preserving animal skins, with which he made caricatures of his enemies.[3] Edward Wortley Montagu, 1713–76, was an English author who traveled widely in the West Indies and the Middle East. His mother Lady Mary, herself an eccentric, described him as "an excellent linguist, a thorough liar, and so weak-minded as to be capable of turning 'monk one day, and a Turk 3 days after.'" She knew her son. In Constantinople, Montagu dressed like a wealthy Turk, sporting buttons and buckles worth £2,500, wearing a beard and turban, adopting Turkish manners, and professing to be a Muslim. This "spectacular 'going native' was the culmination of a career of conspicuous social nonconformity." He presented a narrative of his travels to the Royal Society, of which he was a member. His scholarship is minor, "but he claims a place in any gallery of notable eccentrics."[4] Thomas Manning, 1772–1840, devoted himself to the study of China. Determined to penetrate to the center of the Celestial Empire, he succeeded after many adventures in glimpsing Peking, and he was the first Englishman to visit Lhasa in Tibet and to interview the Dalai Lama. Although he published nothing, he is credited with being the first lay scholar of China in Europe. Upon returning from his travels in Asia, he retired to a house he never furnished, filling it instead with his library of Chinese books, said to be the largest in Europe. Taking "delight in the most monstrous paradoxes," a fellow interpreter said, he "did everything in his own odd and eccentric way," giving as an example Manning howling like a dog and cursing in Chinese to get attention; the Chinese thinking him mad took him across the river, which action was all he wanted.[5] What these travelers had in common was a practice of carrying their eccentricities to "unusual extremes." Cavendish, as we will see, was like them an "eccentric traveler," though his eccentricity was less extreme than theirs, and less amusing.

10.1 Travel Writings in the Eighteenth Century

Lawrence Sterne, novelist and traveler, wrote, "The love of variety, or curiosity of seeing new things, which is the same, or at least a sister passion to it, – seems woven into the frame of every son and daughter of Adam." A scholar comments on Sterne's

[3] Foote, "Waterton," 574. Anon., "Charles Waterton," http://en.wikipedia.org/wiki/Charles_Waterton.

[4] Isobel Grundy, "Motagu, Edward Wortley (1713–1776)," *DNB* 38: 714–16, on 716. Anon., "Edward Wortley Montagu," http://en.wikipedia.org/wiki/Edward_Wortley_Montagu.

[5] Keay, *EccentricTravellers*, 70. Elizabeth Baigent, "Manning, Thomas (1772–1840)," *DNB* 36: 509–10, on 510. Anon., "Thomas Manning," http://en.wikipedia.org/wiki/Thomas_Manning_(sinologist).

quotation: "A person who travels subjects the structures of his personality, his mind and his emotions to a new process of experience, which may, in extreme cases, destabilize the traveller's previous world-view."[6] Cavendish was Sterne's curious traveler, who subjected his mind and emotions to new experiences, but as we should expect, he emerged with his worldview intact and confirmed.

Cavendish's main destinations were sites of British industry at the beginning of what we know as the Industrial Revolution. He and Blagden inspected mines, furnaces, and machinery, and they recorded what they saw in their journals. By this time, the keeping of journals was commonplace. Before we begin the account of Cavendish's travels, we take up the subject of writing about travels.

Travel writing can never be completely objective, for writers bring with them preconceptions and anticipations, but if the object of their travels is scientific, the subjective element is minimized.[7] Cavendish's and Blagden's travel writing is scientific. The entries in their journals are informative, the writers nearly invisible. We might compare them to laboratory notes, the laboratory happening to be movable. As such, we might wonder if it is meaningful to talk about their journals as travel writing.

It is expected of travel writing that it narrates a journey. Even where the writing is severely factual, there should be a narrative flow,[8] and there is little sense of it in Cavendish's and Blagden's journals. They describe what drew their eye at one location, for example, forges, then they do the same at the next location; they say nothing about the journey between stops, or at most mention points of geological interest observed in passing. On this count too, the journals would seem barely to qualify as travel writing.

Yet they do qualify. Like most travel books, their journals had a theme, certain things Cavendish wanted to see or do, which his traveling companion Blagden conveyed to persons he contacted before they set off. To carry it through, Cavendish had a travel plan; his journeys were not a disorderly wandering. To journal entries, he and Blagden assigned dates and locations, and at every location they saw what they came to see, things relating to the theme. Because the journals were not intended to be published, they needed to satisfy no one but their authors. The narrative character of the journals was left largely implicit, but it was there. The departures of the journals from other travel writing lie in emphasis not in nature.

Improvements in roads and carriages encouraged travel in the eighteenth century and with it a curiosity about the world. There was a corresponding market for first-hand accounts of travels. Travel writers took readers to unfamiliar sights, and provided practical information for readers planning to travel themselves, recommending places to see and where to stay.[9] The time had passed when travelers' accounts

6 Barbara Korte, *English Travel Writing from Pilgrimages to Postcolonial Explorations*, trans. K. Matthias (New York: St. Martin's Press, 2000), 44.

7 Ibid., 6.

8 Ibid., 9.

9 Charles L. Batten, Jr., *Pleasurable Instruction: Form and Convention in Eighteenth-Century Travel Literature* (Berkeley, Los Angeles, London: University of California Press, 1978), 3.

freely mixed fact and fiction and spread extraordinary fables. Tastes had changed, people had come to demand believable accounts of the world, and travel writers were expected to report faithfully what they had seen. "Truth and sound knowledge" were the proper subject of travel writers, the travel writer and merchant Nathaniel Wraxall wrote in 1775. "In the travels of a Philosopher," the *Monthly Review* said, we have "the discoveries of science, the improvements of art, the extension of knowledge." Major travel books reported journeys undertaken in the spirit of the Royal Society, and the proper form of writing, according to the *Monthly Review*, paid homage to the ideal of the Royal Society, "a plain unornamented style." The English editor of a Swedish travel book wrote that "every authentic and well-written book of voyages and travels is, in fact, a treatise of experimental philosophy." The traveler was foremost a researcher.[10] Natural philosophers, as we would expect, appeared among the new groups of enthusiastic travelers. In the year Cavendish made his first journey, 1785, the London chemist William Higgins made a journey to visit British factories, following an earlier similar journey by another London chemist William Lewis.[11]

Factualness and plain writing in the spirit of the Royal Society were one ideal of travel writing in the eighteenth century. There were others. Authors often said that they kept a record only for their own use or that of their friends, but this claim could be ingenuous. To appeal to readers and critics, they found they needed to do more than truthfully describe their journeys: they needed to write about them in a pleasing and entertaining way. This need gave rise to a distinctive kind of travel writing, one which blended description with artistic style. Authors of some of the most popular books of travels were literary men such as Sterne, Daniel Defoe, and Samuel Johnson.[12]

Readers of travel books desired novelty above all else. To satisfy the desire, in the last quarter of the eighteenth century, travel writers included new categories of observations such as "men and manners" and "character and manners," which dealt with the habits, laws, religions, and national characters of the inhabitants. Another and more significant trend in travel writing was a change in sensibility: readers' attention was directed to the beauty and grandeur of the landscape, subjective responses and feelings taking precedence over facts. The new writers were known as "picturesque" and "sentimental" travelers.[13]

Let us see what books of travels published around the time of Cavendish's journeys looked like. In 1778–83, Thomas Pennant, a Welsh naturalist and member of the Royal Society, brought out a handsomely illustrated two-volume account of his travels in his own land. In the advertisement of the book he promised readers what the "world justly loves," "reality." He also promised "wild and romantic scenery," but overall his writing is densely factual, with a strong antiquarian emphasis. He closely described the terrain, history, natural history, monuments, tombs, and

---

[10] Ibid., 5, 7, 39, 44, 72.

[11] A. E. Musson and E. Robinson, *Science and Technology in the Industrial Revolution* (Toronto: University of Toronto Press, 1969), 122.

[12] Batten, *Pleasurable Instruction*, 3, 45.

[13] Ibid., 75, 79, 91, 96–97, 109.

churches of Wales. He also gave some attention to the mining of coal, iron, lead, zinc, copper, and silver, and to the associated mineral tracks, though little attention to methods and machinery.[14] Another form of travel book was letters written in the course of a journey. An example is the account of a tour of England, Scotland and Wales in 1778 by Richard Joseph Sullivan, a writer born of Irish parents, and a member of the Royal Society. His observations were whimsical, ecstatic about nature, and without discernible order. He toured Chatsworth, the country home of the duke of Devonshire, which disappointed him, finding the house full of old furniture and the grounds filled with dead walls, spouting horse fountains, and other clutter. He hoped that in time the estate would be restored to its former reputation. Industry was not his subject, though in Birmingham he visited Matthew Boulton's manufactory, a familiar tourist stop. To describe its productions, he said, would require "a quicker degree of comprehension" than his; he found the garden there more interesting than the manufactured buttons and buckles. While he was there he was told about the benefits of canals, and these "stupendous works" caught his attention: "the spirit of industry will do any thing," he said. He believed that manufactures like all the other arts of civilization were connected through travel. "Of all the active employments of civil society, that of travelling seems the most important. The mind, restless and eager in its pursuits, pants after novelty,"[15] a justification of most travel books.

We turn to two travel books with a different perspective, combining a serious concern with industry and a feeling for landscape. Edward Daniel Clarke, world traveler, mineralogist, and chemist, toured England, Wales, and Ireland in 1791. The rugged country of North Wales had an "angry grandeur," he said, and he appreciated "wild scenes" and "picturesque landscapes," but he drew the line at the "affected refinement of a modern sentimentalist." He was interested in the national character of the peoples, in their manners, dress, and features, in ruins, churches, and houses, and in "fine prospects." He included incidents and vignettes, both entertaining and informative. In Cornwall, he described in some detail the extraction and smelting of tin ore, and in Wales the mining and smelting of copper. He responded to the human side of mining. Descending to a depth of 80 fathoms in a coal mine, he was impressed by the endurance of the miners and dismayed by the "cutters, as they are called, a troop of poor miserable black devils, working away their very lives amidst sulfur, smoke, and darkness."[16] Arthur Young, a Scottish agricultural improver and writer, toured South Wales and South Midlands in 1776. Drawn to fine prospects and "rural beauties," he gave a poetic description of Drislan Castle and its setting: "Upon the whole I think this spot the most picturesque residence I have seen in England. Hill and dale, and wood and water necessarily unite to form many beautiful scenes: they are the notes which must every where give the harmony of the landscape, but they are here

[14] Thomas Pennant, *A Tour in Wales MDCCLXX*, 2 vols. (London, 1778–83) 1: i, 415–24.

[15] Richard Joseph Sullivan, *Observations Made During a Tour through Parts of England, Scotland, and Wales. In a Series of Letters* (London, 1780), 141–42, 171–72.

[16] Clarke, *Tour Through the South of England, Wales, and Part of Ireland,* 80–89, 198–201, 274, 293–97.

accompanied with their richest melody." He also described in detail, often quantitatively, fields of cabbages and turnips and herds of cows and sheep. He gave the rents of farmland, the price of hay, the yield of wheat, barley, and oats in bushels per acre, the quantity and cost of lime spread over the fields, the number of calves per cow, and the pay of laborers. He described the land and the towns, how the people lived, how they dressed, and what they looked like. Agriculture was his primary interest, but he was interested in industry too. In Birmingham, he came to an edged-tool manufactory, where bloomery cinders, refuse from previous iron-ore smelting, were re-melted and used. "Such is the superior skill of the present age over the preceding, that they get almost as much iron from them, as from fresh ore." Because of recent inventions, buttons were better made, woolen cloth was better woven, canal locks were better constructed. "What trains of thought, what a spirit of exertion, what a mass and power of effort have sprung in every path of life, from the works of such men as Brindley, Watt, Priestley, Harrison, Arkwright.... In what path of life can a man be found that will not animate his pursuit from seeing the steam engine of Watt?" Young was a perceptive traveler, responsive to progress in all its varieties. His travel account was informative, filled with objective observations and sound judgments. It also showed aesthetic appreciation and feeling.[17] To Young, but to other travel writers too, industry with its new technology was inspiring and worth a look.

Cavendish had a choice of ways to keep a journal. Of the writers above, he was closest to Young. Young distinguished two forms of travel reporting: a diary or autobiographical register of the journey, and a summary of results of the journey. He recognized that almost all travel books were of the first kind; they had greater credibility, though at the cost of repetition and digressions. The advantage of travel books of the second kind was brevity and forcefulness. Young combined both kinds in his travel writing, designating the first part as "Journal" and the second part as scientific "General Observations." James Boswell in his book of travels in Corsica in 1768 used the same organization as Young, only reversing the order of the two parts.[18] To some degree, Cavendish followed their plan. For each of his three journeys in 1785–87, he or Blagden or both kept a journal, and later he wrote up the geological results in a separate text, summarizing his observations on strata.[19] Cavendish's interest, however, was narrower and more focused than Young's.

It would appear that his model came from the Royal Society, which in the second half of the eighteenth century directed a number of expeditions to the far corners of the world. In the decade before his own journeys, Cavendish served on a committee

---

[17] In 1768, Arthur Young published *A Six Weeks Tour through the Southern Counties of England and Wales*. He also published his account of this and other tours in the periodical he edited, *Annals of Agriculture*. The quotations above are from the latter source, not paginated. Arthur Young, *Tours in England and Wales, Selected from the Annals of Agriculture*, http://www.visionofbritain.org.uk/text/contents_page.jsp?t_id=Young.

[18] Batten, *Pleasurable Instruction*, 32–33.

[19] Henry Cavendish, untitled, 21-page paper on observations of strata, Cavendish Scientific Manuscripts, Devonshire Collections, Chatsworth.

of the Royal Society ordered to draw up scientific directions for a voyage to the North Pole by C. J. Phipps, the principal object of which was to determine how far navigation was possible in the direction of the North Pole. The secondary object was to carry out scientific investigations. Phipps took with him a range of scientific instruments, including a thermometer invented by Charles Cavendish, which was used to take the temperature of the sea down to 780 fathoms. Henry Cavendish gave Phipps an explanation of this thermometer including a correction to be made of the readings based on John Canton's demonstration of the compressibility of water and other liquids, which his father had confirmed for the Royal Society. Phipps's account, *Voyage towards the North Pole,* consists of a narrative of the voyage followed by appendices containing observations and experiments, a variant on Young's plan of a travel journal. The narrative was factual, in keeping with the object of the voyage, but at the same time it recounted an adventure. At one point the ships were hemmed in by ice, and the crew was resigned to abandoning them, but in the end the ships broke free. There were stunning prospects, recorded by an artist of the crew. Great bodies of ice filled valleys between high mountains, breaking off from vertical faces at the sea to form icebergs. "The black mountains, white snow, and beautiful colour [light green] of the ice, make a very romantick and uncommon picture."[20] This was a rare picturesque digression; for the object of Phipps's travel book was not to please but to record scientific results useful to navigation, geography, and natural history.[21]

Toward the end of the century, it became fashionable for the upper classes to travel through England, Wales, and Scotland. The internal tour became a competitor to the Grand Tour on the Continent, the traditional introduction of largely upper class youth to the cultural roots and fashionable society of Europe. By then the Grand Tour had been largely written out anyway. In his account of a tour through Wales, Thomas Pennant said that Britain had as much to offer as the Continent in the way of natural history and art.[22] If he had included industry, he would also have been correct. When Cavendish set out on a journey to the mines and metal works in Wales in 1785, he was in step with the times.

10.2 The Journeys

Like other travel writers, Cavendish and Blagden wrote in the first-person, in their case, "we." They used the personal pronoun sparingly, following the practice of travel writers wishing to avoid the appearance of egotism. The new trends in travel

[20] Constantine John Phipps, *A Voyage towards the North Pole, Undertaken by His Majesty's Command, 1773* (London, 1774), 27, 32–33, 70, 142, 145. Henry Cavendish, "Rules for Therm. for Heat of Sea," Cavendish Scientific Manuscripts, Devonshire Collections, Chatsworth, III(a),7.

[21] Batten, *Pleasurable Instruction*, 28.

[22] Ibid., 92–94.

writing did not appear in their journals. The object of the picturesque or sentimental traveler was to entertain or arouse feeling. The object of the scientific traveler was to *see*; seeing was associated with knowledge,[23] Cavendish's first interest, always.

The journals are written in three different hands. One is Cavendish's, hastily written on the trips. Another is a copyist's. The copyist left blank spaces where he could not make out the words in the original; in the copy, missing and misread words are supplied in Cavendish's handwriting. The third hand is Blagden's. The Blagden papers in the Beinecke Library at Yale contain a journal of the second half of Cavendish and Blagden's journey of 1787, written by Blagden. A comparison between it and the copyist's journal in Cavendish's papers at Chatsworth shows that Blagden's journal is the source of the copyist's. Words crossed out in Blagden's journal do not appear in the copy, and illegible words are misread or left out. Cavendish kept a journal of the first half of the 1787 journey. He and Blagden wrote about the same kinds of things and in the same plain, factual style.[24] They agreed on what to record, which would have been Cavendish's to decide.

A person who helped make arrangements for one of their journeys called it a "philosophical tour,"[25] which it was, though they called it simply a "journey." Their journeys had a twofold purpose: to observe the heights and geology of the island and to observe the machinery and processes of industry. The journeys took place on three successive years, always in the summer months when the roads were good. The first, in 1785, took them to the west of London, to South Wales; the second to the north of London, as far as Yorkshire; and the third to the southwest of London, to Cornwall.

Cavendish had long been a vicarious traveler. In addition to drawing up scientific instructions to observers who undertook voyages of discovery, he was an avid buyer and reader of books by travelers, and he invited travelers to meetings of the Royal Society and to dinners at the Royal Society Club. He recommended ten travelers for membership in the Royal Society; because he did not recommend many persons, the number shows a strong interest.[26] He valued the information that travelers brought back with them, and in the 1780s he himself became a traveler bringing back first-hand information.

Cavendish had made a few early trips. He traveled to Cambridge to pursue his education, and he traveled to the Continent with his brother. In 1778, he made a 6-day circuit through Oxford, Birmingham, Towchester, St. Ives, Ely, and back to London, making trials of Edward Nairne's magnetic dipping needle at every stop, evidently the point of the journey; he had recently reported on the Royal Society's

---

[23] Barbara Maria Stafford, *A Voyage into Substance: Art, Science, Nature, and the Illustrated Travel Account, 1760–1840* (Cambridge and London: MIT Press, 1984), 52.

[24] Charles Blagden, untitled journal of the second half of the journey of 1787, Yale, box 1, folder 2. Copy of the same, and Cavendish's journal of the second half of the journey, Cavendish Scientific Manuscripts, Chatsworth, X(a), 6 and 7.

[25] George Hunt to Mr. Hext, 23 January 1787, Blagden Papers, Yale, box 1, folder 4.

[26] The travelers Cavendish recommended are identified in Jungnickel and McCormmach, *Cavendish* (1999), 261–65.

meteorological instruments including its dipping needle made by Nairne.[27] A short journey from London to Folkstone on the English Channel, passing through Maidstone, Ashford, and Hythe, can be mapped from a list of stones he brought home to examine.[28] In the journal of his journey with Blagden in 1787, Cavendish wrote that the road from Bath to Devises had been changed since he was there "on a former journey,"[29] about which we have no further information. He may have made other trips for which he did not leave a record, but the journeys of the 1780s were unusual for their range and objectives.

What do we make of Cavendish's adventurous turn in midlife? On first thought, there seems to be a contradiction here. Travel entails an interruption of routine, and this man of inflexible habits voluntarily makes a sudden appearance as a repeating tourist. The explanation lies in part in the nature of the journeys: intended as an extension of his studies of the physical world, they continued the essential routines of his life. It lies in part too in the kind of tourist he was. He minimized the disruption of his life by bringing its essentials with him. He traveled with the man who was with him on a daily basis, Blagden; he brought a servant; he brought instruments and a portable laboratory and pen and paper to record his observations. It helped too that the journeys were planned well in advance. Persons who provided extended hospitality were not complete strangers. For example, in preparation for Cavendish and Blagden's journey to Cornwall in 1787, an acquaintance of theirs in the Royal Society, Roger Wilbraham, wrote to George Hunt, a Whig Member of Parliament from Cornwall, to introduce Cavendish and Blagden, both "well known in the literary world," to Mr. Hext in Cornwall who was asked to be of "service to them in the mineral way." If Mr. Hext were to take them to Lanhydrock, Hunt's home, he should get them a good mutton dinner and produce the marked "old" port and sherry from his cellars.[30]

On the journeys, Cavendish was the person we recognize from Wilson's description: from his monk cell's "narrow window he saw as much of the Universe as he cared to see."[31] His window was narrow, but less narrow than we might imagine. His interest remained the physical world, but we learn that this embraced a wide range of industrial processes and machinery. The extension of his interest could be direct. When C. J. Phipps, Lord Mulgrave invited Cavendish to visit his alum works, Cavendish was receptive to the idea, "having formerly made experiments himself on the crystallization of alum."[32]

[27] Henry Cavendish, "Trials of Nairne's Needle in Different Parts of England," Cavendish Scientific Manuscripts, Devonshire Collections, Chatsworth, IX, 11.

[28] Henry Cavendish, "List of Stones with Their Examination," Cavendish Scientific Manuscripts, Devonshire Collections, Chatsworth.

[29] Cavendish, Journal of 1787, p. 6.

[30] Hunt to Hext, 23 January 1787.

[31] Wilson, *Cavendish*, 186.

[32] Charles Blagden to Lord Mulgrave, 2 August 1786, draft, Blagden Letterbook, Royal Society 7: 17.

In his daily life, Cavendish preferred continuity to novelty, but he was open to new vistas on his travels. Blagden told a friend in Bristol that he would try to persuade Cavendish to stop there on their way home from a journey, but he was not hopeful. Because Cavendish had already been to Bristol, he would want to return by a "new road."[33] As in the laboratory, where Cavendish was drawn to novel behaviors of matter as a matter of course, on the journeys he was drawn to terrains, machines, materials, and processes that were novel to him. These he found sufficiently interesting to draw him repeatedly from the sameness of his London life to life on the road.

After industry, the next most common subject of Cavendish's travel journals is geology, the first we learn of his active interest in this science. The leading problem of the day in geology was the succession of strata in Britain,[34]and his journals contain extensive observations of strata. At least part of the reason he took up geology was his recent interaction with John Michell, the natural philosopher most knowledgeable about British strata. Cavendish made a point of visiting him on one of his journeys.[35]

Very occasionally, entries in the journals stray from the objects of the journeys. Cavendish and Blagden learned that a canal had not yet paid any dividends, and that the rate for bringing coals up it was three shillings a ton.[36] This lonely economic fact appears in the journal for 1785. The town of Bridgenorth [Bridgnorth], which rose high above the River Severn, received all of its water from a wheel-work in the river that turned with the current, and there was no fall in the river there other than that produced by stopping the current in that part.[37] They passed through many towns in 1785, but this is the only town that is commented on (Fig. 10.1).

The same journal contains two passages in striking contrast to the factual observations surrounding them. At an ironworks, the scoria was "beautifully veined with white and blue." This was definitely an aesthetic appreciation, for when the scoria looked like this, the iron was extremely bad.[38] The second passage is: "The Terras-walk commands a remarkable scene, from the singular appearance of these rocks all around, but especially on the opposite side of the River Severn, the Eastern, and from the fine view of the River underneath. The remains of the old Castle, battered by Oliver Cromwell, exhibit a remarkable instance of a leaning Tower, or ruin, which produces a fine effect."[39] When Wilson read this passage, he paused. To this point in the journal, he said, Cavendish had not said a word about "the grandeur of natural scenery – the changing aspects of the skies – the striking differences between

[33] Charles Blagden to William Lewis, 11 July 1787, draft, Blagden Letters, Royal Society 7: 338.

[34] Roy Porter, *The Making of Geology: Earth Science in Britain, 1660–1815* (Cambridge: Cambridge University Press, 1977), 119.

[35] Charles Blagden to John Blagden Hale, 14 September 1786, draft, Blagden Letters, Royal Society 7: 33.

[36] The journal is in a wrapper labeled in Cavendish's hand, "Computations & Observations in Journey 1785," Cavendish Scientific Manuscripts, Devonshire Collections, X(a), 4, p. 8. Hereafter cited as Journal of 1785.

[37] Cavendish, Journal of 1785, 56–57.

[38] Ibid., 55.

[39] Ibid., 56.

Fig. 10.1 Bridgnorth. Photograph by Pam Brophy. The Shropshire town in 1998 (Wikimedia Commons)

Fig. 10.2 Bridgnorth Castle. Photograph by Philip McCavity in 2008. The twelfth-century castle was a Royalist stronghold in the Civil War. Cromwell took the castle and ordered it destroyed. Part of the great tower now leans at an angle of fifteen degrees, four times the angle of the leaning tower of Pisa (Wikimedia Commons)

the inhabitants of different parts of a region – the historical associations inseparable from certain localities, and much else, on which Saussure, Humboldt, Dalton, Darwin, Forbes, and other scientific pilgrims, expatiate so largely." The singular passage above, Wilson said, "infuses vitality and a human interest into these formal diaries; and shows that their writer had deep down in his nature, the common sympathies of humanity" (Fig. 10.2).[40] Wilson assumed that the journal was written by Cavendish, though he must have recognized that it was not in his handwriting.

Before we accept Wilson's conclusion about Cavendish's "common sympathies of humanity," we need to consider his traveling companion, Blagden, who we know was the actual writer of the journal. On their journey the next year, Cavendish and Blagden each kept a journal. Blagden's, which was in the form of a diary, was not copied out for Cavendish. Both Cavendish's and Blagden's journals contain observations of stones, but there the similarity ends. In his journal, Blagden commented freely on the natural beauty and the people of the regions they passed through. From the road near Thornhill, they had a "most beautiful" prospect both ways. On the road from Keighly to Halifax, the country looked "dull and gloomy," but on the approach to Halifax, there were "many fine prospects" again. Around Halifax, the people had good complexion and spoke the "most difficult language to understand" he had heard.[41] To Banks, Blagden wrote that in the Lake District, "Winander Meer [Windermere] is a most beautiful piece of water, & the scenery about it both magnificent & beautiful, according to the different positions." By contrast with Lake Windermere, the largest natural lake in England, Derwent Water looked "diminutive among the lofty mountains."[42] Between Windermere and Kendal, Cavendish wrote in his journal, the "prevailing stone was slate but with limestone in places," the limestone lasting all the way to Settle in north Yorkshire. Near Buttermeer Lake, wherever strata of slate were visible, the slate was "nearly perpendicular, and the hills were steep and regular in slope, covered with grass."[43] That was what Cavendish saw in the Lake District worth recording. His description gives the reader a picture without the subjective commentary.

Fifteen years later, Blagden came as close as he ever did to criticizing Cavendish in writing. To a friend, he wrote, "When I went to the lakes it was in company with Mr. Cavendish, who had no curiosity for several things which it would have given me great pleasure to have seen. Winander More struck me as the *prettiest* piece of

---

[40] Wilson, *Cavendish*, 179.

[41] 1 September 1786, Blagden Diary, Osborn Shelves, Yale, fc16.

[42] Charles Blagden to Joseph Banks, 4 September 1786, British Library, Add Mss 33272.

[43] Henry Cavendish, "Computations & Observations in Journey 1786," Cavendish Scientific Manuscripts, Devonshire Collections, Chatsworth, X(a), 3, p. 6. This part of the journal is in Cavendish's handwriting. The other part of the journal is in the copyist's handwriting and deals with Cavendish and Blagden's visit to Lord Mulgrave's alum works. This part does not appear in these footnotes.

water I had ever beheld."[44] For Blagden, the scientific objects in the scene were part of a larger view of the landscape, which included natural beauty. For Cavendish, the scientific objects were the only view of interest and the point of the journey. Not long after Cavendish and Blagden visited the Lake District, the poet William Wordsworth settled there to work on his theory of poetry, which would go in the preface of an edition of *Lyrical Ballads*, usually considered the beginning of the English Romantic movement in literature. In their responses to nature in the setting of the lakes, Cavendish with his strict objectivity is at one end of a spectrum, Wordsworth with his individual personal emotion is at the other end, and Blagden with his inclusive interests spans the middle. The passage in the 1785 journal about the fine view of the river and the fine effect of the castle could have arisen from a conversation on the trip, Blagden remarking on the prospect before their eyes, and Cavendish agreeing. In support of this possibility, Blagden said of their return journey from Cornwall by the north of Devonshire in 1787 that he took "great pleasure in shewing to Mr. Cavendish" the "grand beauties of that remarkable coast."[45]

There had long been interest in developing a reliable method of measuring heights of mountains using the variation of atmospheric pressure with elevation, as recorded by a barometer, but it was not until the 1770s that this method of measuring heights promised any degree of accuracy. The method was tried on Phipps's voyage in 1773. Heights were measured two ways: geometrically, using a theodolite, and barometrically, making use of André De Luc's rules for corrections. As measured by the theodolite, the height of a mountain was found to be 1,503 ft above sea level; as measured by the barometer, 1,588 ft. The discrepancy was considered large, and Phipps could not account for it.[46] There were alternative rules to De Luc's, and Cavendish compared them, drawing on his father's experiments on the specific gravity of air at different temperatures and pressures.[47] During the period of Cavendish's journeys, barometric heights were repeatedly taken on Mont Blanc, the highest mountain in the Alps.[48] They suggested to Cavendish an experiment on temperature corrections to barometric readings, and on his journey in 1787 he set up a meteorological station on Dartmoor, arranging for local persons to record instrument readings over a period of time after he left. Cavendish's interest in the barometric method was an extension of his long interest in meteorological instruments, and it bore on his new interest, geology, which deals with relief features of the

[44] Charles Blagden to Lord Palmerston, 25 November 1800, Blagden Letters, Yale.

[45] Charles Blagden to Mrs. Grey, 28 August 1787, draft, Blagden Letters, Royal Society 7: 351.

[46] Phipps, *Voyage to the North Pole*, Appendix.

[47] Henry Cavendish, "Rule for Taking Heights of Barometers," Cavendish Scientific Manuscripts, Devonshire Collections, Chatsworth, VIII, 12.

[48] From observations made by others, Cavendish calculated the height of the mountain. Charles Blagden to Joseph Banks, 5 October 1786, British Library, Add Mss 33272, pp. 19–20.

earth.[49] The barometric method of measuring heights became a major object of study for Cavendish for several years. The expectation was that if the right rules could be found, measuring heights by the barometer would become a practical and reasonably accurate method.

Cavendish and Blagden regularly made another kind of measurement. Cavendish had long suspected that the mean temperatures of different climates could be determined by taking the heat of springs and wells. If the rule held up, it would provide a convenient method of comparing the average temperatures of different climates not just in England but around the world.[50] In 1775, he advised Blagden to make observations with that view in mind in North America, where he was posted as an army surgeon during the American War of Independence. Blagden and Cavendish measured the temperature of wells and springs on their journeys in 1785–87. In 1788, a paper was presented to the Royal Society reporting the temperature of a well slightly above what was judged to be the mean heat of that part of Britain. In discussing the discrepancy with Blagden, Cavendish thought that the explanation lay in the stratum of clay in which the well was dug and on which London is built. To determine if London wells were too warm, Cavendish built a bucket that could be lowered to any depth, and he and Blagden tried it on several wells. To understand the facts better, Blagden studied the strata of Great Britain in a paper on the heat of springs and wells found among his scientific manuscripts.[51] Like the height of mountains, the heat of wells and springs related to geology, a leading interest of theirs on their journeys and after.

On all three journeys, Cavendish and Blagden brought with them a portable barometer for measuring heights. On their second journey, something went wrong with it, placing in jeopardy a main object of the journey. They appealed by letter to Banks, who asked the instrument maker Jessie Ramsden to send them one of his

[49] Charles Blagden to William Farr, 12 June and 3 July 1787, drafts, Blagden Letters, Royal Society 7: 67 and 7: 335.

[50] "In England the heat of the water in deep wells or quick springs is very nearly equal to the mean heat of the air, and it seems well deserving inquiry whether it is the same in other countries; for if it is so, it would afford the readiest way of comparing the mean heat of different climates." "Meteorological Observations at Madras," extract of a letter by Cavendish, in Cavendish, *Scientific Papers* 2: 394. John Hunter, who became Cavendish's physician at some stage, was sent to Jamaica in 1782 to superintend a military hospital, and Cavendish suggested to him, as he had to Blagden when he journeyed to North America, that he observe the heat of springs and wells while he was there. Hunter published his observations in the *Philosophical Transactions* in 1788, where he gave a full account of Cavendish's hypothesis: assuming that the heat of the earth comes entirely from the sun, not from the earth's interior, measurements of temperatures deep enough inside the earth to remain constant through the seasons ought to yield the mean temperatures of different climates. A few observations of this nature would tell as much about average climates as years of meteorological observations. John Hunter, "Some Observations on the Heat of Wells and Springs in the Island of Jamaica, and on the Temperature of the Earth Below the Surface in Different Climates," *PT* 78 (1788): 53–65, on 53, 58, 65.

[51] Draft of a long paper by Blagden beginning, "The idea of determining the mean temperature of different climates by the heat of the springs and wells." Blagden Papers, Yale, box 6, folder 26.

barometers. When they received it, it was poorly packed, air had got into it, and it was a week late, causing Blagden to exclaim, "Ramsden, as usual, told a bounding lye, & that of the most impudent kind."[52] None of this actually mattered, since in the meantime they had fixed their own barometer. They took the normal travelers' inconveniences in stride, but a faulty instrument for measuring heights was serious. Their daily readings of the barometer and thermometer do not appear in the journals, but they are recorded and analyzed on separate sheets.

Their travels were physically demanding. Midway through their first journey, Blagden reported to Banks that they had measured the "highest mountains" in four counties, having climbed them with their barometer.[53] On their second journey, they were shown a lead mine by the proprietors, who were quite willing to take them further than they wished to go. They found climbing up and down the ladders "very fatiguing work," especially Blagden whose legs were "scarcely long enough to straddle across some of the wider turns."[54] These and other physical discomforts of their travels are never mentioned in the journals.

Reports from senses other than sight are rare in the journals. At a tinplate works, they saw how rust was beaten off iron plates, "struck with great violence by a man on an iron Anvil."[55] The journal does not mention the sound of that activity or that of the hammers and other machinery, which would have been tremendous. They saw finers at their forges exposed to searing temperatures. Unaccustomed to the intense heat, Cavendish and Blagden would have kept their distance, but the journal makes no mention of the heat on their skin. The journal is limited to the matter-of-fact recounting of how things worked and looked and occasionally smelled.

Cavendish was fortunate in his companion, a committed traveler. Blagden had made journeys all his life and had kept a record of them. At age 17, he traveled from London to Scotland to study at the University of Edinburgh. In a letter to a relative, he described the roads on that journey, gave the distances traveled each day, reported the weather, and mentioned the sights: a hose manufactory, a castle, a river, and prospects, beautiful and barren.[56] At age 18, he took a 6-day vacation from his studies to take a journey on foot of some 80 miles, "undertaken chiefly with a view to philosophical curiosities," but which included cobalt mines and the "finest ironworks in Scotland."[57] Several months later he took a "short jaunt" in the Highlands.[58] Everything was interesting to the serious young traveler, who had disdain for people

[52] Charles Blagden to Joseph Banks, 13 and 19 August, 4 September, 1786, British Library, Add Mss 33272, pp. 1–2, 7–8.

[53] Blagden to Banks, 31 July 1785.

[54] Charles Blagden to Joseph Banks, 17 September 1786, British Library, Add Mss 33272, pp. 9–10.

[55] Cavendish, Journal of 1785, 19.

[56] Charles Blagden to Sarah Nelmes, 1 November 1765, Blagden Letters, Royal Society, B.159. Blagden's great aunt, Sarah Nelmes was a spinster living in Bristol.

[57] Charles Blagden to Sarah Nelmes, 26 March 1767, Blagden Letters, Royal Society, B.161.

[58] Charles Blagden to Sarah Nelmes, 19 September 1767, Blagden Letters, Royal Society, B.161a.

who were satisfied with their "little world," incurious about what lay beyond it.[59] Journeys were an opportunity to observe the greater world. William Cullen, Blagden's professor in Edinburgh, warmly recommended him, noting he had "more knowledge than most of his standing."[60] Hannah More after meeting Blagden at a small select party called him "a new bluestocking, and a very agreeable one," who is "so modest, so sensible, and so knowing, that he exemplifies Pope's line, 'Willing to teach, and yet not too proud to know.'"[61] He had a remarkable and accurate memory. James Boswell and Samuel Johnson talked about his "copiousness and precision."[62] He had the right qualifications for Cavendish's daily associate. To the point here, he was interested in the industrial scene.

Blagden was interested in people too. Upon his return from North America, in 1781 he settled in London, where he sought out the company of men of science. The following year, as we know, he became Cavendish's associate, and 3 years after that he became Cavendish's traveling companion. He readily adapted to persons of different stations and temperaments. In his youth he advised a friend to read Jean-Jacques Rousseau, "the most eloquent & feeling of men,"[63] and in middle age he paid homage to the most romantic scene of Rousseau's novel *La Nouvelle Heloïse*,[64] and he went on an extended journey to observe ironworks with a man he described as having no affections. He willingly accepted Cavendish's ways and deferred to his desires. There is no reason to think they had serious collisions, and harmony being the condition of successful travels, we suppose that they pursued their goals without personal distractions. Evidence of this is that after their first journey, they took two more.

The goals of the journeys were Cavendish's, and he chose well. Grounded in his interests, the goals were realistic, and within the limitations of the weather they were all met. On their departure from Wales in 1785, Blagden wrote to Banks that Cavendish "bears the journey remarkably well."[65] When they returned from their journey to Cornwall in 1787, Blagden wrote to Banks that Cavendish "is very well, & I think looks the better for his journey."[66]

When Blagden said that Cavendish always knew what was right for him, he may have had the journeys in mind, among other things. Around the time of the journeys, Cavendish made several major changes in his life, which are discussed in Chap. 8. We briefly recall them here to add to them a major change in his scientific work.

---

[59] Charles Blagden, "Memorandum of a Tour Taken for Four Days Beginning August 18, 1771," Blagden Papers, Yale, box 1, folder 3.

[60] John Thomson, *An Account of the Life, Lectures, and Writings of William Cullen, M.D.*, vol. 1 (Edinburgh and London, 1859), 555–56.

[61] Hannah More in 1788, quoted in Crowther, *Scientists*, 333.

[62] Johnson found Blagden a "delightful person." James Boswell, *The Life of Samuel Johnson, LL.D. …*, rev. ed., 5 vols. (London, 1821) 4: 309.

[63] Charles Blagden to Thomas Curtis, 26 July 1771, draft, Blagden Letters, Royal Society, B.162.

[64] Blagden, "Diary of Sir Charles Blagden," 77.

[65] Blagden to Banks, 31 July 1785.

[66] Charles Blagden to Joseph Banks, 14 August 1787, British Library, Add Mss 33272, p. 35.

The first change was his willingness to take on an associate. Blagden was not one of his young assistants but a well-educated physician and officer of the Royal Society who made a scientific career for himself. The second was location. In 1782, the year he began his association with Blagden, he took a house in Hampstead. The next year his father died. In 1784, he bought a townhouse on Bedford Square, which he largely converted into a library. In 1785, the year of his and Blagden's first journey, he moved into a substantial villa near Clapham, which he largely converted into a laboratory and observatory, and where he would remain to the end of his life. A third major change was the direction of his scientific work. Except for an answer to criticism in 1788, he finished publishing his work in the chemistry of airs (gases) in 1785. Not long after that, he drafted a mechanical theory of heat, which brought together his extensive researches on the subject. His experiments on heat, which had paralleled his work on chemistry from the start of his career, came to an end with publications in 1786 and 1788. His work on electricity, the third field in which he had done important work, had come to an end somewhat sooner, in the early 1780s. In his 50s, having nearly completed his work in the major fields, he was open to a new direction, and his journeys, themselves a major change, pointed the way. He was at an age when men often undergo life changes, but if that was the case with Cavendish, the significant changes were contained within his life of natural philosophy.

In 1786, the second year of his journeys, he began recording his chemical experiments in a book of blank pages; this was the permanent record, transcribed and rewritten from the rough "minutes" of the experiments. He labeled it "White Book, No. 1," suggesting an open-ended subject. In it, he referred to a "2d book," perhaps "White Book No. 2," since lost. The White Book contains some experiments on air, but this was no longer a focus of his work.[67] The new format signaled a change in

[67] After 1786, the year Cavendish opened his White Book, he added some more pages of experiments to a massive bundle of pages, "Experiments on Air," the record of a sustained series of interconnected chemical researches, which contain his experiments on the production of water from the detonation of oxygen and hydrogen, the nature of nitric acid, and the composition of the atmosphere. The later experiments did not have a comparable direction, and there is no indication that he had publication in mind for any of them. They do include an impressive investigation of the expansion of different kinds of air with heat, which determined that within the limits of accuracy of the experiments the coefficient of expansion is the same for all of them, but these experiments are out of place in the bundle, being physical not chemical. They are undated, but as they are recorded on paper bearing a 1797 watermark, they could not have been performed before that year. They appear to be Cavendish's last important research on airs. There are later pages containing small investigations having to do with airs. In light of the great changes chemistry underwent in the late eighteenth century, the most interesting of these is a reconsideration. Writing on paper watermarked 1800, Cavendish returned to experiments he had carried out probably in late 1783 on the nature of charcoal produced by burning willow wood. He carried out computations on the earlier experiments and drew a conclusion, which he expressed partly in the new language of chemistry: "either that charcoal contains hydrogen as well as carbon & water or else that the charcoal after distillation contained some oxygen. There is no reason to think the charcoal yielded any phl[ogisticated] air." The subject of the experiments and their timing point to a likely reason for his uncharacteristic use of the new nomenclature of chemistry in 1800. At the end of his paper, "Experiments on Air," read to the Royal Society in January 1784, where he says it would be very

the direction of his research, one which continued the contacts he had made in the course of his journeys in the country beyond the metropolis. Many of the entries could be called geological and industrial experiments, though mineralogical experiments would be appropriate too, since in the eighteenth century ores and stones were both considered mineral. Many experiments dealt with specimens from industrial processes and mines, which Cavendish either brought home from the journeys, or were sent to him from persons he met on the journeys, or were given to him by colleagues. The specimens included industrial discards such as kish from iron furnaces, slag from the purification of copper, cinder from fineries, and dust from lead smelting furnaces. Byproducts of industrial processes, they held chemical clues to what went on in the furnaces and were interesting scientific objects in themselves. On his second journey, Cavendish made side trips in search of kish, a black shiny matter that came down into the hearths of iron furnaces and was thought to reduce the quantity of iron and also to give iron certain qualities when mixed with it. William Lewis, whom he and Blagden visited, and other ironmasters had not mentioned it, and on his return Cavendish requested information about it and samples to be sent to him.[68] Through his travels, Cavendish widened his active field of interest in ways he would follow up at home.

Wilson was disappointed that Cavendish's travel journals lacked human interest. Cavendish's editors excluded them because they contained nothing of geological importance, and "as narratives of travel they have no particular value," nor do they "afford any special information" about technical and manufacturing operations. They also excluded Cavendish's paper on strata, which drew on observations he made on his journeys, because it had no geological importance and would not enhance his reputation as a geologist.[69] We look at them differently, for what they tell us about Cavendish. They tell us that he journeyed through unfamiliar physical and human surroundings, seeing and recording objects that were new to him. They confirm what this natural philosopher asked of life: to provide him with meaningful physical objects to study.

Crowther saw value in Cavendish's travel journals. Given Cavendish's social background and his "reserved London life," he found them "remarkable," revealing "a keen and penetrating interest in the applied science of the new industry" and a

---

difficult to decide between the chemistry of phlogiston and the new chemistry from France, he says that he prefers the old chemistry because of plants, which seem to him to be more compounded than the charcoal or ash to which they are reduced upon being burnt. According to the new chemistry, a plant is deprived of dephlogisticated air (oxygen), and upon burning it acquires it. The plant is simpler than the ash. According to the old chemistry, the plant contains phlogiston (hydrogen, or related to hydrogen), which is given up when it is burnt, making it more compounded than the ash. "Experiments on Air," Scientific Manuscripts, Devonshire Collections, Chatsworth, II, 5: 251–65, 386–90, quotation on 390. "Experiments on Air," *Scientific Papers* 2: 181.

[68] Charles Blagden to William Lewis, 26 September 1786, Blagden Letters, Royal Society 7: 38. Kish is a form of graphite that separates from some iron in smelting. Plumbago and black lead are other names for the graphite. In 1779, Scheele showed that plumbago is essentially carbon, perhaps the reason for Cavendish's interest.

[69] Evaluations by Thorpe and Archibald Geikie in Cavendish, *Scientific Papers* 2: 431–32.

watchful interest in geological formations of potential interest to mining.[70] To better understand what drew Cavendish to ironworks and mines, we look at him in his familiar setting, the scientific society in London at the time. Many of its members had a strong interest in applications of natural philosophy and chemistry to industry. We see this in the record of a club that met in a London coffee house in 1780–87, the period of Cavendish's journeys: the [Chapter] Coffee House Philosophical Society. Cavendish did not belong to the Society, but he knew many of those who did. They included his customary instrument maker Nairne; Priestley, with whom he had a chemical correspondence; the chemist and scientific polymath Kirwan, with whom he had a public disagreement about a serious point in chemistry; the inventor Watt, an honorary member whom he visited on his journey in 1785; John Hunter, who was Cavendish's physician; the chemist James Lewis Macie (James Smithson), who worked as Cavendish's assistant for a time; and a number of other well-known men of science, medicine, and industry such as the manufacturer and engineer Boulton, the industrial chemist James Keir, the ceramics manufacturer Wedgwood, and the geologist and clockmaker John Whitehurst. The majority of members were fellows of the Royal Society.[71] In 1785, the year of his first journey, Cavendish recommended Watt and Kier for membership in the Royal Society, clear indication that he wanted the Royal Society to recognize the industrial development of science, as he himself did.[72]

The Coffee House Philosophical Society is rare among scientific clubs in leaving a detailed record of its meetings. At its founding, the Society resolved that its conversations "shall be confined to Natural Philosophy, in its most extensive signification." From the minutes of the Society, we see that its conversations were, as intended, about natural philosophy, chemistry, industry and applied science, meteorology, ballooning, and a few other subjects. Among the topics that came up at the meetings were a device for measuring the expansive power of steam, a machine for welding masses of iron together by pressure, chemical differences between the forms of iron and steel, and hammers driven by steam engines in ironworks. Chemistry was the leading scientific subject of the discussions, and in 1785 the question of the existence of phlogiston, the central idea of the older chemistry, then under attack by French chemists, was first raised. During the years of his journeys, Cavendish's interests in industry and chemistry paralleled those of members of the Society, and his interest in geology paralleled discussions of ores in the Society.[73]

---

[70] Crowther, *Scientists in the Industrial Revolution*, 320.

[71] G. L'E. Turner, "The Membership," in *Discussing Chemistry and Steam: The Minutes of a Coffee House Philosophical Society, 1780–1787*, eds. T. H. Levere and G. L'E. Turner (Oxford, New York: Oxford University Press, 2002), 17–44.

[72] Priestley, Smeaton, and Cavendish were the first signers of the certificates for James Watt, elected 24 November 1785, and for James Keir, elected 5 December 1785. Certificates, Royal Society, vol. 5.

[73] T. H. Levere, "Introduction," in *Discussing Chemistry and Steam*, 1–15. Jan Golinski, "Conversations on Chemistry: Talk about Phlogiston in the Coffee House Society, 1780–1787," ibid., 191–205, on 193–97.

When Cavendish visited the industrial Midlands on his journey in 1785, he met several members of the Coffee House Philosophical Society on their home ground. They were also members the Lunar Society of Birmingham, a small club which met monthly at individual houses including Boulton's house at Soho.[74] It differed from the London club in having no written rules or records of meetings or formal membership. Its members were a mix of natural philosophers, inventors, manufacturers, and doctors, several of whose names appear in Cavendish's travel journals: William Withering, Watt, Wedgwood, and Whitehurst. The productivity of the club lay in the collaboration of men with strong talents, joined in the belief that science, industry, and progressive ideas would change the world for the better. The dates of the beginning and end of the Lunar Society are uncertain, but its most productive years were 1780–89, bracketing the years of Cavendish's journeys. The compositions and purposes of the London and Birmingham clubs show that there was nothing unusual in the interest Cavendish took in industry and geology on his journeys. His thinking was aligned with that of the most capable men of the age.

The list below is what Cavendish and Blagden chose to see and record on their journeys:

Air pumps.
Alum mine.
Alum production.
Bellows powered by waterwheel.
Blast furnace.
Brass making.
Button manufacture.
Canals.
Chafery.
Circular motion given to steam engines.
Clay pits.
Cloth manufacture and dyeing.
Coal pits and mines.
Coal tar production.
Coal, coke, and charcoal.
Coke production.
Copper and tin mines.
Copper, iron, and tin ores.
Copper smelting houses.
Cylinder boring mill.
Drilling mill.
File making.

[74] The society is the subject of Robert E. Schofield, *The Lunar Society of Birmingham: A Social History of Provincial Science and Industry in Eighteenth-Century England* (Oxford: Clarendon Press, 1963).

Filigree work.
Finery.
Flatting mills.
Forges.
Hammers operated by steam engines.
Iron mines.
Iron ore.
Ironworks.
Japanning copper.
Lead mine.
Limestone quarry.
Machine for twisting handles of horsewhips.
Nail manufacture.
Needle manufacture.
Parallel motion given to steam engines.
Pitch.
Quarries.
Reverberatory furnace.
Rolling mills.
Shingling.
Silver plate manufacture.
Slitting mills.
Stamping and Potting.
Steam engines.
Stones.
Strata.
Tides.
Tin and copper mines.
Tinplating.
Tunnel cutting for canals.
Water pumps operated by steam engines.
Waterwheels.
Wells.
Wire manufacture.

10.3 Working Iron

Iron was the most important natural element in the Industrial Revolution. In Britain, its production was concentrated in a small number of places, Welsh valleys prominent among them. Before joining Cavendish and Blagden on their journey into one of the valleys, for readers who are unfamiliar with the history of iron, it will be helpful to describe how it was produced in Britain at the time, drawing on today's chemical knowledge and language.[75] Iron is plentiful, but in nature it rarely appears as pure iron. In the form of ore, iron is combined with oxygen in various proportions, and usually other elements are present in small quantities. The ore was reduced to the metal in an operation called "smelting," an old term Cavendish did not use, which required a high heat and a reducing agent containing carbon; in practice, the reducing agent was at the same time the source of the heat, either charcoal or coke. The carbon removed oxygen from the ore by combining with it to produce gases, carbon monoxide and carbon dioxide, which escaped through the top of the furnace. Left behind were elemental iron, some carbon, and chemical impurities such as sulfur, phosphorous, and silicon. The residual carbon had a useful function: for strength iron needed to contain a small quantity of carbon, the proportion of which determined whether the iron was suitable for making cast iron, wrought iron, or steel. Iron had to be separated from impurities as well as from oxygen, for which purpose limestone, called a "flux," was heated together with the ore. The flux also increased the fluidity of a partially vitreous byproduct of smelting, "scoria," or slag, a mix of impurities, cinders, and the added flux, enabling the scoria to be tapped from the bottom of the hearth; the term "flux" derives from this fluidity.

Two kinds of furnaces were used in smelting, the "blast furnace" and the "forge." Iron ore together with limestone and fuel were placed in the hearth of a blast furnace and heated to a high temperature, while air was blown through it by a bellows or an air pump, driven by a waterwheel or steam engine. Molten iron tapped from the corner of the hearth was called "pig iron," the name deriving from the traditional method of forming manageable pieces of solid iron: molten iron from the blast furnace flowed into depressions in sand, forming paired rows of iron ingots at right angles to a central channel, suggesting a litter of suckling piglets. Pig iron could be used to make "cast iron," or iron poured into shaped molds, but because it contained excess slag and carbon it could not be used directly to make malleable or "wrought iron." Pig iron could be shaped by hammering when hot, but once it cooled hammering would crack or break it. Because it had a crystal structure, it was strong in compression, and so it was useful in making bridges and other structures.

Wrought iron had a low carbon content and a fibrous structure, making it tough and malleable, capable of being hammered or rolled. Today wrought iron is not made commercially, and what is called wrought iron is really steel that is handworked, but in the eighteenth century wrought iron was the main form of iron, valued because it could be shaped.[76] To make it, the pig iron underwent a second round

[75] Much of this and the following five paragraphs on iron production in Britain are based on H. R. Schubert, "Extraction and Production of Metals: Iron and Steel," *History of Technology*, vol. 4: *The Industrial Revolution, c1750 to c1850*, ed. C. Singer, E. J. Holmyard, A. R. Hall, and T. I. William (New York and London: Oxford University Press, 1958), 99–117. Laurence Ince, *The South Wales Iron Industry 1750–1885* (Merton: Merton Priory Press, 1993). Richard Hayman, "The Shropshire Wrought-Iron Industry c 1600–1900: A Study of Technological Change," PhD thesis, University of Birmingham 2003.

[76] Wrought iron remained the usual kind of malleable iron until the late nineteenth century, when mild steel became practical. The steel that was made in the eighteenth century had limited uses such as files.

of smelting in a "finery forge," which was a small furnace for refining the impure iron. In the forge, the pig iron was placed on a hearth containing fuel, reheated, and stirred by a finer handling a long iron bar, while air was blown on it. The fuel was charcoal, because a mineral fuel, coal or coke, in contact with the molten iron would reintroduce impurities. The operation removed much of the remaining carbon, silicon, and other impurities by oxidation. The product was a porous lump of iron known as the "half-bloom," its pores being filled with slag. To make the lump more compact, it was placed on an anvil under a massive trip hammer, powered by water or steam. Then it was usually returned to the finery forge, and reheated. To drive the molten slag from its remaining pores, and also to knock off slag adhering to the surface, the reheated half-bloom was beaten with a hammer. This consolidation of the iron by hammering was called "shingling." The hammer-man's work was potentially dangerous, since with each blow of the hammer white hot slag flew across the room. Finally, the iron was taken to a second forge, a "chafery," where it was reheated until it was just malleable, and it was then hammered into a bar of wrought iron, known as "bar iron," ready for use. Because the iron in the chafery was not heated to the melting point, coal could be used as fuel without danger of its impurities entering the iron. The finery and chafery forges were usually housed in a large wooden shed, along with trip hammers, hand hammers, anvils, tongs, carts, and other tools for moving hot, heavy pieces of iron.[77] The words in quotation marks above are ones used in the journal.

The production of iron underwent several major changes in the second half of the eighteenth century. One was a partial change of fuel. Charcoal, a fuel obtained by heating wood, was almost pure carbon, excellent for reducing iron ore to the pure metal. A problem with it was that British forests were being depleted to make charcoal for blast furnaces, which consumed it in prodigious quantities. Naturally there was an interest in finding a practical substitute. Coal, another good source of carbon, was plentiful and relatively cheap, and attempts were made to use it in place of charcoal, though its impurities resulted in a weak iron. Coke, obtained by heating soft coal, was cleaner. Introduced in Britain at the beginning of the eighteenth century, coke came into wide use for blast furnaces in the second half of the century.[78] Coke was less friable than charcoal, and accordingly furnaces could hold larger charges of ore and stronger blasts could be used, resulting in greater production. The disadvantage of coke was that it contained more impurities than charcoal, for which reason charcoal continued to be preferred for most finery forges. A second important change was the replacement of water power by steam power for some purposes. The first use of the steam engine for a purpose other than pumping water out of mines was to drive an air pump at John Wilkinson's blast furnace at New Willey in 1776. It was first used to raise water to turn a waterwheel for operating a

[77] Anon., "Forge, Furnace, What's the Difference?" http://www.engr.psu.edu/mtah/essays/forge_furnace.htm.

[78] Five years after Cavendish and Blagden's journey, 81 out of 106 blast furnaces in Britain were fueled by coke.

forge hammer at Wilkinson's ironworks at Bradley in 1782.[79] Cavendish and Blagden visited both of Wilkinson's sites, and they also visited the designer of Wilkinson's engines Watt at the Soho Manufactory in Birmingham. A third change was an increase in the production of cast iron for construction uses. On their journey, Cavendish and Blagden saw the Iron Bridge [Ironbridge], an early instance of cast-iron construction.

The final important changes were two inventions by Henry Cort, both leading to increased production. The first, in 1783, was grooved rollers for making iron bars, replacing slitting mills; the second, in 1784, was a process called "puddling," which converted pig iron into wrought iron in a reverberatory furnace, described below, without the use of charcoal; the molten pig iron in the furnace was stirred by long iron bars, which assisted the removal of carbon and impurities by oxidation. Over time, puddling replaced the finery process for making wrought iron. On their journey, Cavendish and Blagden saw neither of Cort's innovations, though they saw reverberatory furnaces, and South Wales was one of the earliest locations for puddling furnaces.[80] Later Cavendish took an interest in Cort's process for working pig iron. We know about it from notes he took of a conversation with the engineer James Cockshutt about the advantages and disadvantages of two variants of the process, one requiring a common coke finery in addition to a Cort's furnace, the other using only a Cort's furnace. Cockshutt gave Cavendish finery cinders corresponding to the two variants, which Cavendish subjected to chemical tests with his customary thoroughness.[81]

10.4 Journeys and Science

When Cavendish ventured into the wider world of industrial Britain, he saw a reflection of his own small world on Clapham Common. He visited forges, which differed from the forge at home mainly in scale. In the fiery hearths of ironworks, he saw chemical reactions releasing gases, which differed from what he saw in his laboratory mainly in scale. He brought the wider world back with him in the way of specimens from furnaces, forges, mines, and geological formations. A month after they returned from their second journey, Blagden wrote that Cavendish was busy making experiments on stones and on specimens from an alum works.[82]

In the time of his journeys, Cavendish's most important scientific work was in chemistry, and on his journeys the most important actions he observed were chemical. In 1785, 2 months before Cavendish and Blagden began their first journey,

[79] Harris, "Wilkinson," 1010–11.

[80] Schubert, "Extraction and Production of Metals," 99–107.

[81] The notes, dated spring 1793, are entered loosely in Henry Cavendish, "White Book No. 1," Cavendish Scientific Manuscripts, Devonshire Collections, Chatsworth. His experiments are on pp. 84–90.

[82] Charles Blagden to Joseph Banks, 8 [?] October 1786, British Library, Add Mss 33272, pp. 15–16.

Blagden wrote to a French chemist that the British had not yet given up on phlogiston but that with Cavendish it was a "doubtful point." The question of which hypothesis was right, that of phlogiston or that of Lavoisier's, "cannot remain long undecided."[83] In 1787, 2 months after their last journey, Blagden wrote to the same chemist that the main objections of the partisans of the doctrine of phlogiston had been answered, and that the explanations of Lavoisier's hypothesis were "so much clearer and more pointed" that the "combat must soon be at an end."[84] Blagden wrote as one convinced of the imminent victory of the new chemistry coming out of France. He was in regular correspondence with French proponents of the new chemistry, and he was in the regular company of a principal chemist of the old chemistry, Cavendish, who at the time was in serious doubt. We might expect Blagden's opinion on the great question of the day in chemistry to be aligned with Cavendish's. In the spring of 1787, before Cavendish and Blagden began their last journey, the Irish chemist and warm advocate of phlogiston Richard Kirwan wrote to a French chemist, "Mr. Cavendish has renounced phlogiston."[85] By whatever route Kirwan learned this, we have no reason to doubt him. Given what Cavendish wrote about the two chemistries in 1784 and what Blagden had told colleagues about Cavendish's views since then, Kirwan's news comes as little surprise. With reasonable confidence, we can say that Cavendish was the first British chemist to abandon the old chemistry centered on the concept of phlogiston.

Over the 2 years following Cavendish's last journey with Blagden, the French advocates of the new chemistry were elected foreign members of the Royal Society. Cavendish signed the certificates for all of them, and in the case of their leader, Lavoisier, he was the first to sign.[86] What was called "dephlogisticated air" in the old chemistry was renamed "oxygen" in the new chemistry, where it was regarded it as an elementary substance. The "phlogiston" of the old chemistry did not appear in the new chemistry. Of interest to us in this chapter is the timing. Cavendish's questioning and renunciation of phlogiston coincided with his journeys into an industrial heartland, where he observed chemical processes explained either by phlogiston or oxygen.

The processes of iron production he observed may have played a supporting role in his change of mind on the foundations of chemistry. According to the new chemistry, the principal object of smelting was to remove oxygen from the iron in the ore; according to the old chemistry, it was to add something, phlogiston. In the ironworks Cavendish visited, and in his laboratory where he experimented on specimens from the ironworks, he had frequent occasions to consider the question of

[83] Charles Blagden to Claude Louis Berthollet, 21 and 24 May 1785, drafts, Blagden Letterbook, Yale.

[84] Charles Blagden to Claude Louis Berthollet, 17 November 1787, draft, ibid.

[85] Richard Kirwan to Guyton de Morveau, 2 April 1787, in *A Scientific Correspondence during the Chemical Revolution: Louis-Bernard Guyton de Morveau and Richard Kirwan, 1782–1802*, ed. E. Grison, M. Sadoun-Goupil, and P. Bret (Berkeley: Office for History of Science and Technology, University of California at Berkeley, 1994), 165–67.

[86] Certificates, Royal Society, 5 (3 April 1788).

oxygen or phlogiston. What he saw at the ironworks did not answer the question any more than did his experiments on air, where he had addressed the question in the year before his first journey, but it directed his thoughts to it over a sustained period. Evidence of his thinking on the question is an undated draft of a paper, inserted loosely in the White Book, which Cavendish prepared for the engineer Cockshutt who had sent him specimens of iron and coal to examine. Cavendish wrote out for him a chemical analysis of the production of iron, with recommendations for practice. He addressed the variables in the operation of smelting – cast and wrought iron, charcoal and coal fuels, and hot and moderate temperatures – which he related to materials that enter the smelting process, slag, carbon, and oxygen, though he did not use these terms. He described the chemical difference between cast iron and wrought iron in both the old and the new chemistry: according to "Bergman & the partisans of phlogiston," cast iron contained less phlogiston than wrought iron, and according to "favourers of the new system," it contained some oxygen.[87] He did not say which explanation he preferred, but he acknowledged that the question of phlogiston and the two chemistries came up in the explanation of smelting, the point being made here.

The journeys may also have been a stimulus to Cavendish's work on heat. At the ironworks, he witnessed high heats produced by chemical reactions of the fuels, and he witnessed chemical reactions of the ores caused by the heats; and at some of the ironworks he witnessed steam engines in operation. From early in his career, he was convinced that heat is the vis viva (energy of motion) of the vibrations of the particles constituting bodies, but it was not until the late 1780s, the time of his journeys, that he worked out a theory of heat based on this understanding. The theory contained a derivation of the complete principle of the conservation of energy including heat. He wrote up the theory as a paper with publication in mind, but for unknown reasons he did not follow through with it. It would be 60 years before physics acquired a comparable mathematical-theoretical formulation of energy and heat. In his paper, Cavendish applied the theory to the heats of chemical reactions, heats of a kind he observed in the hearths of furnaces and forges. Where steam engines powered massive hammers and air blasts, he saw instances of the conversion of heat into mechanical work, as explained by his theory. In the working of steam engines, he saw the interconversion of water and steam, the heat of which was explained by his theory. He had worked out the parts of the theory much

[87] Henry Cavendish, "Paper Given to Cockshutt," loose insert in "White Book No. 1." He was probably the engineer James Cockshutt who was in charge of the forges at the Wortley Ironworks near Sheffield, and who co-managed the Cyfarthfa forge and mill for making cannon at the time Cavendish and Blagden visited there. About then, around 1787, Cockshutt together with a partner introduced Cort's puddling and grooved rolling processes at both Wortley and Cyfarthfa. In 1804 Cavendish recommended Cockshutt for membership in the Royal Society. Ince, *South Wales Iron Industry*, 60. *Biographical Dictionary of Civil Engineers in Great Britain and Ireland*, vol. 1: *1500–1830*, ed. A. W. Skempton (London: Thomas Telford, 2002), 156. Sheffield Trades Historical Society, "A Welcome to Wortley Ironworks," http://www.topforge.co.uk/1955%20Guide.htm. Wortley Top Forge, "History of Iron Making at Wortley," http://www.topforge.co.uk/History%20 Notes.htm.

earlier. He said nothing about why he brought them together when he did. He did not refer to any contemporary work that might have prompted it, leaving us with the possibility that his tours of ironworks played a part in his motivation. He was intrigued by the industrial operations he saw on his journeys, and he was prepared to understand them scientifically. They may have stimulated him to complete and draw up a comprehensive theory of heat, and in any event they produced copious examples of it. By bringing together his fundamental researches in chemistry, heat, and mechanics with iron smelting, he would have found a place for industrial Britain in his life of natural philosophy.

An objection might be acknowledged here. The Marxist historian of science James G. Crowther would probably disagree with the suggestion of a connection between Cavendish's journeys and the completion of his mechanical theory of heat. He says that in the second half of the eighteenth century, there arose a new direction in science, one related to industrial society, which emphasized physics, especially heat, and chemistry rather than the astronomy of the solar system, the emphasis in the mercantilist society a century earlier, in Newton's time. Crowther develops this idea in *Scientists in the Industrial Revolution*, a study of Cavendish, Black, Watt, and Priestley, whom he groups with several other men he equally well could have chosen: Boulton, Hutton, Wedgwood, and Erasmus Darwin. Most of them we have encountered as members of the Coffee House Philosophical Society and the Lunar Society of Birmingham. Born in the same decade, 1726–36, they were concentrated in the new industrial districts. The exception was Cavendish, a Londoner, who was exceptional in another way too, owing to his class. Because of his "social outlook" together with his "psychological constitution," Crowther says, he did not hold an "industrial ideology." He together with his associates Banks and Blagden were "attached to the old ruling class," in light of which his interest in industry on his journeys in the 1780s was unexpected. He was curious about Watt's inventions, including Watt's claim of priority in an invention having to do with the steam engine, indicating an appreciation of the inventor's point of view, and he talked with Watt about heat experiments. Crowther doubts that Cavendish brought up his law of energy conservation with Watt, thinking it too abstract to be interesting or to have a bearing on the steam engine. Having derived the law from Newtonian mechanics, the science he was educated in, Cavendish would have regarded it as of only intellectual interest, detached from the industrial sciences of chemistry and heat and from the new technology of power. Cavendish's failure would have come out of the "cultural outlook of the social class to which he belonged," the outlook that prevailed in the Royal Society, Cavendish's principal social connection with science.[88] Cavendish probably did not bring up his derivation of the energy law on his visit to Birmingham as Crowther says, but I doubt that it was because he thought of mechanics as detached from heat. Indeed, it was his interpretation of heat as mechanical energy that separated his and Watt's understandings of the composition of water, the

[88] Crowther, *Scientists of the Industrial Revolution*, 2–3, 321–22. With his approach, Crowther makes some interesting observations about Cavendish, but for the same reason, he can be overly schematic.

core of the water controversy. I think it more likely that he was encouraged by what he observed on his journeys to draw up his theory of heat as mechanical energy, together with the conservation of energy law.

10.5 Journey of 1785

To become acquainted with Cavendish the traveler, we single out one of his journeys, the first, in 1785. In what follows, we depend primarily on the journal for that year and secondarily on letters Blagden wrote before, during, and after the journeys together with a range of published and unpublished works. In the biography of Cavendish I wrote with Christa Jungnickel, I did not fully understand his journeys and their place in his life. In this respect, as in the matter of his personality, this book complements the biography.

Descriptions of industrial processes in the 1785 journal are precise, explanatory, and clear. Technical terms that were used at the site are defined. Quantities refer to a small number of objects: mainly to dimensions of blast furnaces and forges, lengths and times of piston strokes and diameters of pistons, weights of hammers, weights of coke to make a ton pig iron, weights of pig iron to make a ton of refined iron, and times of heating and cooling. Personal names in the diary are few.

The idea of travel almost certainly originated with Blagden. In early 1785, he tried to interest Cavendish in going to Thornhill in Yorkshire to see the progress John Michell had made in the great telescope he was building. Blagden was unsuccessful at first, and by the time Michell extended a formal invitation, he and Cavendish had already set out in another direction.[89] The journey to Wales was Blagden's second plan. He "proposed the scheme one day," he said, of inspecting the ironworks near Cardiff, and when he described them, Cavendish became "very curious" and agreed to make the trip (Fig. 10.3).[90]

They traveled by carriage, but we do not have a description of what kind. Their carriage would not have been a stagecoach, which was unwieldy, uncomfortable, and capable of traveling no more than 3 or 4 miles per hour on good roads. Mail-coaches traveled twice that fast. Among the notes from their journey, there is mention of a "chaise,"[91] which was a lightweight carriage with a folding top, seating two persons side-by-side, having either two or four wheels. If that is what was meant, the travelers would have rented it on their journey for making a side trip, affording them a panoramic view of the land. If a "post-chaise" was meant, it would have been their principal mode of transportation. A post-chaise had four wheels and was drawn by two or four horses, with fresh horses provided at each post along the way.

[89] Charles Blagden to John Michell, 25 April 1785 and 13 September 1785, drafts, Blagden Letterbook, Yale; in McCormmach, *Weighing the World*, 379–82, on 380; 395–400, on 399.

[90] Charles Blagden to William Lewis, 20 June 1785, draft, Blagden Letterbook, Yale.

[91] "Chaise" appears in an untitled document containing barometer and thermometer readings made on the 1785 journey. Cavendish Scientific Manuscripts, Devonshire Collections, Chatsworth.

Fig. 10.3 Map of 1785 Journey. Several places where Cavendish stopped are shown on the map, indicating his itinerary 1. London. 2. Alderley. 3. Cardiff. 4. Merthyr. 5. Colebrook Dale. 6. Birmingham

It accommodated two to four persons, and there could be an accommodation behind for servants. By the standards of the time, it was comfortable, equipped with leaf springs and upholstered seats, and through large front and side windows it too offered a wide view. The post-chaise was popular among travelers in the eighteenth century, and it or the chaise was the typical carriage of families who were respectable but not wealthy. However, because Cavendish owned his own carriage and horses, it seems likely that he would have chosen that means, as it was the usual preference of well-to-do travelers. In a porter's account, his carriage was referred to as a "chariot," a light four-wheeled carriage.[92]

Roads were rough, but improving. Turnpike trusts were set up by which users of roads paid for their maintenance. Every 5 miles or so, roads were blocked by tall white gates. Upon hearing the approach of travelers, turnpike men emerged from little houses beside the gates to take tolls for each vehicle they let pass. Signposts

[92] Rosamond Bayne-Powell, *Travelers in Eighteenth-Century England* (London: John Murray, 1951), 8–9, 19. Anon., "Georgian Index-Carriages," http://www.georgianindex.net/horse_and_carriage/carriages.html. Anon., "Chaise," http://en.wikipedia.org/wiki/Chaise. History World, "History of Transport and Travel," http://www.historyworld.net/wrldhis/PlainTextHistories. Repair of "chariot" was billed by a coach maker: porter's accounts, March and December 1809, "Bedford Square. James Fuller's Account with the Exec. of Hen: Cavendish Esq. Balance 37.6.4. Settled 30 August 1810."

along roads were rare, but many had milestones.[93] Cavendish and Blagden read milestones and carried maps, and their carriage was probably fitted with Cavendish's way-wiser, or mileage counter.

Except when they stayed with Blagden's relatives, they stayed at inns. Inns were prominent buildings in the towns and villages they passed through. They were easy to see, identified by huge signs with colorful if obscure names such as Bull and Mouth. A wide archway opened onto a yard, which accommodated carriages. The proprietor of an inn, who was a substantial person in the vicinity, would welcome travelers in person. Inns were plentiful; travelers usually had a choice, and they chose inns that were meant for their kind. There was a hierarchy of inns. At the top were posting houses for travelers who arrived in post-chaises or their own carriages. These inns might not accept passengers from mail-coaches, and inns that accepted the latter might not accept passengers from stagecoaches. At the bottom were persons arriving in wagons or on foot, who might not be accepted anywhere, or if they were they might be put up in the kitchen and fed remains.[94] Cavendish and Blagden were welcome at the best inns.

On a recommendation, Cavendish and Blagden might stop at a particular inn where they would be assured an agreeable visit, though it did not always work that way. On their travel to Yorkshire in 1786, they stayed at an inn recommended by their colleague and host John Michell, after which Blagden complained that it was "the vilest house at which I had ever the misfortune to put up."[95] Michell apologized, explaining that the inn was new and he had not stayed there but had heard good reports; he would not repeat the mistake. By and large, English inns were of good quality. Gentlemen who arrived in post-chaises or private carriages were always given private rooms. Rooms were large, clean, and furnished with mahogany furniture and four-post beds heaped with feather mattresses, requiring two steps to mount. Wax candles were provided. The stairs and landings were carpeted, so the traveler would not be disturbed by other lodgers. There was a common dining room, called the "coffee room." Food was usually excellent, distinguished by the quantity and variety of meat, eaten with steel two-pronged forks and knives: mutton, beef, ham, fowl, and fish. There were vegetables, cheese, cakes and pies, and beer and wine. Dinner was followed by tea in the evening.[96]

The historical setting of Cavendish and Blagden's journey into Wales is the Industrial Revolution. Eric Hobsbawm begins his history of the event by calling it the "most fundamental transformation of human life in the history of the world recorded in written documents."[97] It began in Britain with a conjunction of favorable circumstances, which included an expanding supply of labor, capital, and land; participation

[93] Bayne-Powell, *Travelers*, 29–31.

[94] Ibid., 41–42.

[95] Charles Blagden to John Michell, 19 September 1786, draft; in McCormmach, *Weighing the World*, 409–12, on 410.

[96] Bayne-Powell, *Travelers*, 43–47.

[97] Eric Hobsbawm, *Industry and Empire: The Birth of the Industrial Revolution* (New York: The New Press, 1999), xi.

in an international economy; incentives of low interest rates and expectation of high profits; and sources of water, coal, and steam.[98] In the journal of the trip, sources of power and mechanisms for transmitting power to do useful work are recorded, but nothing is said about the social and economic aspects of industrialism. Cavendish's interest in the Industrial Revolution was selective, in keeping with the interest he brought with him on his journeys.

Cotton textiles are synonymous with the early Industrial Revolution,[99] their manufacture having undergone rapid change through a series of technological inventions. Cavendish was not especially attracted to textile sites, however. He was more interested in the mining of ores and the extraction of metals from the ores, subjects with clear connections to his scientific work. The larger part of the journal of 1785 is taken up with descriptions of ironworks.

A Welsh guide book described the production of bar iron. "The impression on the spectator, to whom this scene is novel, is one of extreme interest; he finds himself in the midst of machinery of a peculiar character, and of extraordinary magnitude, remarkable for its massiveness and its weight, and its effect is astonishing."[100] Our spectators were Cavendish and Blagden, and the impression the scene made on them is hard to judge, but in a few places in the journal there is a hint that they were impressed. Because iron production was carried out with intense heat, fiery chemical reactions, and heavy mechanical violence, scenes described in the journal unavoidably appeared Dantesque. Under the hammer, balls of iron removed from the furnace "strike off sparks, some of which fly to a great distance, and a few have the brilliant appearance of steel dust in fireworks. There comes besides a white flame from different parts of the mass, and at times a different flame from certain spots, of a light bluish colour, like that from burning Sulphur."[101] Coalfields in the vicinity of the ironworks suggested an underground inferno. Cavendish and Blagden passed a coal pit that had been burning many years. "From some places close by the road, a strong flame was now issuing, and the earth seen through the crevices and apertures in many places was red hot or even white hot. All about the places actually burning, lay the cinders of old conflagrations."[102]

Blagden's family connections eased their journey. They stayed several nights at each of two locations. The first was Alderly (Alderley), an estate in Gloucester, the birthplace of Matthew Hale, lord chief justice of England at the time of Oliver Cromwell. They were put up there by "Mr Hale," who was John Blagden Hale, Charles Blagden's older brother, who had changed his name only the year before. Having married into the Hale family, John Blagden assumed the name of Hale upon succeeding to the estate of Alderley and the arms of Hale at the death of a member of the family, a later Matthew Hale. The second location was Pentyrch,

[98] Ibid., 32. T. S. Ashton, *The Industrial Revolution 1760–1830* (Oxford, New York: Oxford University Press, 1996), 17.

[99] Hobsbawm, *Industry and Empire*, 34.

[100] T. E. Clarke, *Guide to Merthyr-Tydfil* (London, 1848), 33.

[101] Cavendish, "Journal 1785," 53.

[102] Ibid., 57.

with "Mr Lewis," who was William Lewis, ironmaster of Pentyrch Ironworks and Blagden's brother-in-law.[103] The month before setting out, Blagden wrote to Lewis about their plans and asked if he and Cavendish could stay with him or if not if he could recommend a place for them to stay in Cardiff. Lewis replied that his house had plenty of rooms, but they were small and low, and his house was close to the hammers. If they wished, they could stay at another house at a distance from the pounding. He promised to show them ores and ironworks.[104]

From the seventeenth century, the prolific Lewis family had played a major role in the rise of iron production in South Wales. William Lewis was an important man in the area. In addition to Pentyrch, his enterprises included iron and coal mines, a warehouse near the Town Quay at Cardiff where he conducted his shipping business, and part interest in Dowlais Ironworks at Merthyr.[105] He owned turnpike gates, farms, and a corn mill. Lewis's letters to his brother-in-law John Blagden Hale give an idea of what it was like to be an ironmaster in South Wales at the time of Cavendish and Blagden's visit. One problem after another required Lewis's attention. Despite his many properties, in the month before Cavendish and Blagden visited him, he wrote to John that he was anxious for the repayment of £200 owned him by a certain party. "I never was so short of Cash as I am," he complained. "During the dry weather we employed our Finers to make Blooms, but very little was drawn out for want of Water, therefore we have a large stock in hand, and the quantity of Charcoal brought in daily is very great w[ch] takes away the Cash." In addition, besides his regular monthly payments, he had several big sums to pay for

---

[103] Gloucestershire County Council, "Gloucestershire Archives: Online Catalog," http://www.gloucestershire.gov.uk. In the description of the Hale family papers, the catalog guide identifies William Lewis as John Blagden Hale's brother-in-law, and therefore Charles Blagden's brother-in-law. In the correspondence between William Lewis and John Blagden Hale around the time of Cavendish and Blagden's journey, Lewis refers to his wife Mary, presumably Blagden's sister Mary. In the correspondence, Lewis speaks of moving to Alderley. Edgar Chappell writes that in later years Lewis "married a Miss Hale of Alderley and took up his residence at that place." Edgar L. Chappell, *Historic Melingriffith: An Account of Pentyrch Iron Works and Melingriffith Tinplate Works*, 2d ed. (Merton: Merton Priory Press, 1995), 25. The Blagden, Hale, and Lewis families were closely connected by marriage and business.

[104] Blagden to Lewis, 20 June 1785. William Lewis to Charles Blagden, 25 June 1785, Blagden Letters, Royal Society, L.46. William Lewis wrote to Blagden that "we shall with great pleasure receive you & your Friend M[r] Cavendish, in the best manner our small House will admit of." He said the Joy House, previously occupied by Mr. Curtis, was a quieter alternative to his house, and when the river was low only a 3-min drive. In an earlier letter to John Blagden Hale, he referred to the alternative house for Blagden and Cavendish as the Ivy House. 7 June 1785, Gloucestershire County Archives, D1086/F116. Ivy House is thought also to be near the site of another furnace, Tongwynlais Furnace, which had been abandoned by the beginning of the eighteenth century. The house is associated with the Price family, which had close business ties with the Lewis family. It was Nicholas Price father and son and Thomas Lewis who built the Pentyarch Ironworks. We are not told which house Cavendish and Blagden stayed at, but we know they had an obliging host. Philip Riden, "Early Ironworks in the Lower Taff Valley," *Morganwg* 36 (1992): 69–93, on 86; http://welshjournals.llgc.org.uk/browse/viewpage/llgc-id:1169834. Ince, *South Wales Iron Industry*, 145.

[105] Chappell, *Historic Melingriffith*, 23, 26, 38.

cordwood, which he needed for making charcoal. He had a steady order from a buyer of 40 tons of iron a quarter, which he was glad for, but which also troubled him. He thought that the buyer was running his operation at a loss, trading with other people's capital, and that eventually he would fail. He feared he was dealing with an unsound business, and he wanted John's opinion. He considered how to punish a father and his sons who overcharged him, having put ordinary stones and earth in the middle of the heaps of ore in order to be paid for it all as ironstones. He worried about the lease; Billy Price promised them a lease for another year after he came of age. While he was away from Pentyarch, the workers used too much wet charcoal and wet Lancashire ore, which nearly cost him his furnace. One of their partners, he believed, set fire to the rope and all the wood construction in a coal pit, and further threatened to burn it all down. This partner had formerly on two occasions stopped workers from bringing up coal for several weeks, and Lewis thought he should be sued for damages. Winter floods shut down the forges completely for a time, and because of a lack of water to power the furnace, work went very slowly. Lacking workmen, the forges made little iron for a period, and the furnace did not make the usual quantity of iron because the quality of Lancashire ore was poor and the heap of charcoal had gone out. From time to time, Lewis could report good news. He told his brother-in-law that he was about to build an "Air Furnace," a term then used for a reverberatory furnace, employing a flow of heated gases. A month later, he told him that the air furnace was finished, and soon he was melting down old castings and making new ones with it. On Christmas, 1785, he told his brother-in-law that the "Furnace & Forges go on very well," and he had "given the Workers & their Wives plenty of Ale today." In the new year, owing to the death of an owner at Merthyr, he had hopes of buying a coke furnace "cheap."[106] In the midst of his busy life, with its cycle of trials and rewards, Lewis welcomed into his home his brother-in-law Charles Blagden and his friend Cavendish, who had come to be shown how an ironworks works.

The journal of 1785, because of its sparseness and lack of fresh observations of manners and men, makes for dry reading as an account of travels, however informative it may be as technical reportage. As a source of information about Cavendish, however, it is revealing and consistent with what we know about him. We may take it as further evidence of Cavendish's view of the world.

10.6 Synopsis of Journal of 1785

Because the journal was kept as a record for the two travelers only, much was taken for granted that is unfamiliar to readers today. For this reason, I provide a synopsis. It includes information about the journey not given in the journal; it outlines the

[106] William Lewis to John Blagden Hale, 12 March, 7, 22 June, 24 August, 18 October, 25, 31 December 1785; 13, 23 January 1786; 12 January 1787, Gloucestershire County Archives, D1086/116.

industrial setting; and it briefly describes what is new to the travelers at each stop. It also refers to researches Cavendish carried out at home that relate to what he saw on his journeys, pointing to a continuity of interest in new settings.

Day 1. Sunday, July 17. Reading, Crown Inn.[107] Their working day began at 6:30 AM, when they took barometer and thermometer readings up one pair of stairs at the Crown Inn, where they spent their first night. On the road, at 10:23 AM and again at 10:40 AM, they took readings of the same kind on the ground floor of the George & Pelican in Speenham. Through the day, they took readings at inns and stops along the road, a total of nine locations. The last readings were taken on the ground floor of the house in Alderley belonging to Blagden's brother, John Blagden Hale. It was 11 PM; it had been a long day.

Throughout the journey, they took barometer and thermometer readings to determine the heights of places above the sea, and the differences in height between places. They used maps of the region for locations. On their journey to Wales, they recorded their movement through three dimensions.

The barometer and thermometer readings are recorded on sheets separate from the journal, and they are in Blagden's handwriting, so there may have been a division of labor. It was Cavendish in any case who made the extensive calculations for determining heights from the readings.[108] Because with a couple of exceptions they took barometer and thermometer readings every day of their journey, there is no need to bring them up regularly in this synopsis.

Day 2. July 18. Alderlery, Gloucestershire, John Blagden Hale's house. On their journey, at two locations, in Hale's garden on this day and on 25 July at the Pentyrch Ironworks, they determined their true latitude from observations of the sun, using a small Hadley's quadrant and an artificial horizon as a circular spirit level. Cavendish and Blagden both made observations, again on sheets separate from the journal. Cavendish made all the computations, which were considerable, entailing corrections for temperature and time.[109]

Day 3. July 19. Alderley, Hale's house. They observed limestone in the vicinity. Off a turnpike and down a lane, they talked to farmers about blue limestone embedded in local blue clay. They tested coarse-grained limestone with acid. They estimated the angle to the horizon of a stratum of limestone. They passed quarries in which the limestone lay in flat stones. In addition to limestone, they saw blue, red, and white clay, granite, sand, slate, and other rock formations along the way. Here

[107] The inns where they stayed overnight are not included in the journal for 1785, but they are in a document by Charles Blagden, cited in the next note. On any one day, I give the inn where they started out from.

[108] Charles Blagden, untitled document containing barometer and thermometer readings, journey of 1785. Henry Cavendish, untitled document containing computations of barometric heights, journey of 1785. Both documents are in Henry Cavendish, Scientific Manuscripts, Devonshire Collections, Chatsworth.

[109] Charles Blagden, "Observations with Mr Auberts Small Hadley's Quadrant of 3 Inches Radius by Ramden & an Artificial Horizon as a Circular Spirit Level," ibid. Henry Cavendish, untitled document containing computations of true latitude, ibid.

as elsewhere on the journey, they regularly observed strata, rocks, and pebbles surfacing the road.

Day 4. July 20. Alderley, Hale's house. On this day, they began their visits to industrial sites, which they continued throughout the rest of the journey. They witnessed the process of dyeing cloths in a copper boiler heated by a furnace; the journal gives a detailed account of the boiler, the heat, the chemicals, the procedures, and the handling of cloth. Blagden carried out a small chemical experiment on making dye by dissolving tin in aqua fortis (nitric acid), which he bought. The solution had a yellowish color and the smell of aqua regia (nitric and hydrochloric acid). Early in his career, Cavendish had made a study of the solution of metals and acids, and he had worked out a theory to explain it. The process of dyeing resembled what Cavendish did in his chemical laboratory.

Day 5. July 21. Alderley, Hale's house. They measured the temperature of a spring and of water brought up in a bucket from a deep well. Further along in their journey, in Wales, Cavendish took the temperature of springs while climbing a mountain, and he stopped by a warm spring to test for the kind of air that rose in bubbles,[110] an observation related to his studies of mineral water and of the chemical identity of gases.

Day 6. July 22. Alderley, Hale's house. They visited the Frombridge [Fromebridge] Mill, near Frampton-on-Severn, one of the largest producers of wire in the country.[111] Cavendish and Blagden were shown how wire was made from brass, iron, and steel, the main uses of which were for fishhooks and for teasing and aligning woolen fibers in the cloth trade. Cavendish was interested in wires for their mechanical strength. At home he made experiments on the twisting of wires, and after a meeting with the industrialist Matthew Boulton several years after this journey, he brought back with him a list he had copied out of the weight needed to break wires made of brass, iron, and five other metals.[112] Cavendish and Blagden were shown how brass was made, beginning with melted copper.

Fromebridge also produced pig and bar iron. Cavendish and Blagden saw slitting mills for cutting bars of metal into rods, and they saw flatting, or rolling, mills for producing metal sheets, the mechanical alternative to hammering. All of the mills were powered by water.

At Oakley Wood, they stopped to look at shafts dropping into a tunnel cut for a canal. The shafts, as deep as 35 yards, were vents for circulating air in the Sapperton Canal Tunnel in Gloucestershire. The canal was the Thames and Severn Canal linking the two rivers, intended to supply London with cheap coal from the northern and

[110] Charles Blagden to Joseph Banks, 9 October 1785, Banks Correspondence, Royal Botanic Gardens, Kew 1.210.

[111] Stephen Mills, "Fromebridge Mill, Frampton-on-Severn, Gloucestershire," *Gloucestershire Society for Industrial Archaeology* (1998): 6–22, on 8–9.

[112] Henry Cavendish, "Weight in pounds required to break wires of metal 1/10 inch diam. Taken from Boulton," Journal of 1793, Cavendish Scientific Manuscripts, Devonshire Collections, Chatsworth, X(a), 2:9. Experiments on twisting of glass tubes and springing of wires, ibid., VI(b), 21. Experiments on twisting of wire tried by the time of vibration, ibid., VI(b), 22.

northwestern collieries. Only about 40 yards of the tunnel had been completed when Cavendish and Blagden saw it, but when it was finished it would stretch 2.2 miles, making it the longest tunnel in the world at the time. This canal did not meet expectations, but generally canals played an essential part in the British Industrial Revolution by providing an economic and reliable means of transportation of materials and products. Cavendish and Blagden's journeys fell in the period known as the "Golden Age" of British canals, 1770s to 1830s. At Oakley Wood, they were interested both in the canal and shafts as engineering works and as an opportunity to study exposed strata.[113]

Day 7. July 23. Gloucester, King's Head. Near Gloucester, where the River Severn divides into two channels, they observed the tide at several locations, timing it and comparing times with local accounts. Cavendish's manuscripts contain a paper about the tides and the earth's motion.

Days 8–10. July 24–26. They arrived at Lewis's house at Pentyrch Ironworks, on the western bank of the River Taff, about 6 miles north of Cardiff. It is where the River cuts through the southern rim of the South Wales Coalfield, and where iron, coal, limestone, and water are found in close proximity. Because of its resources, the region had been home to iron production since Elizabethan times. A furnace and forge were set up at Pentyrch in 1740. On their first day there, Cavendish and Blagden took barometer and thermometer readings in the ironmaster William Lewis's parlor.[114]

Blast furnaces were usually made of stone blocks and lined with firebrick. They narrowed toward the top, and they could be very tall. The furnace at Pentyrch, not one of the tallest, reached 26 ft with a funnel that rose higher. At ground level, blast furnaces were square, holding a hearth with access in the front for tapping the molten iron and slag. Built of stones, the new hearth at Pentyrch was 2 ft on a side and 4 ft deep (Fig. 10.4). Furnaces were built into hills for ease of loading, the ore, fuel, and limestone being alternately introduced from the top.[115] Once going, the red-hot charge reached to the top of a furnace, which might operate continuously for weeks or months. From partway up the furnace, molten iron dripped down through the fuel onto the hearth, where the fusion of the metal took place.

A constant companion of a blast furnace was a means of producing a blast of air, which by increasing the flow of oxygen raised the temperature of the furnace high enough to melt the materials. The blast usually entered the furnace through one or two side openings, or "tuyères." Traditionally the blast was produced by leather bellows operated by cams from a waterwheel. In the second half of the century, in many places bellows were replaced by cast-iron cylinders and pistons, which had greater force. These again were operated by waterwheels, sometimes augmented by

---

[113] Candida Lycett Green, "Sapperton Canal Tunnel, Gloucestershire," http://www.candidalycettgreen.co.uk.

[114] Chappell, *Historic Melingriffith*, 23, 26, 38. The book which was originally published in 1940 is updated in an introduction to the second edition by Philip Riden, pp. iii–xii. The starting year at Pentyrch is given by Riden, p. v.

[115] Ince, *South Wales Iron Industry*, 9.

Fig. 10.4 Blast Furnace. The drawing shows a typical blast furnace for smelting iron. At the *bottom* is the hearth, which when tapped releases molten iron. Two ways of collecting it are shown, one a flat sand casting, the other a series of troughs. From the *top*, the furnace is fed iron ore, coal, and flux. A leather bellows powered by waterwheel provides the blast to produce a high temperature (Courtesy of Ironbridge Gorge Museum)

Newcomen steam engines that returned water from the downstream to the upstream side of the wheel. This machinery for producing blast was fairly standard in Welsh ironworks, though occasionally Watt steam engines were used for the purpose.[116]

The River Taff was not navigable, but there was a turnpike route along the valley for transporting iron.[117] From the quay at Cardiff, iron was shipped to Bristol, the commercial center for South Wales iron production. From Bristol William Lewis wrote to his brother-in-law John Blagden Hale that he sent 50 tons of bar iron from the warehouse there by boat for Harford & Co,[118] a typical business operation. Partnerships consisting of merchants from Bristol and other major commercial centers invested heavily in South Wales. To secure their interests they gained control of the sources of supply by engaging in both the production and the selling of iron. If this broader setting of the ironworks was talked about, it did not enter the journal.[119]

[116] Ibid., 9–11.

[117] The road, following a turnpike act of 1771, instigated by the prominent ironmaster Anthony Bacon, ended at Tongwynlais, 4 miles north of Cardiff. Richard Hayman, *Working Iron in Merthyr Tydfil* (Merthyr Tydfil: Merthyr Tydfil Heritage Trust, n.d.), 8. Riden writes that "there was an easy road journey, improved by turnpiking in 1767, down the valley to the quay in Cardiff." "Introduction," Chappell, *Historic Melingriffith,* iv.

[118] William Lewis to John Blagden Hale, n.d. Gloucestershire County Archives, D1086/F116.

[119] Riden, "Introduction," Chappell, *Historic Melingriffith,* iv, vi.

Cavendish was interested in the smelting processes at blast furnaces and forge furnaces, where raw earth was made to yield its metal. Pentyrch had both kinds of furnaces, but because just then the hearth for the blast furnace was being replaced, he could only see the forges in operation. He was told that when the furnace was working, it used charcoal for fuel, as there was still ample woodland in South Wales for making it.[120] On the day after his and Blagden's arrival, they were shown the forges, one a finery forge and one a chafery. The hearth of the finery was a pit formed of heavy iron plates, with dimensions 2 ft 4 in. in length, 1½ft in width, and 7 in. deep. The nozzle of a bellows rested on the edge of a side plate of the hearth. "Firework sparks fly off in all parts of the operations in the Finery," the journal records. The chafery hearth differed from the finery chiefly in having no hole in front for the scoria to run off; instead the scoria cohered in masses that were lifted out. The finery used coal as well as charcoal, though the latter worked better. The coal came from Pentyrch mines.

At the finery, a pig of iron was placed on burning charcoal with a blast from a bellows directed at it and workers stirring it constantly. The hot iron lump was removed from the bottom of the hearth with iron pincers, or tongs, and beaten with iron rods to remove scales. Then it was struck with a massive hammer, forcing white hot scoria out of cracks on the surface. The resulting oblong mass or "half-bloom," measuring about 20 in. long and 4 in. thick, was reheated and struck by the hammer again and then beaten into a long bar with knobs at each end, affording a grip for pincers. The iron was moved to the chafery, which was heated by coal, where the knobs were drawn out and the iron was fashioned into utensils. Cavendish and Blagden were shown three kinds of iron, distinguished by color, and having different qualities. Bright gray was the best color, indicating that the iron was made with the right proportion of ore to charcoal in the furnace. The ironworks was set on the bank of the river, but it was powered by water from a tributary, the usual arrangement.[121]

Cavendish and Blagden explored the hills and coal pits around the village of Pentyrch, looking at strata, testing stones with acid, and measuring elevations. The dominant feature of the land there is Garth Mountain, or the Garth, a high, steep ridge formed of sandstone separated by a hollow from another ridge marking the edge of the South Wales Coal Field. Where the River Taff cut through this ridge, it left a hill on its west bank known as the Lower, or Lesser, Garth, a valuable source of iron ore and limestone for ironworks. William Lewis owned an iron-ore mine on the Lower Garth, a big pit on top of the hill. Ore was raised from the pit to the surface and hauled in baskets on the backs of mules and horses along a zigzag path to the Pentyrch Ironworks. Limestone was quarried on the Garth for use as flux in the blast furnace, and coal was mined in levels, slants, and pits on the sides of the Garth and in the valley.[122] On July 26, Cavendish and Blagden climbed to the top of the

[120] Ibid., vi, viii.

[121] Riden, "Ironworks in the Lower Taff Valley," 87.

[122] Chappell, *Historic Mellingriffith*, 38. Riden, "Early Ironworks in the Lower Taff Valley," 69. Anon., "Garth Hill," http://en.wikipedia.org/wiki/Garth_Hill.

Fig. 10.5 Garth Mountain. On the *lower left*, we can see a furnace (Courtesy of Cardiff Central Library)

Garth from Pentyrch carrying their barometer. They may have been dissatisfied or else they wanted a second observation, for they repeated the climb the next day. Cavendish used the two measurements from the second try in calculating the height: He put the top of the Garth at 1,038 and 1,045 ft above sea level.[123] The elevation given on maps today is 1,007 ft (Figs. 10.5, 10.6, and 10.7). After the journey, Blagden wrote to Lewis that their measurements of heights in Glamorganshire were "very coarse, not only because the method with the barometer is not capable of great exactness but also because we had only one."[124]

Day 11. July 27. Pentyrch, Mellengryffyth [Melingriffith] Tinplate Works, or Tin and Iron Works. A forge was established at Melingriffith around 1749, and tinplating was added later. The works was conveniently located 2 miles from Pentyrch along the River Taff, and because Pentyrch smelted hermatite ore, a tough iron

[123] Henry Cavendish, untitled document containing computations of barometric heights, Cavendish Scientific Manuscripts, Devonshire Collections, Chatsworth.

[124] Charles Blagden to William Lewis, 16 February 1790, Blagden Letters, Royal Society 7: 404.

Fig. 10.6 Portable Barometer. (Photograph by the author at Chatsworth). This is probably the barometer that Cavendish took on his journey in 1785 and that he carried to the top of Garth Mountain to measure its height. When folded into its mahogany case, the barometer measures 43½ inches. The instrument is suspended in gimbals. At the bottom, near the wooden cistern, there is a

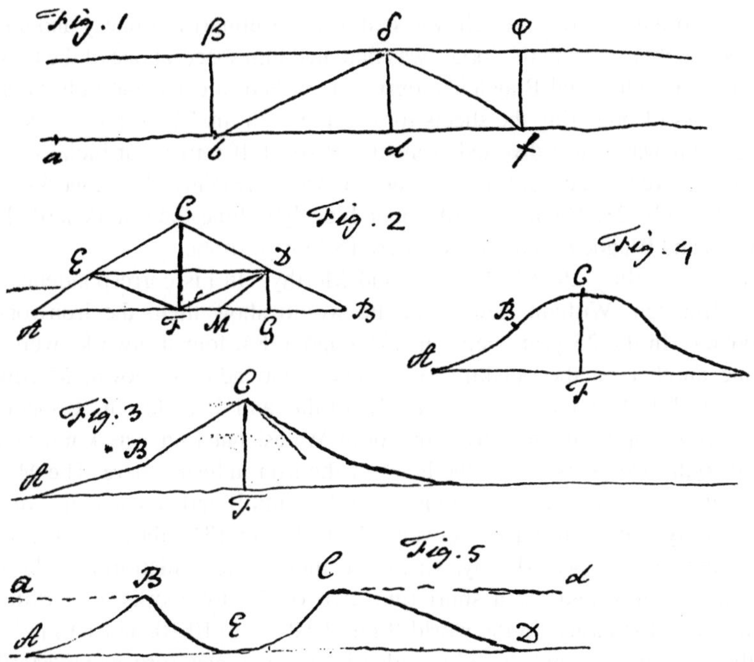

Fig. 10.7 Cavendish's Drawings of Mountains. On his journey of 1785, Cavendish measured the height of mountains. Twelve years earlier, he helped with another measure of mountains, their bulk. This was in connection with an experiment by the Royal Society to measure the gravitational attraction of mountains as a means for determining the average density of the earth. Cavendish drew up rules for selecting an actual mountain for the experiment, and for calculating the attraction of the mountain. He considered a number of shapes of mountains for the purpose ("Mr Cavendish's Rules for Computing the Attraction of Mountains on Plumblines," Cavendish Scientific Manuscripts VI (b), 2. Courtesy of the Chatsworth Settlement Trustees)

well-suited for tinplating, the two plants worked together from the beginning. For much of the time they had common ownership. Melingriffith was the first and most important of a number of tin-plating works in South Wales, the center the industry until the end of the Second World War.[125]

Fig. 10.6 (continued) thermometer with a corrections scale. William Roy, with whom Cavendish collaborated on experiments with barometers, used a portable barometer almost identical to this one for taking heights of mountains. Although the Chatsworth barometer is unsigned, we know from Roy that this kind of barometer was made by Jessie Ramsden. It was highly accurate. The height of the mercury column was read to one-five hundredths part of an inch by means of a nonius moved by rack work. Roy used two such instruments in his experiments, finding them to agree within a few thousandths of an inch (Roy, "Experiments and Observations Made in Britain, in Order to Obtain a Rule for Measuring Heights with a Barometer," *Philosophical Transactions* 67[1777]: 653–787, facing 658. The photograph is reproduced by permission of the Chatsworth Settlement Trustees)

[125] Riden, "Introduction," in Chappell, *Historic Melingriffith,* iv–v.

Pig iron produced by Pentyrch was sent to the Melingriffith finery and chafery forges. Wrought iron from the forges was flattened into sheets at one of the two rolling mills. Cavendish and Blagden's journal described in detail the rather complicated process of preparing the sheets for coating with tin. The source of power at Mellingriffith was water diverted from the River Taff into an artificial channel, which also served as a canal bringing raw iron from the Pentyrch Ironworks.[126]

Day 12. July 28. Having heard about blue lyas limestone at Penarth Point, Cavendish and Blagden spent the day there looking at strata.

Days 13–14. July 29–30. Pentyrch and Merthyr. In his correspondence with Charles Blagden, William Lewis said he would show them the ironworks at Merthyr too. In the 25 years between 1759 and 1784, four ironworks were built near one another on the outskirts of Wales's first industrial town, Merthyr or Merthyr Tydfil: Dowlais, Plymouth, Cyfarthfa, and Penydarren. Copper and tinplate were important industries in South Wales, but iron was king, and the Merthyr-Tydfil ironworks were the heart of the iron industry there. The Merthyr works were still modest in size when Cavendish and Blagden saw them. In 1784 they had only 4 blast furnaces between them, but in 1812 they had 18, and in 1850, at their height, 41. Merthyr was the center of the British iron trade in the nineteenth century, and for a short time two of the ironworks, Cyfarthfa and Dowlais, were the largest in the world (Figs. 10.8, 10.9, 10.10, and 10.11).[127]

In the morning, Cavendish and Blagden inspected a new furnace at "Merthyr." Their description indicates that they probably visited Penydarren Ironworks, the latest and final ironworks in the vicinity of Merthyr. In the journal, Cavendish wrote of a furnace "just erected" by "Messrs Humphries." Penydarren Ironworks was begun in 1784, the year before Cavendish and Blagden's visit, by the Homfray brothers, Samuel, Jeremiah, and Thomas, and the first blast of the furnace was in 1785. On their visit to this ironworks, Cavendish and Blagden spoke with "the Engineer, whose Name is Birch." He would have been James Birch, engineering partner and manager of Penydarren until 1801.[128] The new furnace burned coke, the first mention of this fuel in the journal, though it was not the first coke-fired furnace of the Merthyr group of ironworks.[129] The coke was made on the spot, from local coal, and it was made in an open pile rather than in a covered oven, a method common in the second half of the eighteenth century. The use of coke in furnaces made South Wales, with its abundant carbon-rich coal, attractive to the iron industry, though coke supplemented rather than replaced charcoal.

---

[126] Friends of Melingriffith Water Pump, "Melingriffith Tin Works," http://friendsofmelingriffith-waterpump.org.

[127] Lewis was part owner of Dowlais. Hayman, *Working Iron in Merthyr Tydfil*, 3, 8–10. John A. Owen, "Merthyr Tydfil – Iron Metropolis 1790–1860," published in the series *Merthyr Historian* 1 (1976): http://himedo.net/TheHopkinThomasProject/TimeLine/Wales/MerthyrTydfil/OwenMerthyrTydfil.

[128] Ince, *South Wales Iron Industry*, 36, 57, 80. Anon., "Samuel Homfray," http://en.wikipedia.org/w/index.php?title=Samuel_Homfray&oldid=513293991. If by "Merthyr" Cavendish and Blagden did not mean Penydarren, the ironworks would likely have been Dowlais, of which William Lewis was part owner.

[129] The first was Cyfarthfa.

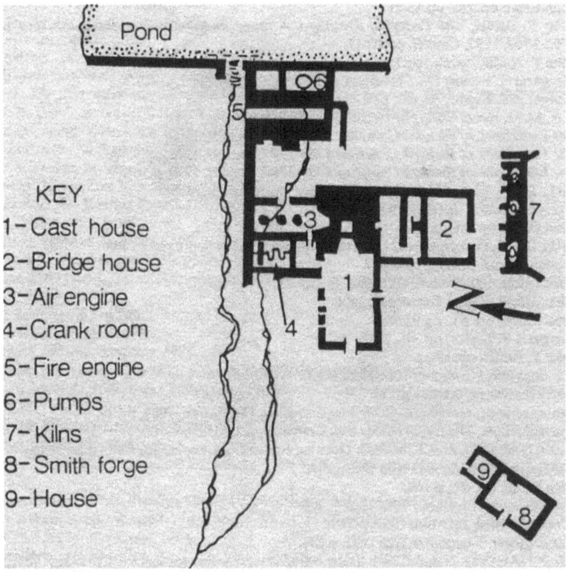

Fig. 10.8 Plan of Early Ironworks. Dowlais furnace 1763. Dowlais Ironworks was founded in 1759, the first of the four Merthyr Ironworks. The so-called "Merthir Furnace" was the first modern furnace in the area. Compared with later ironworks at Merthyr, the one shown on the plan is small, about 150 ft across; its largest building "cast house" is about 35 by 85 ft, not much larger than a large cottage (Bruce Thomas, "Merthyr Tydfil and Early Ironworks in South Wales," in *The Company Town: Architecture and Society in the Early Industrial Age*, ed. J. S. Garner [Oxford, New York, Toronto: Oxford University Press, 1992], 20–21. Ince, *South Wales Iron Industry*, 1, 68)

Fig. 10.9 Working Iron at Merthyr Tydfil. Watercolor by J. C. Ibbetson in 1792. A mass of hot iron is being struck by a trip hammer to remove slag (Courtesy of Cyfarthfa Castle Museum)

Fig. 10.10 Cyfarthfa Works and Waterwheel. Drawing by William Pamplin between 1791 and 1800. In the *center* we see a wonder of the early Industrial Revolution, an overshot waterwheel 50 ft across and 6 ft wide, made of cast iron, weighing approximately 100 British tons, reputed to be the largest in the world at the time. It was named Eolus after the Greek god of the winds. Water to turn it was conveyed from the surrounding hills by a raised stone double aqueduct. On each side of the waterwheel was a secondary wheel which operated the bellows that supplied blasts at Cyfarthfa's four furnaces and four fineries. Behind the Ironworks, we see coals being burnt to produce charcoal for use in the furnaces. (Thomas, "Merthyr Tydfil," 20. Courtesy of Cyfarthfa Castle Museum)

Fig. 10.11 Cyfarthfa House and Works. Drawing by William Pamplin between 1791 and 1800. The house of the ironmaster is next to the works. In the foreground we see the head of the Glamorganshire Canal. In 1790, the ironmasters of the four autonomous Ironworks around Merthyr agreed to build a canal. Until then, transport of materials between Merthyr and the port of Cardiff was by mountain road and pack animal, slow and expensive. The canal, which was completed in 1794, was 25 miles long, laid along the west bank of the River Taff. The drop was over 500 ft, requiring 51 locks. The canal was an immediate success. A barge could travel from Merthr to Cardiff in 20 h. More than any other project, it was responsible for the large-scale industrial development of Merthyr. (Thomas, "Merthyr Tydefil," 22. Courtesy of Cyfarthfa Castle Museum)

Because coke was harder to burn than charcoal, wherever coke replaced charcoal in blast furnaces a powerful blast was needed. The blowing cylinder of the coke blast furnace at Merthyr was 6 ft across and powered by “a fire engine on the old construction.” “Fire engine” meant steam engine in the eighteenth century, and the reference to “old construction” would have been to a Newcomen steam engine, the first appearance of a steam engine in the journal. In this case, the blowing cylinder was operated directly by a steam engine rather than indirectly through a waterwheel.

The new furnace at Merthyr rose 60 ft. Tall as it was, Cavendish and Blagden could see a strong flame issuing from the top. The proportion of width to height of this furnace differed from that of the one at Pentyrch, the Penydarren furnace being higher and narrower making for a better draft and hotter fire. The castings from the iron produced by the Merthyr furnace on Cavendish and Blagden’s visit were mainly iron plates used for copper plating. The ore for smelting came entirely from local hills.

They next visited Cyfarthfa Forges, about 1 mile from Penydarren.[130] Charcoal finery forges had low output. To increase it, a process known as “stamping and potting” was introduced in the 1760s, one of a number of inventions that made use of coal or coke rather than charcoal. At Cyfarthfa Ironworks, a relative of one of the owners was employed to build a forge for potting and stamping, for which he and his brother had a patent. In potting and stamping, pig iron from a coke blast furnace was first melted and stirred in the forge until it became a lump, as in the charcoal process only using coal or coke as fuel. The lump was hammered into plates, which were broken up, or stamped, with hammers. The pieces together with the small bits of iron broken off were placed in uncovered cylindrical clay pots, insulating the iron from impurities in the coal. The small bits served as flux, which assisted in welding the plates together, and they also entered into the composition of the iron. The pots were heated but not so much that the iron melted. This process produced the bloom, which was laid under the hammer again. Water from the River Taff powered the blast at this forge, as it did at the chafery forge. It also powered the hammers, stampers, and a clay mill for making pots used in stamping and potting.[131] Two years after Cavendish and Blagden’s visit, Cyfarthfa adopted the superior Cort puddling process for making wrought iron, one of the first ironworks to do so.

They looked only at the forges. The journal does not mention the furnace for making pig iron or the boring mill for making cannon. Like many ironworks, Cyfarthfa had secured good contracts with the Board of Ordnance for cannon during

---

130 The journal mentions that Mr. Bacon leased the forges to Mr. Tauper. The second name is unfamiliar. Anthony Bacon and William Brownrigg started Cyfarthfa Ironworks in 1765. In 1771 Brownrigg retired, and in 1782 Bacon leased part of the ironworks, which included a forge, to Francis Homfray, who held it for 2 years. In 1784, it passed to David Tanner, who was owner when Cavendish and Blagden visited.

131 Anon., “Cyfarthfa Ironworks,” http://en.wikipedia.org/wiki/Cyfarthfa_Ironworks. Anon., “Potting and Stamping,” http://en.wikipedia.org/wiki/Potting_and_stamping. Hayman, “Shropshire Wrought-Iron Industry,” 79.

the recent American War of Independence.[132] As before, Cavendish and Blagden studied the strata of the hills around the ironworks.

Day 15. July 31. Brecknock, Golden Lion. They traveled this day, noting lime kilns on the way, and gathering a bounty of heights as measured by the barometer. From Monmouth, Blagden sent Banks a positive report of their journey so far. They had seen cloth and iron manufactures in "great perfection," and in measuring the highest mountains in four counties, they had been "perfectly successful." They were going to Birmingham, with plans to measure the Malvern Hills.[133]

Day 16. August 1. Monmouth, Beaufort Arms. They had left Wales and were back in England, still on the road.

Days 17–18. August 2–3. Worcester, Hop-Pole. Birmingham, Swan Inn. At Birmingham, they visited Matthew Boulton's Soho Manufactory. Birmingham had long been a center of the ironworking industry, and when Cavendish and Blagden arrived it was also a center of the steam-engine business. Steam belonged to a branch of natural philosophy on which Cavendish was an expert, heat. Several years before his journey, he had made a thorough experimental investigation of the elastic force of steam, which he carefully wrote out but did not publish. The editor of his papers supposed that Cavendish took up the subject both for its scientific interest and for its connection with the steam engine, then growing in importance and in need of good empirical data.[134]

The inefficiency and other deficiencies of the early Newcomen steam engine were answered by Watt, who patented an improved steam engine incorporating a separate chamber for condensing steam at the end of a cycle. Watt entered into partnership with the Birmingham businessman Boulton to market the engines, and in 1776 they installed two engines, one at a colliery and one at Wilkinson's ironworks at New Willey, mentioned above. The engines were soon in demand, and Watt continued to improve them (Figs. 10.12, 10.13, 10.14, 10.15, 10.16, 10.17, 10.18, 10.19, 10.20, 10.21, 10.22, 10.23, 10.24, 10.25, 10.26, 10.27, 10.28, 10.29, and 10.30).

Cavendish and Blagden's journey coincided with a turning point in the development of steam power. In the early 1780s Watt patented three major improvements. The first translated the reciprocal motion of the steam engine into a rotary motion, useful in manufacturing of many kinds. Watt's machinery for accomplishing this is known as a "sun-and-planet gear." Cavendish described the gear and also the alternative, a crank and wheel patented in 1780 by one of Watt's former employees. The second improvement was the "double-acting" rotative steam engine, which enabled the same engine to deliver twice the amount of power. The third was an application of the pantograph principle to give the piston the motion it needed for a double-acting engine. The journal contains Cavendish's drawing of the pantograph,

[132] Ince, *South Wales Iron Industry*, 60–61. Hayman, *Working Iron in Merthyr Tydfil*, 8. John Lloyd, *The Early History of the Old South Wales Ironworks 1760 to 1840* (London: Bedford Press 1906), 51, 93–94.

[133] Charles Blagden to Joseph Banks, 31 July 1785, Banks Correspondence, Royal Botanic Gardens, Kew 1.199.

[134] Henry Cavendish, "Tension of Aqueous Vapour," *Scientific Papers* 2: 362–72.

Industrial Tourist

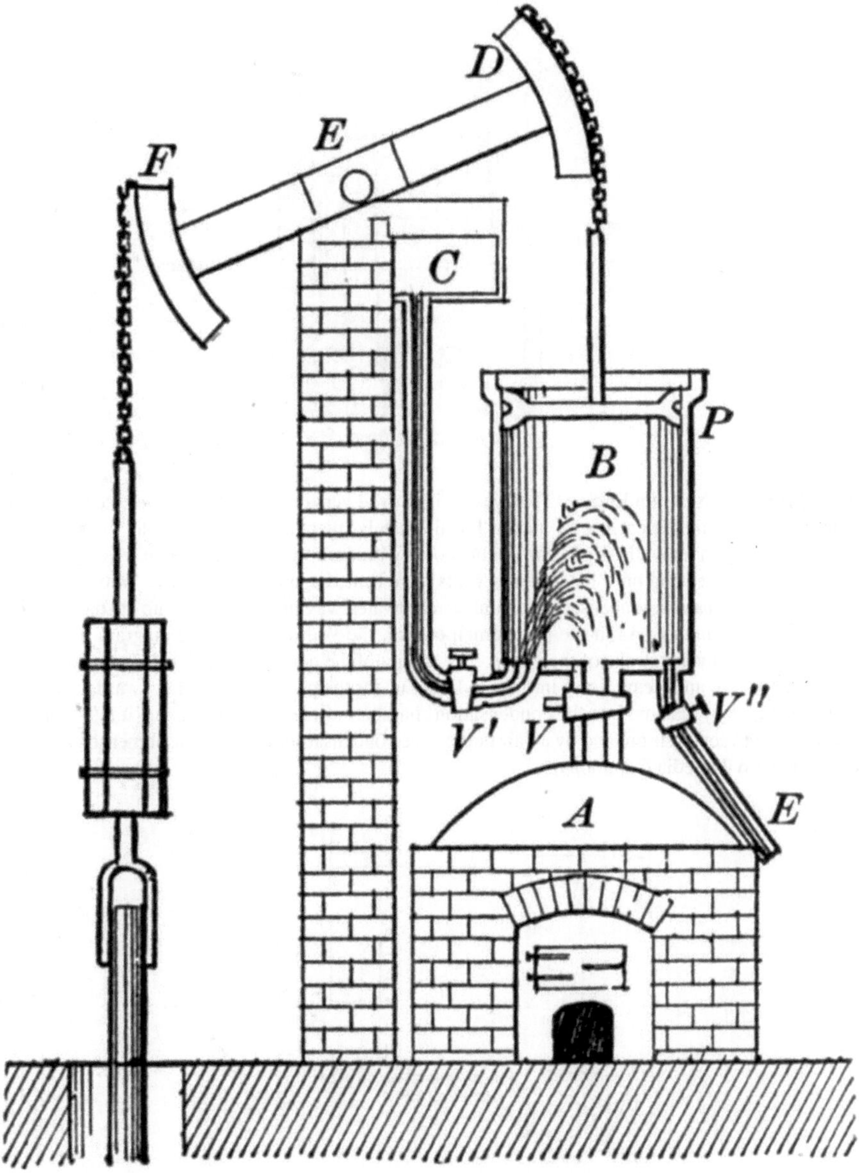

Fig. 10.12 Newcomen Steam Engine. Also called the "atmospheric engine", the Newcomen was the first steam engine that could raise water from deep mines. It was invented by Thomas Newcomen in 1712, and through the eighteenth century it was installed at British mines. It was also used to lift water to drive waterwheels at ironworks. In the diagram, the pump is on the *left* outside the engine house. FED is a rocking arm. Steam from the boiler enters the cylinder lifting the piston. Cold water from the water tank C is injected into the cylinder condensing the steam and creating a vacuum. Atmospheric pressure drives the piston down, rocking the beam and working the pump (Wikimedia Commons)

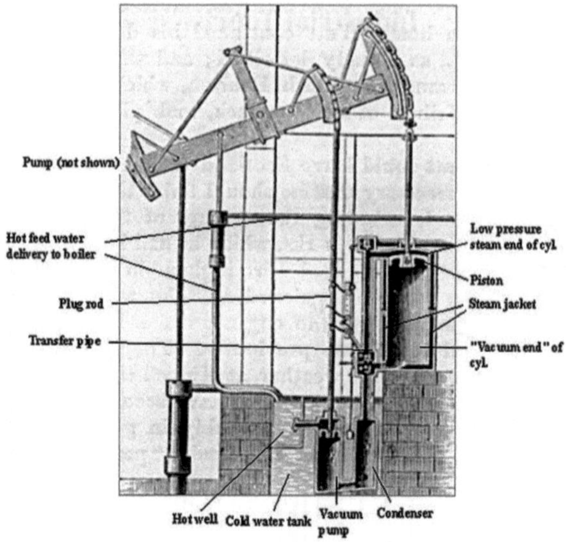

Fig. 10.13 Watt's Steam Engine. This was an improved version of the Newcomen steam engine. The latter made inefficient use of steam. By alternately injecting steam and cold water into the cylinder, the cylinder walls were heated and cooled with each stroke; when steam was introduced, it continues condensing until the cylinder walls were heated again, wasting heat. Watt's new idea was to add a separate condensation chamber; coldwater was injected only into it, the working cylinder remaining hot. As in the Newcomen engine, the Watt engine in the diagram does work only on the downward stroke of the piston. The weight of the pump, the counterpoise on the beam, raises the piston, and steam flows into the cylinder under the piston. On the downward stroke, the steam is drawn by vacuum into the condensation chamber, whereupon condensing, it maintains the vacuum. The vacuum is created by an air pump. The condensation chamber is submerged in a cold water bath (Wikimedia Commons)

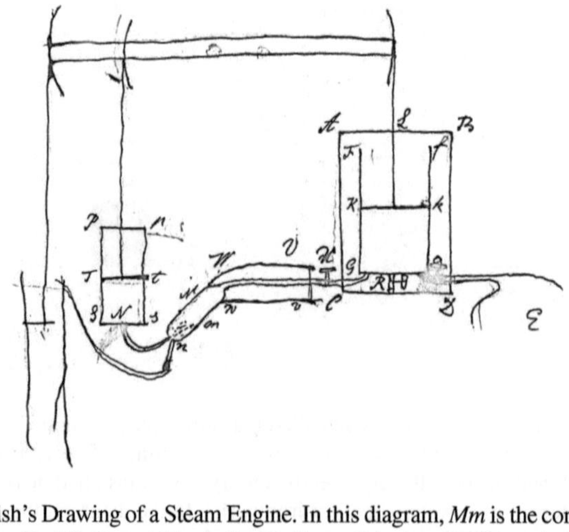

Fig. 10.14 Cavendish's Drawing of a Steam Engine. In this diagram, *Mm* is the condensation chamber, *Pp* is the air pump, and is *Ff* is the working cylinder. Cavendish gives the dimensions and the strokes per minute of the engine he describes, and he notes its advantage: "In common engine [Newcomen engine] as much steam condensed on sides as is used to fill the cylinder." (Cavendish Scientific Manuscripts, Devonshire Collections, Misc. Reproduced by permission of the Chatsworth Settlement Trustees)

Fig. 10.15 Old Bess Steam Engine. An early beam steam engine built by Boulton and Watt, 1777. Preserved in the Science Museum, London (Wikimedia Commons)

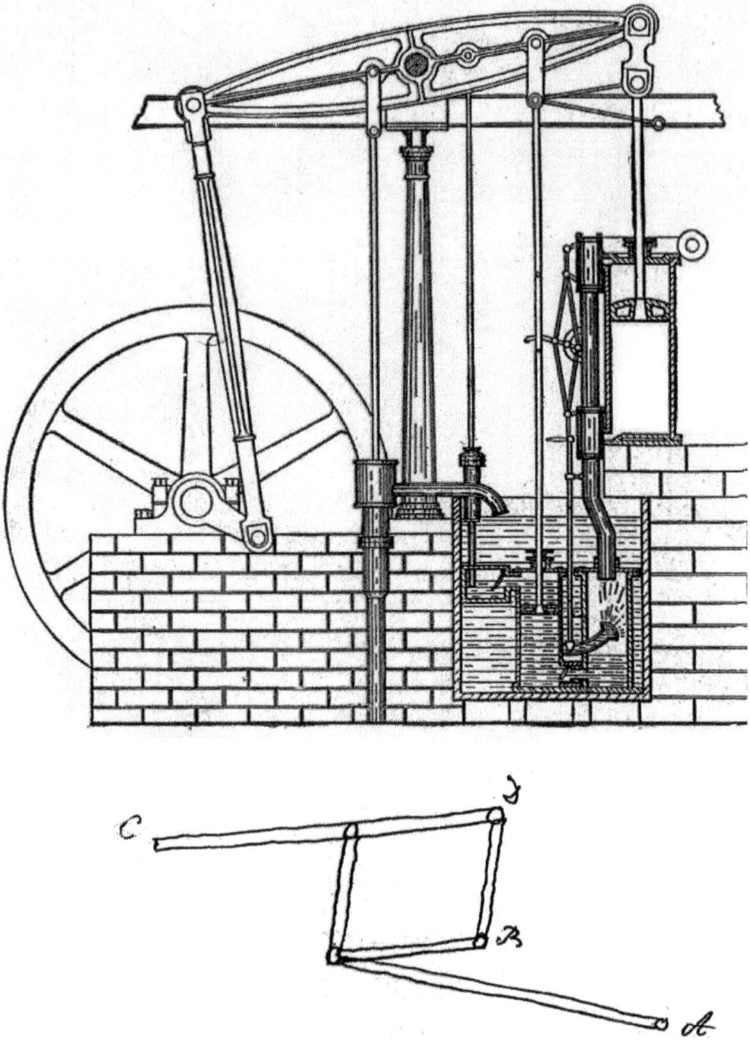

Fig. 10.16 Parallel Motion. In the early Watt engines, the piston was connected to the beam by a chain. By replacing the chain with a rod, it was possible to develop power on the upward as well as downward stroke, to push as well as pull, doubling the action of the engine. There was a problem, however. The piston rod moved vertically, while the beam moved circularly. Watt solved the problem with a four bar linkage between the rod and the beam. This took the form of a familiar pantograph, which produces parallel lines; in this case, parallel motion. A piston moving vertically up and down transmitted force in both directions to a circularly moving beam. Watt took out a patent on his "parallel motion" in 1784. The device on the steam engine is seen on the *upper right*. Cavendish drew a picture of the linkage in his 1785 journal; it is shown at the *bottom* of this illustration (Wikimedia Commons)

Fig. 10.17 Sun and Planet Gear. The gear is shown on a Boulton and Watt engine of 1788. There was a related problem to parallel motion, the translation of the motion of the beam to the turning of a wheel. This can be done simply by a crank, but there was already a patent; a Watt engine with a crank is shown in the previous illustration. To solve the problem, Watt adapted an idea proposed by one of his employees an epicyclic sun and planet gear. It involves two cogwheels. One is fixed to a rod descending from one end of the beam, the other is attached to the axle of the wheel to be turned. Watt patented the gear in 1781. Cavendish and Blagden referred to it in their journal (Wikimedia Commons)

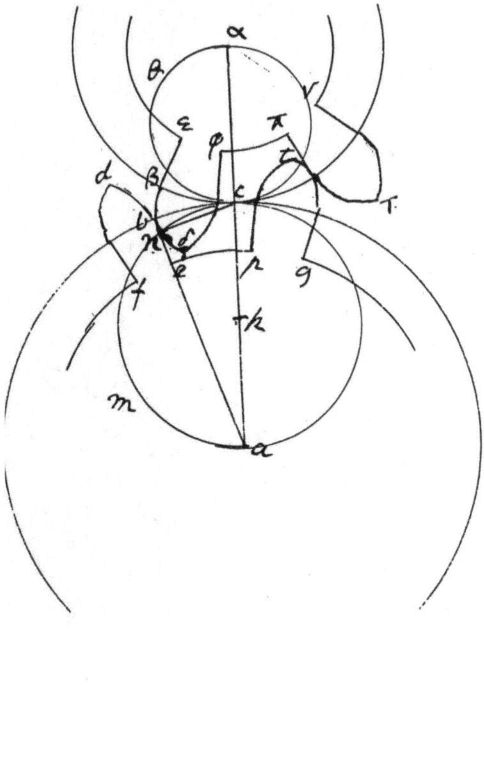

Fig. 10.18 Cavendish's Drawing of Rackwork. In their journal of 1785, Cavendish and Blagden described actions of machinery, including gears, or rackwork. For his course of lectures on natural philosophy at the Royal Institution, Thomas Young asked Cavendish's help with a problem of the teeth of gear wheels. The drawing here was part of Cavendish's response. (Cavendish Scientific Manuscripts VI (b), 31. Reproduced by permission of the Chatsworth Settlement Trustees)

Fig. 10.19 Soho Manufactory. Aquatint over etching by Francis Eginton in 1773. Near Birmingham. Founded by Matthew Boulton and John Fothergill, and completed in 1766, the factory had a Palladian front, bays for loading and unloading, and quarters for clerks and managers. Other buildings contained workshops with advanced metalworking equipment. The industrial complex was much admired and frequently visited (Wikimedia Commons)

Fig. 10.20 Matthew Boulton. By Carl Frederick von Breda in 1792. Manufacturer and business partner of James Watt. Boulton & Watt steam engines were installed by the hundreds in the late eighteenth century (Wikimedia Commons)

Fig. 10.21 James Watt. By Carl Frederick von Breda in 1792. Engineer and inventor. Best known for his improvement of the Newcomen steam engine (Wikipedia Commons)

Fig. 10.22 John Wilkinson. By Lemuel Francis Abott. Ironmaster and inventor (Wikimedia Commons)

Fig. 10.23 Smelting House at Broseley. Engraving by Wilson Lowry, after George Robertson, 1788. In 1757, John Wilkinson with partners erected a blast furnace at Willey, near Broseley in Shropshire. He later built another ironworks at New Willey, which Cavendish visited. In this picture, we see molten iron flowing from the furnace into a sand pig bed (Courtesy of the Ironbridge Gorge Museum Trust)

Fig. 10.24 Reverberatory Furnace. The fire is at *a*, the iron at *b*. (http://www.probertencylopaedia.co/photolib/misc)

Fig. 10.25 Ironbridge. "The Cast Iron Bridge near Coalbrookdale," oil painting by William Williams about 1780. Opened in 1781, the 100-ft Iron Bridge spans the River Severn in Shropshire. John Wilkinson was instrumental in initiating the construction of the first bridge to be made of cast-iron. Abraham Darby III's ironworks in Coalbrookdale produced cast-iron at a reasonable cost, making a bridge feasible. The bridge is another wonder of the early Industrial Revolution (Courtesy of Ironbridge Gorge Museum)

Fig. 10.26 An Iron Work for Casting Cannon. Engraving in 1788 by Wilson Lowry, after George Robertson. Shown is Alexander Brody's cannon foundry on the south bank of the River Severn, a half-mile downstream from the Iron Bridge. Cannon-making was big business in the Gorge (Courtesy of the Ironbridge Gorge Museum)

Fig. 10.27 Upper Works at Coalbrookdale. Hand-colored engraving in 1758 by François Vivares, after Thomas Smith of Derby. In the foreground, a wagon drawn by horses transports a cast-iron cylinder for a Newcomen steam engine. Piles of coke can be seen on the far *right* and the square top of a furnace in the *center*. Houses belonging to the succession of Darby ironmasters can be seen on the hill (Courtesy of the Ironbridge Gorge Museum)

Fig. 10.28 Coalbrookdale by Night. Oil painting in 1801 by Philip James de Loutherbourg. The source of the fiery glow is molten iron tapped from the Madeley Wood, or Bedlam, Furnaces on the right. The large building on the *left* is Bedlam Hall, which has nothing to do with furnaces. In the foreground, we see horses pulling a wagon on a plateway alongside some large castings. From 1776 to 1796, the furnaces belonged to Darby's Coalbrookdale Company. Built in the 1750s, they were probably where iron was cast for the Iron Bridge, a half-mile upstream (Courtesy of the Ironbridge Gorge Museum)

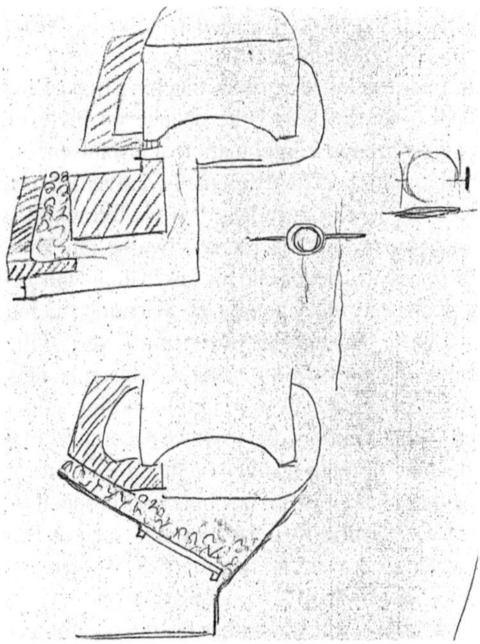

Fig. 10.29 Cavendish's Drawing of Watt's Furnace for Burning Smoke. In 1785, Watt patented a smoke-consuming furnace, which had two sources of heat. On a grate, there was a regular fire. Where the fire was drawn into a flue or chimney, there was a second grate containing red-hot coals that had ceased to smoke, and there the smoke of the first fire was consumed (Cavendish Scientific Manuscripts, Devonshire Collections, Chatsworth, Miscellaneous)

Fig. 10.30 Albion Mills. Cavendish may have drawn Watts smoke-consuming furnace in Birmingham on his journey in the summer of 1785, or he may have observed it at Albion Mills, located on the Surrey side of Blackfriars Bridge. Built-in 1783–86, Albion Mills was the largest and technologically most up-to-date flour mill of the time (Wikimedia Commons)

considered by Watt his masterwork. Watt's double-acting steam engine would be the standard engine to the end of the century.

Industry in Britain had developed without much assistance from science, but that would change. Cavendish foreshadowed the change on his visit to Watt, who had improved the steam engine through a combination of inventiveness and applied science. Cavendish and Watt both studied heat experimentally, and at their meeting in Birmingham Watt described his experiments to Cavendish.[135] If we follow Wilson in calling on the ancient concept of microcosm and macrocosm, we see in Watt's experiments and Cavendish's preoccupation with science and his fascination with technology a foreshadowing of the dynamic society set in motion by the Scientific and Industrial Revolutions.

Day 19. August 4. Near Broseley in Shropshire, New Willey Ironworks. New Willey was the logical next stop, its ironmaster Wilkinson having aided Watt materially in his improvement of the steam engine. His boring mill for making cannon turned out to be exactly what Watt needed to make accurate cylinders for his engines, correcting their inefficiency through leakage of steam.[136] Powered directly by a Watt steam engine, the cylinder and piston machine for producing the blast for New Willey's two furnaces was similar to what they had seen at Penydarren. A second steam engine lifted water to power the hammers. The finery used coke instead of charcoal, and the method was basically the same as the one at Cyfarthfa, stamping and potting. The resulting wrought iron was rather brittle, but good enough for nails, the main product of the New Willey finery forge. Wilkinson blamed the brittleness not on the coke but on the nature of the ore. All of the iron produced at New Willey was made with coke, which was got from local coal. The iron ore too was local. Cavendish and Blagden discussed ores and local strata with Wilkinson (Figs. 10.22 and 10.23).

Day 20. August 5. Coalbrooke Dale [Coalbrookdale], Rathbone's Works; Broseley, New Willey Ironworks. In 1709, Abraham Darby rebuilt a blast furnace at Coalbrookdale, made notable for its use of coke instead of charcoal as fuel; that year he started Coalbrookdale Company, destined to prosper as a family venture. His pots and other products were all made of cast iron. In 1778 Abraham Darby III, third-generation head of the company, made castings for Iron Bridge (Ironbridge), the first cast-iron bridge in the world and the first major structural use of cast iron. The still-standing 100-ft, semi-circular bridge spanned the River Severn, linking ironworks at Coalbrookdale with industrial sites across the river.[137] It happened that two Rathbones of Liverpool

---

[135] In 1793, Cavendish took another journey, this time without Blagden, initiated by the president of the Royal Society, Banks. A new steam engine had been installed at the Gregory lead mine in Derbyshire, in which both the Banks family and the Cavendish family had interests. Banks wanted Cavendish to meet him there to view it, and he wanted Watt or Boulton to join them. The notes Cavendish kept of the journey describe an experiment Watt made with a steam engine to determine the density of steam. Henry Cavendish, Journal of 1793. Joseph Banks to Matthew Boulton, 6 and 18 July, 10 August 1793, Birmingham Assay Office.

[136] Schubert, "Extraction and Production of Metals," 103–4.

[137] S. B. Hamilton, "Building and Civil Engineering Construction," *History of Technology*, vol. 4: *The Industrial Revolution, c1750 to c1850*, ed. C. Singer, E. J. Holmyard, A. R. Hall, and T. I. William (New York and London: Oxford University Press, 1958), 442–519, on 455–56.

married into the Darby family; Cavendish and Blagden called the ironworks they visited Rathbone's Works (Figs. 10.24, 10.25, 10.26, and 10.27).[138]

The plant was large, a quarter mile in length. The blast for its two furnaces was delivered by two cylinders powered by water raised by a steam engine of Watt's design. The year of Cavendish and Blagden's visit a steam engine was installed to blow air at two fineries located outside the building.[139] There was the expected mill for boring cylinders. In a plant belonging to the same company, they saw tar being made from coal.

For lack of water for power, the operation of shingling had been abandoned at Coalbrookdale, and Cavendish and Blagden returned to the New Willey Ironworks to observe it there. The operation differed from what they had seen principally in the use of a reverberatory ("wind") furnace (Fig. 10.28). In this type of furnace, the iron did not touch the contaminating fuel as it did in a blast furnace or in a finery but was heated by hot gases that flowed over it and also by the radiant heat reflected from the roof of the furnace; "reverberatory" refers to reflection. The initial steps were ones we have seen before in stamping and potting. The pig iron was heated in a coke finery to turn it into tough iron, which was then stamped and broken into pieces, and they together with small bits knocked off by the stamping were placed in cylindrical clay pots. The pots were then set in a reverberatory furnace fueled by coal and left there until the iron was at its melting point but still solid. Under the intense heat, the pots cracked, scoria flowed out, and the iron retained the cylindrical form of the pots, called "balls." The balls were dragged by pincers to a large forge hammer and beaten. Finally the iron, now oblong in shape, was drawn out into bars in the chafery forge.

Day 21. August 6. Birmingham, Swan Inn. Returning to Birmingham, they turned off the road at Bradley Furnace and Forges, a new ironworks built by Wilkinson. A steam engine of Watt's construction powered the blast at the furnace. Nothing there was cast directly. Instead all iron pigs from the furnace were carried to a foundry, equipped with reverberatory furnaces and the machinery of cranes. Blooms from the forge hearths were hammered to give them the proper thickness and then they were cut in lengths. These were passed through a rolling mill several times, producing long bars, which were then sent to a slitting mill, producing rods from which nails were made. A steam engine turned two wheels on separate axes, each wheel turning a rolling and a slitting mill.

Cavendish and Blagden learned that Watt had invented a furnace to burn smoke, which he intended to apply to the steam engine. Cavendish drew a sketch of it. In

[138] Joseph Rathbone married Mary Darby, daughter of Abraham Darby II, and the year after Cavendish and Blagden's visit William Rathbone married Hannah Mary Darby. Relatives of the founder owned most of the shares, some of which were mortgaged to William and Joseph Rathbone, who were large investors. Alan Birch, *The Economic History of the British Iron and Steel Industry, 1784–1879* (1967; reprinted Abingdon, Oxon: Routledge, 2006), 62. The name "Rathbone's Works" may have been an alternative: in 1781 a pumping engine "was built for Joseph Rathbone & Co, better known as the Coalbrookdale Co." Handsworth, "1780s Pumping Engine for the Coalbrook Dale Company," http://www.search.digitalhandsworth.org.uk/engine

[139] Hayman, "Shropshire Wrought-Iron Industry," 71.

the fall of 1785, Watt came to Albion Mills, located on the Surrey side of Blackfriar's Bridge in London, where his steam engines were to be installed. These were his advanced double-acting, rotative engines, proper for turning mills, and they were to be worked by his newly invented smoke-consuming furnaces (Figs. 10.29 and 10.30).[140]

Their industrial visits ended here, at Bradley. They began their return journey to London on this day, making geological observations as they traveled.

Day 22. August 7. This day was spent traveling, making frequent stops to measure elevations with the barometer.

Day 23. August 8. London, Bedford Square. This day began at the Star Inn in Oxford and ended in Cavendish's townhouse in London. The last barometer reading was made in Cavendish's library there. Four days after arriving home, Cavendish measured the height of his library above the hall of the Royal Society apartments in Somerset Place, and he compared the portable barometer he used on the journey with his stationary barometer at Bedford Square and also with a second portable barometer at his country house at Clapham Common. The traveling barometer examined, the journey was over.

10.7 Summary of Part II

During three consecutive years in the mid-1780s, Cavendish together with his associate Blagden made extensive journeys to different parts of Britain. In broad terms, the journeys took Cavendish on roads and to places he had not been to before, where he made geological observations and where he was shown mines and industrial plants in operation. Midway through his first journey, he saw Ironbridge cross the River Severn at Coalbrookdale, the first bridge to be made of cast iron. His father devoted years of conscientious administrative work to the building of Westminster Bridge, the second bridge across the River Thames after the centuries-old London Bridge. Westminster Bridge was going to be built of wood, but then it was decided to use stone, a more expensive but more durable material. In the iron and coal valleys of South Wales, the material foundation was beginning to be laid for new ways of construction and of living. If Cavendish sensed this when he saw Ironbridge, he kept it to himself, but he definitely would have realized that his world already was not his father's, and that it held new possibilities.

---

[140] Initially there were problems with the piston rod and the sun-and-planet gear of Watt's engine, but by early 1786 the repairs had been made. In 1789, a second engine was installed. In 1791, Albion Mills burned down. It bears on Cavendish's interest that in 1791 he together with Blagden, Banks, and the engineer John Smeaton were invited to inspect drawings of a steam engine and a waterwheel at Falcon Stairs, near Blackfriar's Bridge and the site of the former Albion Mills. Charles Blagden to Joseph Banks, 23 October 1785, Banks Correspondence, Royal Botanic Gardens, Kew 1:212. John Maitland to Joseph Banks, 19 December 1791, British Library, Add. Mss. 33979, p. 118.

Neither in the journals nor in any other writings from that time do we find indications of Cavendish's "peculiarities." A possible exception is the narrowness of his interests. If his journals were to be taken simply as scientific memoranda no different in kind from minutes of experiments in his laboratory, their narrow focus would be expected. Yet even if this were so, still it was his decision to confine his travels and journals within his narrow scientific interests, and this reveals an important side of his personality, which to his contemporaries appeared peculiar.

The Aristocrat. Cavendish made his journeys during a time when the British aristocracy was in ascendancy. This was also a time of industrialism. Aristocrats and industrial entrepreneurs alike benefited from the political temper of the period, which favored minimal government interference. Eventually the transfer of wealth to the rising middle class of industrial Britain would challenge the aristocratic order, but this prospect was remote, and it would not have troubled Cavendish. It was common in his day for aristocrats to participate in industrial undertakings, and they profited handsomely if their lands were found to hold mineral wealth. Cavendish inspected the installation of a steam engine at a lead mine in which the duke of Devonshire held part interest. Cavendish did not visit mines and ironworks out of monetary or class interest, but as an aristocrat he would have had no aversion either.[141]

The Natural Philosopher. Cavendish looked at mines, ironworks, and factories through the eyes of a natural philosopher, interested in their workings, in cause and effect. He showed no outward interest in the commercial side of industry or in the conditions of labor.

Strangers. As travelers do, he met new people along the way. They included innkeepers, villagers, farmers, miners, and ironworkers. They were given names if they were owners or engineers. The journals recorded information they provided, but nothing else about them.

Regularity and Routine. Travelers often become interested in something they did not foresee at the start, occasioned by the travel itself or a by a spontaneous impulse. Nothing in the journals suggests that Cavendish was such a traveler. We know that Blagden wished to see more of the Lake Country, but Cavendish was not interested. Before Cavendish set out on his journeys, he had a plan, which he did not significantly depart from. It called for observing the same few kinds of things at different places, noting any variations. He and Blagden took barometer and thermometer readings multiple times every day. This was the routine of the journey of 1785.

Wealth. On a whim, Cavendish took a 3-week journey because he wanted to. He could do it because he was wealthy and his time was free; he had no family or professional obligations to hold him back.

Physical World. Methods of extracting and purifying ore and other industrial processes offered Cavendish another avenue of the physical world to explore. In the machines, fires, and materials of metal production, he observed the interplay

[141] Cannon, *Aristocratic Century,* 6, 178–79.

of forces he was intimately familiar with from his studies in mechanics, heat, and chemistry.

Mathematical Mind. From industrial sites, he took away such measurements as were given him such as the sizes of hearths and the quantities of fuel and ore. They were part of the description, not the starting point of calculations of any length. It was otherwise with the readings of measuring instruments he brought with him– barometer, thermometer, and theodolite – which entered extensive calculations for elevations and latitudes.

Senses. Other than for measurements of elevations and occasional measurements of the angles strata made with the horizon and of the thickness of strata, his study of the earth was qualitative and relied on his sense of sight. His geological observations were discerning, comparable to his chemical observations in his laboratory.

Objectivity. His journals are impersonal, objective. They contain only technical and scientific descriptions and observations.

Accuracy. One of the main objects of his journeys was to measure heights of mountains with a barometer. Like others, he was interested in finding a method that was reliably accurate. In general, he and Blagden took pains always to describe what they saw on their journey accurately.

Knowledge. When Blagden told Cavendish about industry in South Wales, Cavendish was curious. He was curious to see it, to know.

Publication, Writing. He kept journals of his tours. It was a custom of the time, but it was also his practice to keep a written record of observations and explanations. Like most of his papers, he did not publish the journals.

Scientific Society. It was common for natural philosophers to take an interest in industry, and like Cavendish some of them made journeys to see industrial works and to make geological observations.

Way of Life. Cavendish's journals give us an idea of his view of the world. His industrial and geological tours were an extension of his interests, as we know them from his more sedentary activities. His regular reading of meteorological instruments on his journeys followed his practice of daily consulting instruments attached to his house or set out on his grounds on Clapham Common. When he returned from his journeys, he submitted his readings to calculations, deriving physical results. The rocks he collected on the journeys became material for chemical and physical analysis at home, and they probably ended up in museums in his townhouse. His observations of strata were incorporated in a comprehensive paper on the strata of the island, which he wrote after returning. He followed up his observations on industrial technology with requests for specimens from mines and smelting furnaces. His journeys were a stimulus to new directions in his research, our clearest evidence that he incorporated his journeys into his life of natural philosophy. The world beyond his laboratory was a larger laboratory, full of interest to this natural philosopher.

Part III
Some Supplementary Materials on Cavendish's Life

Chapter 11
Journal of 1785

Editorial Note. This journal covers a 3-week journey taken by Cavendish and Blagden, July 17 through August 8, 1785. The manuscript is copy of the original journal written by Blagden and subsequently discarded or not identified. The copyist was unfamiliar with the subject of the journal, as is evident from his occasional misreading of words. Cavendish made corrections in the copy, which are incorporated here without comment. Extra spacing between words denotes places in the manuscript where words were illegible to the copyist or where words or numbers were meant to be added.

Except in the case of a few missing letters in words, which are supplied in square brackets, inconsistencies and errors of spelling and grammar are retained. So too are capitalizations, with one exception. The letter "s," where it appears as the first letter of a word, is almost always capitalized in the manuscript, and since it is the only letter written this way, most capital "S"s are removed. Any illegible letters or editorial additions are indicated by square brackets.

The 62-page journal is carefully written, though not with publication in mind. It is reproduced by permission of the Trustees of the Chatsworth Settlement.

Sunday July 17th

Just to the left of the Road between Woolhampton and Thatcham, some of the Fossilwood Peat had been dug, and was then burning to all appearance for Manure.

Tuesday July 19th

In going down Nind Lane, from Wortley Turnpike, we soon got to the end of the calcareous freestone, and came into the stiff blue clay. The Farmers told us, that in digging, they first found pieces of blue limestone a little rounded, lying scattered

Henry Cavendish, "Computations & Observations in Journey 1785," *Cavendish Scientific Manuscripts*, Devonshire Collections, Chatsworth, X(a), 4.

R. McCormmach, *The Personality of Henry Cavendish - A Great Scientist with Extraordinary Peculiarities*, Archimedes 36, DOI 10.1007/978-3-319-02438-7_11,

irregularly in the blue clay, which they considered as an upper layer of the limestone beds, and called it the crop. Below the crop in digging they find successive layers of this Limestone, which, as they go lower, lies in more close compact beds, always with clays between, but where the Strata are very close, that clay is compacted into a hard Substance almost like Slate.—In ascending from this blue clay up the Fortworth ridge, we soon got upon a sort of Stone, looking like a gritty or coarse grained Limestone, which tried with an Acid, was found to contain not much calcareous matter. The dip of this Stratum was highest towards the Freestone hills, and amounted in some parts to 40° or 50° to the horizon. In descending the Fortworth ridge from Wickwar towards Splats hill, we came toward the bottom to the same stiff blue clay, in which the blue Limestone was lying as about Nind. In the quarries here, the Limestone was lying in the form of flat stones, some very thin, lying upon one another, with hard compacted blue clay between, having somewhat the resemblance of Slaty matter. Through some of the stones there ran veins of yellow matter seemingly mundic. The blue clay continued till we came near Hilsley, and there seemed to be nothing interposed between it and the beginning of the Freestone hills.

Wednesday July 20

In making the solution of Tin, Mr. Blagden used two parts of the Aqua fortis he had purchased to one part of Water. He drops the Tin in carefully, taking particular care, that the effervescence shall not be violent, otherwise a clear solution is never obtained.—If any Straws or pieces of Wood are in the Liquor during the solution, they are apt to make the mixture take fire, which entirely spoils it for dying, the Tin subsiding as a white powder: it also takes fire with the same effect, if the solution be made too violently. The Aqua fortis is prepared by adding to it a small quantity of Common Salt; and the Tin by being poured melted into cold Water, which divides it into thin irregular plates. No yellow fumes were perceived during the solution, nor anything like the smell of nitrous Air; the predominant smell was that of Aqua regia. We were told that when the solution heated much, it yielded yellow fumes, and became proportionably more turbid; and this was the case particularly when it took fire.—The solution, if well made, would keep many days, and even some Weeks, if the Weather was cool: The slower the solution had been made, the longer it would keep; and when the Operation had been nicely performed, the solution spoiled not by dropping any sediment, but by losing its yellowish, reddish or amber colour, and becoming pale and colourless. The powdered Cochineal was thrown in upon the Liquor in the boiler all at once, and floated on the surface, where it was suffered to remain, without being stirred into the Liquor; the solution called Spirits was then sprinkled on the surface of the Liquor at different Times, and in successive periods of the boiling, and the heat was urged at the same time till the surface broke, that is, till it boiled up violently. — It appears that the cloths are dyed without such preparations with a *mordant*, as is described by Hellot, the Tartar or whatever else may be necessary to fix the colours, being put into the boiler with the dying drugs. In dying the scarlet with grain, it is necessary that the cloths should not come into immediate contact with the copper boiler, other wise they are spotted. This is prevented by an osier frame in the boiler, cloths round the edges, &c. Alum entirely spoils the grain colours, turning the composition to a dirty crimson.—The cloths are never dyed crimson with Cochineal alone, but with Spirits in the same manner as for scarlet, but less composition, not much more than half, and then the proper hue is given by means of Archall, which brings it down to the crimson, from a scarlet which is not very deep, or rather from a bright red. In dying Ora[n]ge, they put the Turmeric and Tarter into the copper first, and when the Liquor was boiling (simmering) though not boiling hard, they throw in the powdered grain, and afterwards sprinkled in the composition (Spirits) successively, so that all the ingredients were in the furnace at the same Time.

Thursday July 21

The fine and constant spring called Kinner Well in the road from Mr Hale's house to Monk's Mill, had been previously found to be $50^1/_5$ by the Thermometer with a sliding Scale. We now tried the temperature of a bucket of water, taken up briskly from the well at the Lower house Alderly. By the same Thermometer it appeared to be 50¾; but by a Thermometer with a fixed Scale (supposed to be adjusted for the diameter of the tube) it was only 50¼ consequently supposing this last the most accurate, Kinner Well is only 49¾, and the Lower house well is probably the same, as the water may have heated a little in coming up. We then measured the rope with the bucket of the Well at the Lower house, and found it to be nearly 107. feet. We then measured by the barometer the height of the Lower house well top (as before) above Kinner Well, and it came out feet, consequently they seem both to be nearly the same level.

Friday July 22

At Frombridge we were informed, that in order to draw Steel or Iron wire, after they had annealed it down to the proper temper, they found it necessary to rust the rods or large wires, by means of sour waters, that they might not heat too much in passing through the holes. They first rub off the scales contracted in the annealing by means of a Mill, and then go to the operation for rusting, for which Sour Whey is much employed; still more to prevent the heating, they use much grease in the holes, which makes up into a sort of paste with the rust.

In making brass, they first divide their Copper, by dropping it melted into cold Water, which reduces the Metal into large irregular drops or lumps.—These are mixed up with about an equal weight of charcoal, but a much larger proportion of powder'd Lapis culaminaris, and the mixture is then put into a crucible (of Stourbridge Clay) so as to almost fill it, and then the crucible is put into a furnace, where the heat is immediately raised, so as speedily to reduce the Copper into fusion. In this state it remains 12 hours, and upon cooling the crucible a lump of brass is found at the bottom, about 1/5 of the quantity of the Materials employed. In making both brass and iron to wire, they are obliged to anneal each drawing; but with the brass it is not necessary to rub off the outer coat, produced in the annealing, which must be done with the iron. In making iron wire, they sometimes employ slit rods, but prefer certain rods made by hammering with irregularly lozenged Surfaces, from the Strokes of the hammer, as being of a more uniform texture, and drawing better.

In the slitting mills, both rollers were both turned by rack work, moved by the same axis. In the flatting mills, two rollers were turned by 2. separate water wheels.

At the Tunnel cutting for the canal at Saperton, we observed that the freestone brought up was partly yellow and partly blue, the same piece of stone being often tinged through of a blue colour, not exactly in the direction of the layers, but approaching to it. The different layers of freestone had in many places layers of clay interposed. The shafts sunk down the Tunnel were, we thought, remarkably near one another: the deepest now made, is 35. yards; but nearer Saperton there are to be shafts as deep as 60, we were told. The Tunnel itself was by estimation from 20. to 24. feet high; arched and walled with bricks: only about 40 yards are compleated.

At Stroud we were informed, that the old Canal had not yet paid any Dividend to the Proprietors, and that scarcely any thing but coals were brought up it, at the rate of 3^s/[?] a ton. The new cut extends from Stroud near 5 Miles to Chafford Prince, so as to be navigable; but no further as yet.

Saturday July 23

At Stone Bench the head of Tide passed at 9^h 50′ A.M. by the watch[. L]ast night at Alney, it came up nearly at the same time p.m. At Alney it made a good deal of noise before it came; at Stone Bench but little. The tide passed this Evening at Newnham soon after 8 in the Evening.—It rose with great rapidity, and after the waves caused by the first coming, had entirely subsided, perhaps half an hour or more after the first head, it seemed to rush up again with renewed violence, so as almost to form a head. We were told, that the tide, though earlier than we had expected it, did not come sooner than it ought: that in spring tides, the interval between each tide, seldom exceeded ¼ of an hour, and that it was this Evening about one quarter of an hour later than it had been in the Morning. A person present said, that about 5 or 6 years ago, the Channel of the River, which used to run close to Newnham, had shifted to the other side, and now ran equally close to Arlingham.

Monday July 25

The furnace at Pentyrch not being in blast, we could only examine the forges. They are of 2 kinds, the Finery and the Chafery. In the Finery they make use of Charcoal only. A pig of iron is put on a hearth in the midst of a heap of burning Charcoal, and the blast of the bellows directed upon it, as it softens and partly melts, it runs down into the lowest hollow part of the hearth, the workmen stirring it all the while, till at length it is collected in the hollow as a soft tenacious lump, which can be taken up with the proper pincers. At the front of the hearth is a hole for the scoriae to run out. The lump while hot, is at the proper time taken up with a pair of large iron pincers, out of the Charcoal, and struck with iron rods all over the outside, to beat off all the Scales and adhering foulness which can be separated in that manner. It is then laid under the hammer, weighing 6 or 7. C^t weight, the first strokes of which, besides making a large number of hot scales fly off to a distance, is to make a white hot fluid matter run out of different cracks of the surface, some times in considerable quantities. From some of the cracks also a bluish light flame issues, often to the height of 2. or 3. inches. This flame comes out without any strokes of the hammer, though more when the lump is under the hammer. This lump is then beaten with the hammers till it is reduced to an oblong form, ab[t] 20. inches long and 4 inches thick, called half-blooms. These are then heated again in the same fire on the hearth where the pig is melting, and when white hot, are again brought under the hammer, and reduced, by beating in the middle, to a long bar, about inches in length, terminated at each end by a knob of 8, or 12. inches, being so much remaining of the original half-bloom. In this operation one end of the half-bloom is but little heated, as it sticks out of the fire, and affords a hold for the pincers to move it round in the fire, and at length to take it out. After these blooms are finished they are removed to the Chafery, where the knobs at the end are drawn out, and the Iron is fashioned into various utensils. All this is done with pit coal; but only particular veins of it are fit for this purpose. The knob at the end is heated in a hearth like that of the finery, and when white hot is laid under the hammer, and beaten till it is drawn out even and similar to the original bar of the bloom, and it is found to be in all respects equally good Iron.

The Scoriae of different kinds are heated up again in the Charcoal fire with the pigs into those lumps, of which the half blooms are formed.

The shape of the hearth at the finery is as follows: It consists of an oblong pit, formed of cast iron plates, about 2 inches thick. The pit is 2 feet 4 inches long, 1 foot 6 inches broad, and about 7. inches deep. In the front plate is a hole even with the bottom plate, which they tap by thrusting an iron into it, to let the Scoriae run off; and at the bottom of the front plate is a notch, lower than the under side of the bottom plate, through which an iron is thrust to heave up the bottom plate when necessary. The bottom plate does not come quite home to the front plate, for which no reason was assigned, but that it was unnecessary for it to be close; the distance was about 4 inches.

The weight of each lump, as it melts down from the pig, is about ¾ of an hundred. The pig rests upon the edge of the back plate, and the nozzle of the Bellows on the edge of one of the side plates, much nearer the back plate than the front plate, and just under the projecting end of the pig. As the pig melts, the matter collects at the bottom of the hearth, and does not come into liquid fusion, till it is among the cinders, or Scoriae, which the workmen consider not only as bringing it to the quality of tough wrought iron, but even as necessary to melt it perfectly. After being stirred, and as it were kneaded some time with the Scoriae, it ceases to be in liquid fusion, and becomes that tenacious lump, which they take out with pincers to put under the hammer. An half bloom is commonly about 20. inches long and 4. inches thick. The Blooms are of uncertain length, according as they are drawn out into a longer or flat bar: the former is about 4, the latter only 2 feet, besides the knobs at the end, the longest of which is near a foot, and the shortest about 7 or 8. inches. The firework sparks fly off in all parts of the operations in the Finery.

In the Chafery the hearths have no front holes for the Scoriae to run off by, but the refuse matter collects at the bottom into masses they call *Mosses*, which are taken out every 12, or 24. hours, when they make up the fire afresh. Those in the Chafery hearth, worked with pitcoal, contain much iron, but are not found worth working over again, so that they are absolutely refuse; but when the Chafery hearth is worked with Charcoal, the Mosses contain no iron. The former ones are round

Mosses, of the size of a large Man's head, hollow in the middle, and open at top, very spongy and rusty. In the Chafery when they use Charcoal, it is commonly small and dusty, which they call brain, and this works up the Blooms & with less loss.

At Newnham the Cliffs consist of a red stone partly calcareous, but leaving a good deal of undissolved matter, divided by thin seams of whitish wax-coloured shivery stony matter, which did not dissolve (in Aqua fortis) and appeared to be argillaceous.

When we got among the Hills about Pentyrch Works, they were found to consist of a reddish Lime Stone, laying in Layers which dip from S. to N. the whole Stratum very thick; the angle to the horizon at least 45°.—Proceeding northward we came to a blue shivery clay Stone, in some parts soft, deeper down very hard, in which both the Iron Stone and coal seem to lie; or at least at Pentyrch Church the Lime Stone was as at the cliffs above the works; then we passed a Lane, and the first Stone, as we ascended again toward the Garth, contained no calcareous matter. In the Valley between the Garth and the hills of Pentyrch, were the Coal pits. In the Lane, ascending toward the Garth, where the first Stone not calcareous was found, the layer dipt very much, and apparently in the same direction as the Lime Stone of Pentyrch hill. In digging the Coalpits, they pass through nothing but the blue shivery claystone, which the lower they go, the harder it is, till near the coal it is quite a Stone; but on being brought up, and exposed to the Air, it presently shivers to pieces, and moulders down.

On Caerfilly Common we had the calcareous hills to the S. The common itself consists chiefly of the blue shivery clay Stone, which also becomes harder the deeper they dig, till at length they come to a seam of Iron Stone, from 6. inches to a foot thick, with pellets or nodules, interspersed above and below the seam. In other places of the Common, in digging through the blue clay, they come to coal. Under the seam of Iron Stone; they have the same blue clay stone as above it. They have 5 different seams of Coal with the blue clay stone interspersed. The seams of coal are from 2 to 4. feet thick each.

At the furnace we saw three different kinds of Iron, mottled, dark-grey, and bright-grey. Of these the mottled melts the easiest; bright-grey next easiest, and dark-grey most difficult. The mottled will not make good tough Iron by itself, but either bright- or dark-grey will. Light-grey is always used by itself; but dark-grey is more frequently mixed with a little mottled, because it is very slavish to work, and of difficult fusion. The proportions are one ton of mottled to 3 of dark-grey. The mottled on breaking, looks more like light-grey, but has some white spots, especially toward the edge of the fracture. The general ground seems granulated, but the spots look like crystallized plates broken edgewise. The light and dark-grey are both granulated, but the former exhibits finer grains in the fracture, than the latter. In a Ton of pigs 28 lb is allowed for sand; and then 4 Tons of pigs, make 3 Ton of blooms 21. C^t w^t to the Ton. To produce a Ton of Bar Iron, 21. C^t w^t of blooms is allowed, and 22. C^t of half blooms. The whole produce from 4 Ton of pigs then is 3 Ton of Bars, 20 C^t w^t to the Ton. The greatest advantage is to have the Iron bright-grey, shews that the proper proportion of ore to the charcoal in the furnace is used; if too little ore, it then becomes dark-grey, if too much, it becomes mottled, and when in excess, the furnace cools, and sometimes the mass congeals and spoils the furnace, so that they are obliged to take it down. Producing dark-grey, they call a low furnace, mottled and tending to congeal is raising the furnace high.

At Pentyrch they were putting a new hearth in the furnace. The hearth-part, which is the part in which the ultimate fusion of the metal is effected, is nearly square, about 2 feet in length and breadth, and perhaps 4 feet deep, built very solid. The hole for the blast was on the left hand wall, exactly in the middle between the back and the front, and about 18. inches from the bottom. The front is made by what they call a temp, which reaches from above to about 18. inches from the bottom; and the lower part of the outer front is a plate of Iron 2 inches thick, provided with a large notch in the top edge, and two arms on the right hand side. This Iron plate is put close home to the left hand wall, but leaving space between it and the right hand wall, which in working is stopped up with sand. The arms above mentioned reach to the right hand wall, into which they are fastened. Between them the poker is thrust, when it is necessary to stir the melted matters; and the furnace is tapped, to let off the melted metal, by removing the sand which fills up the interstice between the right hand edge of the front plate and the right hand wall, close to the bottom; when the Metal discharges itself through a gutter made in sand, into a series of parallel side gutters: the middle gutter and Metal congealed in it, being called the sow, and those on the sides each a pig. Immediately above this square part, the furnace spreads on each side at an angle of between 30° and 40° to the perpendicular, and after enlarging in this manner to a certain height, it again contracts to form the Mouth at the Top, by which the Furnace is fed. The parts that rise up immediately above the square hearth are called the *Bosses*.—When the Furnace is working, the whole load of Matter up to the Mouth is red hot, as we were told. After tapping the Furnace, they cease blowing some time, about half an hour, to get out the *Scoriae* and clean the hearth; after which the blast being renewed, the Operation is found to have been in no respect injured by the cessation. Round the hearth they make what is called Air pipes, forming a hollow communication, from the external Air at the Water wheels, about even with the bottom of the hearth, where there is one hole, to the Air in the casting house, where there is another hole in the wall from the right hand side of the hearth above, level with the top of the cast Iron front plate. The use of these we could no otherwise learn, than that the Furnace would not work so well without them. Each hole, and we suppose for the whole length of the communication is about 2. inches square. The hearth is built of a sort of Plumpudding stones, of which we saw great numbers about Pentyrch loose in the channel of the Taaf.

Wednesday July 27th

Visited the Works at Mellengryffyth. The Forges, &c. seemed the same as at Pentyrch. For making plate tin, the Iron is first drawn out into long and broad bars, which are then cut into slips nearly of the breadth that the tin plate is to be. These are then drawn out between rollers. At last they come to be rolled 4 sheets thick, and are then cut at the edges all the four sides of the square, to be made smooth, and then the 4. plates are separated from each other, which requires some force, as the cohesion is considerable. Each plate is next dipped separately in Spirits of Salt, being previously bent in the middle, and laid in a kind of Oven, to be made red hot; they are then taken out, suffered to cool, then bent back as smooth as they will readily become by the hand, and several of them being laid together, they are struck with great violence by a Man on an iron Anvil. By this operation the general scale or coat of rust, formed by the Spirit of Salt and the heating, is beaten off, and the plates rendered of an uniform though roughish surface; in some places discoloured of bluish or yellowish brown, which we were told was not a scale, and had no bad effect in the tinning. Each plate is next passed through a pair of rollers to make it smooth and even, and it is then steeped a proper time, half an hour more or less, in a sour Liquor, made of bran and water, fermented. From this Liquor, it is put in common water, and is thence carried immediately to the melted tin, without being dried. The tin is melted in large boilers of a proper depth to take in the whole plate, and its surface is covered with tallow or hogs Lard; but these being now dear, they were trying fish-oil and found it answer. The iron plate is taken out and turned, and passed again into the Tin several times: so that the whole operation, till the surface be properly covered, takes up about half an hour. The tinned plates are then scoured with bran, till they have acquired a proper polish. At the time the iron plates are steeped in the sour Liquor of fermented bran, they are well scoured with sand, &c. so as to clean the surface perfectly.

At this Work they make bolts for ships by welding together the shavings of the tin plates, by means of a hand hammer, and they consider the Iron as tougher than any solid piece would be. The inside parts are not compacted into one mass, but only cohere, and they are thought less likely to break, than if they were one mass, besides that the iron would be much burnt, before they could be so compacted.

Thursday July 28th

Having heard that the blue Lyas Limestone was at Penarth point, we went thither. After passing the Salt meadows near Cardiff, we came to a hill which they call Lecquith; we found it lying in Strata nearly horizontal divided by thin seams of blue indurated clay; the whole hill seemed composed of these alternate seams, & probably contained as much Limestone as clay. When we came to Penarth Point, we found the cliff consisting of the red Clay stone, with seams or veins of white clay stone, as at Newnham, above which the Lyas with its clay was laid, composing the upper part of the cliff. A[t] the point some of the thin seams between the red Stone were of Gypsum or Alabaster.

Friday July 29th

The Furnace at Pentyrch is 26 feet, but the Funnel reaches much above it, and other Furnaces are much higher. The bottom of the hearth is narrower, and proceeds upwards widening a little, so that the Temp-stone is supported by the inclining sides. In wet weather they can use to work the furnace only one measure of the wet charcoal to 6. of dry; but when the weather is dry, they can use 2. wet to 7. dry. In the forges they do not regard whether it be wet or not, and are frequently wetting it, to prevent the iron from cracking, from the difference of temperature between the parts of the Bloom out of the fire, and those in it.

At Merthyr a new Furnace with Machinery is just erected by the Messrs Humphries. The blast is given by means of a fire engine on the old construction, the cylinder 40. inches diameter. At the further end of the beam, a piston was worked in a cylinder 6. feet in diameter, which forced the air into two cylinders, one on each side: from which went pipes that united in a reservoir, from which the blast was conveyed by proper pipes to the Furnaces. The covers of the side cylinders were loaded with a weight equal to 4 inches of quicksilver. When these covers, which were moveable within the side cylinders, rose to a certain height, by the Air forced in underneath them, they raised a frame which opened a valve, that afforded a discharge for the superfluous Air from the middle cylinder in which the piston works. The Engineer, whose Name is Birch, supposed that half the Air escaped this way; and it is intended to convert this Air to the working of a Forge. The piston of the middle cylinder, which was 6 feet in diameter, made a stroke of 5 feet, 12 times in a minute. The pipe which let the Air into the Furnace, was fastened to the main pipe by flexible leather, so that it could be pushed in or out, and otherwise guided.

The Furnace was like Mr Lewis's at Pentyrch; but the boshes reached up not more than ⅙ or ⅕ of the way from the hearth to the mouth where the Furnace is fed; and the width of this Furnace seemed much less in proportion to its height than Mr Lewis's. The width was 12 feet, the height 60. The Air pipes extended round, under, and up, in all directions, and are intended to prevent the Furnace from bursting, by the Air and Steam, when it comes to be heated. This Furnace had cracked toward the top, which was attributed by many to its having been heated too soon. They use 3 ton of Coak to make 1 ton of pig Iron, and the ore used, is entirely the stone of the surrounding hills. The casting we saw, consisted almost entirely of plates for the Anglesey Copper works. They were oblong, weighed nearly 1 C^t^ w^t^ had 2. holes toward one of the long sides, and a protuberance between them, and were about 1/2 an inch thick. The holes are intended as handles, and the protuberance to keep the plates apart, when they are laid in the Copper water. They were going to cast a hammer beam, that is, the axis by means of which the hammers in working are raised up.

There are several veins of coal in the neighbouring hills, with the blue clay above, beneath, and interposed. The bed of Iron ore, of which there is only one, lies underneath the coal. The veins often failed, and were found again at another level. The seam of Iron ore is attended with pellets and nodules, as in the Caerfilly Seam. They coak the coal by laying it in a large heap, several times broader than deep, and then light it at the top. When the flame has spread something more than half way over, which it does in 6, 9, or 12. hours according to the nature of the coal, they cover it up with ashes; but the heaps or pile being hollow underneath, by a foundation of large coals laid upright and separate from each other, the fire spreads at bottom, after it is smothered on the upper surface, sufficiently to coak the whole heap. The ore is roasted in the same manner as lime is burned. Some of the kilns in which it was roasting afforded a smell of Sulphur, in others scarcely any thing could be distinguished but the common smell of coal.—At the top of the furnace in which the ore was melting, though 60. feet above the hearth, the flame was very strong, Mr Humphries prefers these high and comparatively narrow furnaces, because that construction makes a greater draught of air in the middle parts, and consequently a stronger heat.

The Method of tapping at Mr Humphries's Furnace, was exactly like that described by Mr Lewis, by removing the sand at the side of the front iron plate, close to the bottom of the hearth; except that in his Furnace, the interstice filled up with sand is at the left hand side of the hearth. As soon as all the Metal had run out, they began to clean the hearth, by removing the Scoriae and congealed matter with iron hooks and pincers, and the blast was suspended all this Time, perhaps a quarter of an hour, till the hearth was sufficiently freed of impurities. Mr. Humphries informed us, that the Materials, tho' much heated, and almost ready to melt in the body of the Furnace, do not really melt till they come under the blast; but immediately as they are acted upon by the Air, run down quite fluid. In consequence of the crack, in the top of this Furnace, the flame forced itself out at the holes, that had been left as Air holes near the Top.

Saturday July 30th 1785

This morning we went to Cyfarthfa Forges, Mr Bacon's, now let to Mr Tauper, about a Mile from Merthyr. Here coak iron is reduced to malleable iron, without charcoal. The pigs are first melted down in a hearth[1] and stirred till they become a lump, as in the charcoal works; they are then put under the hammer hot, and reduced to flat plates in general about ½ an inch thick, of an irregular shape, and rough cracked irregular edges. These plates are then broken under a hammer into pretty large lumps, which are laid in a heap, and kept moist by water dropping upon them. We found them blackish on the outside, with spots of rust, and at the fractures of a granulated texture bright light colours, but also in places much rusted. The small stuff that was collected in the hearth, they washed with water, and then laid it by separate. Afterwards the broke lumps are put into cylindrical pots, made of refractory clay near a foot deep, with a certain proportion of the washed small stuff, which served as a flux. There is no other addition, and the surface is not covered. The heat is not urged for the metal to melt, but so that the different pieces may be welded together, which happens by their sinking down into a sort of mass, and contract so as to leave the sides of the pot, which usually cracks and is easily separated, This mass is then formed into a half-bloom, a bloom, &c. as in the common forges.

In proceeding up the Valley of the Taaf, toward the Van, the Strata were not much inclined. At first we had nothing but the same rock as about Merthyr, which seems a sort of Clay-stone, or at least is a hard Stone not calcareous. (Much of the Stone about Merthyr, of the same appearance nearly, is soft and crumbles.) Some Miles from Merthyr we got to Limestone, which composed several of the rising hills, and lasted for some miles but a few miles before we reached Pont-y-Taave. At the bottom of the Van, we came upon a sort of slaty rock, which lasted all up the Van, with some difference of appearance. Toward the Top of the higher hills on each side of the Valley of the Taaf, there were Strata of stone of a different appearance, which Mr Humphries said was Lime Stone, and this seemed to form the upper Strata of stone on these hills.

[1] These hearths are nearly of the same form as those in the charcoal Works, but with some peculiarities, which the Agent told us were known only to the workman.

Sunday July 31st 1785

On the road from Brecknock to Abergavenny, we had lime kilns on the right, near the top of the high hills on the S. side of the Usk. The Sugar loaf mountain seemed to be the same stone as about the Van. Many miles before we reached Monmouth, the soil had the same red appearance as about Newnham and Chepstown.

Tuesday August 2d 1785

In ascending Malvern hill at Wytch from Ledbury, we saw Lime Stone rising up along the hill, to the right, just at the foot of the down part, which they were burning into Lime. The highest point of the hill, is immediately above Malvern village. In ascending we saw rock projecting in several places, which was a sort of coarse Granite, below almost resembling a very fine plum-pudding stone, but near the top composed of much finer parts or grains

At Birmingham we were informed by Mr Withering and Mr Watt, that the part of the road through which we descended off the [--]cky on the northside, when fresh cut, appeared evidently to be a granulated Quartz.

The Machine for twisting the handles of horse whips, is so constructed, that the bobbins can move only one way, as the Machine is turned backwards and forwards; but they slip over one another alternately in succession.—When the bobbins rise up to a certain height, a catch is set loose, which suffers the thread to unwind from the bobbin half a revolution.

The great inconvenience complained of in the needle Manufactory, is their warping in hardening. In that case they must be made straight by a stroke with the hammer; but this leaves an impression, and the needle is always the worse.

In the rolling plate Manufactory at Snowhill, a circular motion is given by means of the steam-engine. A bar of iron is fastened to the beam of the Steam engine, and at its lower end turns a crank, which gives motion to a wheel, communicating with the rest of the Machinery. In order to keep that in motion, when the crank is at its upper and lower points, it moves a wheel by rack-work, which makes its revolution in half the Time; and to one part of this wheel a heavy weight is fastened, which tends to accelerate motion most, when the crank is at the upper and lower points, and retard it, when the crank would give the quickest motion. It was invented, or at least the patent for it obtained, by a Mr Picard, who has sold it to the present proprietor; but Mr Watt claims the original Idea. In order to make the silvered plates draw, it is necessary to forge them, previously heated, before they are first put between the Rollers. In order to fix the plate of silver on the plate of copper, they make the surfaces of both exceedingly clean, and then wet them with a solution of Borax. They are then put into the fire, and the workman watches the time, when the edges of the two plates in contact melts. Whenever that is observed, he takes them out of the fire, and they are found perfectly fixed to one another: not so however as to bear being rolled, till they have been again heated and forged with a hammer.

Wednesday August 3d 1785

At Soho. The pieces of steel which are inlaid into the Bath metal for buttons, are made of a steel very highly converted, so that it can be hardened by a very slight heat. Some steel is so converted, that it can be hardened with a heat scarcely sufficient to make it slightly red hot; but that kind of steel has not much strength.

The fashionable excellence of gilt buttons is that they may look red, much like copper. For this purpose the gilt button scarce polished up, is dipped into a solution of some salts, amounting as Mr Watt said, to a kind of Aqua Regia. It is then set on a part of a stove, which seems to make it obscurely red hot, and then is thrown into cold water. Upon being afterwards burnished, it looks so like burnished copper, as not to be distinguished even by workmen, without applying Aqua fortis.—The very fine black polish is given to silver by burnishing it with Haemalites; nothing else will make it so perfect.—Filigree work is fastened in by strewing sodder with Arsenic over the surface, and melting it with a Blow pipe, when it is absorbed at the parts in contact, without filling up the interstices meant to be left hollow.

Mr Watt's new method of giving a circular motion by the Steam engine, is by making a small wheel fastened at the bottom of the bar, suspended to the beam of the Steam engine, pass round a larger wheel, without revolving at all on its own axis. The centers of these two wheels are connected by a lever, keeping the small one constantly at the circumference of the large one. The effect of this is, that the larger wheel performs always one revolution, and as much more as corresponds with the number of teeth in the smaller wheel, while the latter is passing once round the circumference of the former.

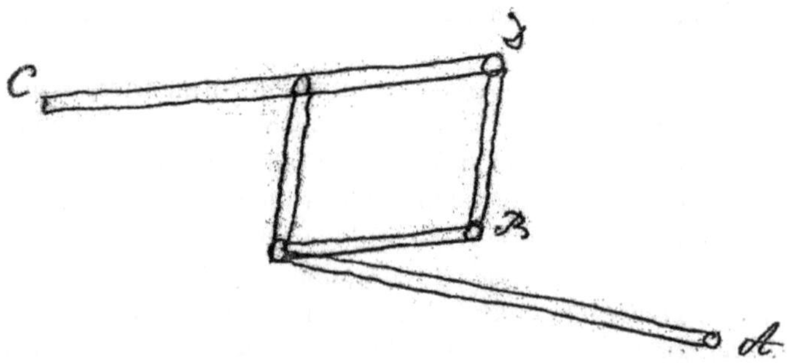

This figure represents the manner of making a rod, moved by the beam of a Steam engine, perform its motion up and down vertically in a straight Line, without the help of a circular arch and chains. CD is the Beam of the Steam engine, C. its center, A. a fixed point, and B the point which is to move vertically in a straight line.

Mr Watt mentioned, that having found that some steam is condensed in the cylinder of the Steam engine, tho' surrounded with steam, he made an experiment to discover what happened. He threw steam into a Glass vessel close at top[2] and found that upon making the Vacuum some of the steam condensed on the sides of the Glass Vessel; and having heated the sides of this Vessel, so that none could condense upon them, he observed the condensation to take place, so as to render the Steam visible in the middle of the Glass vessel; that is, when the sides were much heated, no cloud was perceived; but as they cooled, a cloud began to appear at each Vacuum,

[2] By making it communicate with that part of the cylinder of this fire engine in which there is alternately Steam and Vacuum.

in the middle of the glass vessel; and when any parts of its sides were sufficiently cooled, the steam then condensed on that cooled part.

Mr. Watt thinks to have ascertained by experiment, that the less heat water is converted into steam with, the more latent heat it requires to assume the elastic form; more even than the difference of the sensible heat: for instance, that water converted into steam under a pressure that suffers it to take that form at 100, contains more heat, sensible and latent taken together, than when it is converted into Steam under the whole pressure of the Atmosphere at 212°. He states the difference at 100. or more. Mr Watt considers the heat of steam at 212. both sensible and latent, as near 1160, reckoning from *0* of Farenheits scale; therefore the heat absorbed in the act of conversion from water into steam, is 1160 − 212 = 948. He considers the Density of the steam of boiling water at 212. to be 1800. less than that of water.

Thursday August 4th 1785

In the road between Wednesbury and Bilson, going from Birmingham to Woolverhampton, we saw a coal pit hot and smoking, which not many months before, we were told, had been burning. The Steam had the appearance of coming from the Top, but really came from the bottom, and continued transparent all the way up, being condensed, and so becoming visible, when it came near the top in contact with the cold air. Several other pits in that neighbourhood are in that same state, some of them throwing up fire or flame, which is visible at night many feet above the Top. The road here about is mended with a red substance like potsherds, being the clayey matter about the pits, burned by this internal heat.

Between Wolverhampton and Shifnal, great part of the way we saw lying by the side of the road for the purpose of mending it, both large and small stones of Granate, all rounded in the manner of pebbles, many of them seemed much to resemble the Granate we had found on Malvern hills. The red soil continued with little interruption all the way to the wooden Bridge over the Severn.

The place where Mr Wilkinson's Iron works are carried on, near Broseby, is called Willey.——This Iron is all made with coak, from ore and coal found on the spot. The ore is of 2 kinds, called *pinny* and *flinty* ore: the latter harder than the former; but both he considers as a clay Stone, impregnated with Iron. Both ore and coal lie in clay, the former usually over the latter, when both are found together; but in some places, one in particular he mentioned, they interchange, the Seam of Iron ore running under the coal. Mr Wilkinson described to us a fault in the Seam of coal near his Works, which he called a hog's back. It means an interruption in the Stratum of coal dipping according to the usual direction; but when the Stratum is found again, that part dipping in a contrary direction: so that if a Line were drawn from the bed on each side of the fault according to the angle of inclination, the two lines would meet one another in an angle in the middle, raised above the surface of the earth. He said the regular and usual dipping of the Strata was a little to the Southward of East. He spoke of a place in Staffordshire, where after a fault the order of the seams of coal was inverted, those lying undermost, which before the fault were uppermost. The Limestone he uses as a flux, is of a lightish colour, inclining to yellow, and has very distinct marks of extraneous fossils, shells, coralls, &c.

He has two Steam engines, one for the blast, and the other to raise Water to work the hammers. The cylinder for forcing the Air into the reservoirs, is 72 inches in diameter. The piston in it, is worked up and down by the Beam of the Steam engine, and I believe he said makes a stroke of 7½ feet. From this cylinder pipes go into reservoirs made of brick, open at bottom, and surrounded with a column of water,[3] by the pressure of which it is meant, that the effort of the air to escape, should be uniform. Out of these Reservoirs, for there are two of them communicating together, pipes are led which convey the blast wherever it is wanted. The Air was let into the furnace in the same manner as at Merthyr. The furnace is 25. feet high, it was all flaming on top. He said the height of a Furnace should be determined by the nature of the ore, particularly its more easy or more difficult fusion. He is of opinion that the ore melts and runs down a considerable way above the point where the blast comes into the furnace. He said the blast might be suspended 16 or 18. hours, without spoiling the Furnace.

In the Finery he had a current of water under the bottom plate of the hearth, but conceived it to be of no other use than to keep that plate from melting. The pig is here melted down with coak into a lump or loop, which is beat under the hammer into flat plates about an inch thick, with irregular edges. These are afterwards broken with a hammer into irregular pieces, and then picked, to separate any metal which may not have been properly refined. These pieces, with the small bits broken of[f] from them in the hammering, the latter previously washed, are put into an earthen pot, and kept in a fire, 3, or 4, or 5 hours, in the course of which Time they run together into a mass, tho' without coming into fusion, it is rather as a sort of welding, The pots soon break, and some fluid matter runs off, which is one sort of finery cinders. When this lump is taken out of the fire in the pot, it is again hammered, and another quantity of similar finery cinder is separated from it. The small stuff or bits put into the pots with the large pieces, are what is broke off in hammering out the flat pieces or cakes; for the original lump does not bear the hammer like iron made of charcoal, but breaks and crumbles very much, this forming the small stuff in question. Mr Wilkinson employs 31. or 32 C^{t} w^{t} of coak pigs, to make one Ton of Iron. The first refining reduces the quantity to ab^{t} 25. C^{t} w^{t} and the second that from the pots, reduces the 25. C^{t} w^{t} to 20. The pigs do not come into Nature, as the Term is, till they have discharged a large quantity of finery cinders: after which the mass puffs up and becomes spongy, like Fermenting, and has then assumed the Nature of tough Iron, tho' still as yet but imperfectly.

The forged Iron Mr Wilkinson here makes, is chiefly used for nails; it is not so malleable and tough as the Charcoal Iron made in England; but that Mr Wilkinson imputes to the nature of the ore, and says, that this Coak Iron is better than it could be made by Charcoal with the same ore; the reason he assigns is that more of the original pig is worked away.

Friday August 5th 1785

We went to Mr Rathbone's Works, at Coalbrooke Dale, where the Iron Bridge was cast. They have two Furnaces, one a little below the other. The upper Furnace,

[3] The pressure of this, he said, was equivalent to 2. lb on each square inch.

the smallest, is 25, or 26. feet high, and 12 feet wide at the Top of the boshes. It flamed at Top like the others we had seen. The blast was given by two cylinders, without any regulator, only a comparatively small box of Iron, into which the pipes from the two cylinders united. Hence a pipe conveyed the blast directly to the Furnace, and was furnished with flexible leather as usual.—All the motions were produced by water, raised by a Steam engine from the depth of 101. feet. The Engine was of Watt's construction: but in the little dish which sits upon the Collar in which the shaft of the Piston works, greatest part of the Liquid was water; but with some thickness of oil swimming upon the Top: and the Agent told us, that a small quantity of water getting down into the Cylinder, did not injure the motion; but rather served to keep the piston tight and in good Order. W[h]en the Engine began to lift, it made a short pause, in the manner of a common Engine, and it worked remarkably slow 12. or 9. times in a minute. The works extend near a quarter of a Mile, and the same Water serves all, seeming to have about five successive falls: then from a dam at the end of the works, it is carried by a level to a pit under the Engine just 101. feet deep, whence it is raised by the Engine into the upper reservoir, and so again makes the round of the Works. At the Furnace they employ three sorts of ore, called *white flats*, *pinny*, and *ball*. These lie intermixed with the coal in Strata of clay, and are found from 2 to 3 miles up the country. They have lately dug from near the Iron Bridge, a harder sort of ore, which lies under the coal and all the other measures, and is supposed to be richer than the former ones. They had desisted from shingling at the forges for want of water, but wer[e] trying an experiment of working up the broken pots, kettles and other pieces of cast-Iron, by merely heating them with coal in the shingling Furnace without pots. The Iron so made, did not prove very good, but was manufacturing into stampers, for which purpose, they said, it was good enough. There was a Mill for boring Cylinders. A square Iron box on a long shaft of the same Metal, received an Iron wheel of near the diameter of the Cylinder to be cut. This wheel had four pair of notches, each two near together, but each pair at intervals of a quarter of a circle from each other: Into one of each pair were fastened the 4. cutters employed, and the other of each pair received a piece of wood, of the same width as the cutters, so as to fit exact in the Cylinder, but rather behind the fore edge of the cutter, that it might only touch parts which had been cut. The use of these pieces of wood was, to make the Work go smooth, and prevent jarring.

About half a Mile below the Iron Bridge, is a Manufacture of Coal Tar, belonging to the same Company. The Coal to be distilled, is put into chambers, built of fire brick, in the form of plates, and a fire is made under them, which drives the vapour through communication pipes of brick into a common gallery, or chimney, nearly horizontal, which is conveyed round, and terminates in a Reservoir for the Tar or oil, over which the water swims. As the Tar accumulates, the Water is carried off, till the Reservoir is filled with Tar, when it is raised by pump into another Reservoir, whence it is conveyed to the Alembic, where ⅓ of it is distilled off, and what remains is thought Tar of the proper consistence. The matter distilled off is very pungent, and has been obtained of the colour of Mountain Wine, but they could not make it turn to any account; it was tried for Varnish without success. The water smells and tastes strongly of *hepar sulphuris*, and has besides a peculiar pungency, which seemed to come from the oil; they had never ascertained the quantity of this water produced, and there was no appearance of its containing any volatile Alkali.

They also shewed us some Pitch, exactly like Lord Dundonald's. They said that they could not afford this Tar much cheaper than the foreign, on account of the great expence of casks for packing it up. It is used by the craft upon the Severn; many of the bargemen prefer it to every other kind, but others complain, that it being of a very penetrating Nature, is too soon absorbed by the Wood and Oakum, and exposes the vessels to grow leaky. They do not suppose any Tar comes over from the coal, till the smoke is of a thick yellow colour, the first smoke is chiefly waters. The communication pipes or tunnels between the chambers containing the coal, and the great tunnel for the vapour to be condensed, are furnished with a damper, that is a plate which can be occasionally shut down, when they have occasion to open the door of the chamber. If this door and the damper were up together, the distilling coal would be likely to take fire, and probably blow up the chamber. No elastic fluid was observed to escape; nor have they any apparatus to let it out, if any such should be generated. The Strata where they could be traced dipped down the River, inclining a little towards it. The hill a little above the Iron Bridge consisted almost entirely of Limestone, except a little towards the top, which was what they called dye earth, in which they observe no Iron stone or coal are ever found. This seems to be a clay laid on irregularly, so that the layers could scarcely be traced. In many places the separations between the layers of the Limestone were filled up with clay.

The part of the hill which had slid down into the Severn some years ago, appeared to consist of soft clay, together with some masses which had a stony appearance, but which softened in water, and were found to be clay like the former. That clay and the dye earth, were just of the colour of the Severn mud.

We returned to Mr Wilkinsons Forges at Willey by Broseley, to see the operation of shingling. The pig of Iron is in the finery converted into tough Iron, after which it is stamped in a sort of Iron pan, and then broken. These broken pieces with the small stuff which comes off in stamping, is put into the pots of yellow burnt clay, perhaps 9 or 10. inches in diameter, nearly as much in height, and cylindrical. These are placed in a Wind furnace, made to hold 18. of them at a time in 4. rows: that is 6. in the row behind, 5. in the next row forward, 4. in the next, and 3 in the nearest. Large lumps of coal are put to the side of each row, to prevent the flame from cutting off the Top of the pot, which it is apt to do. The fuel employed is raw coal. The fire in the Wind furnace was chiefly made at the side upon a grate; but there was a large space sideways with a solid floor, on which the pots stood, with a certain quantity of bright coal under them, and the funnel or chimney was continued up, still further sideways from behind the whole. They find a certain proportion of the small stuff absolutely necessary to the success of the operation. They call it cinders, and conceive that it not only assists in welding the different pieces of Iron together, but likewise enters into the Composition, and feeding the Iron. Their idea is, that the more cinders can be brought to unite with the Iron, of the better quality it becomes. When the pots are put into the fire, which they call charging, some of them crack very soon, others not till later, but all are cracked more or less during the operation, not so however but they adhere together, keeping in the Iron in a lump, till the workman at the end of the operation, pulls them off from the lump with a hooked Iron, when they separate very easily, and are brought off in several pieces.

Toward the end of the operation, a hole toward the lower part of the solid floor, of the furnace, on which the pots stand, is opened from the cinders and coal which before stopped it up, and then a quantity of liquid Scoriae is seen running out for a considerable time, composed, as we were informed, partly of vitrified parts of the pot and coal, but partly also, perhaps principally, of matter which had run off from the Iron through the cracks in the pots. To facilitate the discharge of these Scoriae, the floor is made inclining a little from all parts toward the abovementioned hole. When the Furnace is to be drawn, the Term for taking out the pots, which is from 3 to 4. hours after it was charged, they begin by examining the appearance of the pots, and judge that they are fit, by the whole appearing of an uniform white heat. They then pull off the pot in pieces, as before mentioned, leaving the Iron it contained in the form of a cylindrical lump, which they call a ball. At the Chafery then an other workman makes the end of an Iron bar white hot, and when the first brings the white hot ball under the hammer, the latter lays the hot end of the Iron bar upon it, which is welded to it by a few strokes of the hammer, and then serves as a handle to turn and manage it. The hammer weighs about 6¼ C^t w^t. Its first effect is to strike off sparks, some of which fly to a great distance, and a few have the brilliant appearance of steel dust in fire works. There comes besides a white flame from different parts of the mass, and at times a different flame from certain spots, of a light bluish colour, like that from burning Sulphur. As the operation proceeded, some cinder ran down the Anvil, but in small quantity, and as the ball cooled, it assumed under the hammer an oblong angular form, 2 feet, or 18. inches in length, and perhaps 3 or 4. in thickness, and is then called a half bloom; which is afterwards drawn out in the Chafery, into a bar with *Maggot* ends, answering to the Blooms of Glamorganshire, here called an ancony. Some of them however appear to be drawn out in the first instance into uniform bars. Formerly the Iron was made without this operation of shingling; but then it was of much worse quality. The shingling properly means, beating the ball with a hammer; that of putting the iron into the pots to be heated, they call potting it.

By the Steam Engine at Rathbone's work, Coalbrook Dale, we saw a basket lying with several balls of fresh horse-dung: on enquiring the use, we were informed that they put it into the boiler of the Engine, when they observed any thing to leak out, and that it worked into the leak and stopped it.

We were told here and in other places, that they distinguished the goodness of the Iron produced by the Scoriae which come from the Furnace: those that are black rather spongy, and not very perfectly vitrified, indicate the best and toughest Iron. When the Scoriae comes out beautifully veined with white and blue, &c. such as we were shewn some at Willey Forge, the Iron is extremely bad. The general quality of the Iron produced at all these Forges, we were told was cold-short[4]; it is the quality of the ore, or mine to produce such, and they say positively that the same

[4] [Cold-short iron is brittle when cold.]

mine worked with charcoal, would make Iron still worse in this respect; but by the manner in which it crumbles in the operation of stamping, to produce the small stuff for the shingling pots, it appears clearly to be also in some degree at least, red-short.[5] Even under the shingling hammer, a great deal scales off, so that though it seems to be very strong in that state, yet it is not perfectly divested of its red-short quality. The breaking of the stamped Iron is effected by means of a strong iron plate, longer than broad, provided with ribs at proper intervals, set breadthwise: the stamped cake being laid upon [---] ribs, the workman strikes it with a sledge hammer on the part answering to the hollow, or interval between two ribs, when it commonly breaks off short, but sometimes only bends, and then must be laid upon the ribs with the convex side uppermost, when it always gives way to the stroke of the sledge.

About Bridgenorth on both sides of the Severn, we saw principally red rock, with a large proportion of sand: and it seemed to lie in all directions, or consisted of no regular Strata, for the lines or cracks sometimes ran one way, and sometimes another, in pieces that were close together. The Terras-walk commands a remarkable scene, from the singular appearance of these rocks all around, but especially on the opposite side of the River Severn, the Eastern, and from the fine view of the River underneath. The remains of the old Castle, battered by Oliver Cromwell, exhibit a remarkable instance of a leaning Tower, or ruin, which produces a fine effect. The Town is supplied with water from the Severn by means of wheel-work, turned by the current of the River, without any other fall than what is produced by stopping the current in that part and the height of great part of the Town above the River, is very considerable. We were told they had no other water.

Saturday August 6th 1785

In returning from Woolverhampton to Birmingham, about half way between Bilson and Wednesbury, we turned out of the way to the right 3 or 4 hundred yards to Bradley Furnace Forges, lately constructed by Mr Wilkinson. The road went close by the coal pit heath which has been burning many years. From some places close by the road, a strong flame was now issuing, and the earth seen through the crevices and apertures in many places was red hot or even white hot. All about the places actually burning, lay the cinders of old conflagrations, forming that reddish matter with which the road here is mended; and in many parts the ground had given way, forming hollows of greater or smaller diameter and more or less deep, on the edges or brink of which the ground was broken into long cracks of different width, some apparently very recent. The places from which the flame was now issuing, were not old pits, but apertures or cracks formed by the sinking down of the ground. About 3 years ago, the fire seemed approaching the Works at Bradley, and the ground became hot in many places; but it has been stopped by throwing upon the ground, at different times, considerable quantities of Water.

The Steam Engine at Bradley was upon Watt's construction, and made the blast by working the piston in a single cylinder, whence the Air was conveyed to a

[5] [Red-short iron is brittle when hot.]

Water-regulator, like that at Willey, but less. The Furnace here is 45. feet high, and 12. feet wide, at the Top of the boshes: flame issued from the Top of it. They use two kinds of ore, or mine, found on the spot among the coal, some above, some under, and some mixed with it. They do not distinguish the two sorts of Mine by different names, but use some of both, and find them both cold-short. They seem like the others a clay Stone impregnated with Iron, some pieces richly, but others poorly. They do not here cast any large things immediately from the Furnace, but carry the pigs to a foundry where there are several hearths, with Machinery of cranes, &c. They bring the streams from these several hearths to join in one, when the piece to be cast requires so great a quantity of Metal. These hearths were all wind Furnaces, with the fire made at one side.

At Bradley we had a further account of the Finery hearth. The bottom plate from 18 inches to two feet square, serves as the top to a cistern of water beneath the hearth, supplied with a small stream of water, which runs in and out through the cistern. The surface of the water is kept 3 or 4. inches below the bottom of the plate, which is interposed between it and the Metal in the hearth; that is, the bottom plate of the hearth: and its use is to keep that place cool, as they suppose. We observed that the shingling Furnace had a small pit at the hole through which the Scoriae are discharged, toward which pit the floor supporting the pots inclines on all sides. The Steam Engine here worked with the rod of the piston quite dry, and made no pause at beginning to lift. We saw a fly-wheel of Iron lately cast, and observed that the circumference was not continued, though the whole was cast at once, and the center and spokes were one: the separation in the circumference was in the middle between each spoke, and was left that it might not give way of itself in cooling, by the unequal contraction of the spokes and circumference. At the edges of the separation on each side were holes, for joining them together by means of an Iron plate to which each edge was to be fastened: thus the whole was stronger that it would have been likely to be if a single piece.

The rolling and slitting Mill was worked with a Steam Engine, which communicated a circular Motion by means of a crank. There were two wheels of cast Iron turning on separate axes, very near each other: each wheel having round its circumference a number of large holes, into which entered a large Iron pin, which passed through the hole or loop at the end of the rod. Thus the rod moved both these wheels round, keeping between them, & could thereby communicate movement to two different sets of Machinery. Each wheel turned the pinion of a large fly-wheel, by means of which the proper action of the crank was answered. The fly was not loaded on one side. Each of these wheels was made to turn a rolling and slitting Mill; but one set only was finished. The two cylinders of the Rolling Mill were inclined to each other, and adjusted to four gradations, so that the bar was rolled 4 times with the same heating. After being passed through the widest part, it was received by a workman on the opposite side, and conveyed over again to the first workman, who then passed it through the second gradation, and so on for the 3^d^ and 4th gradation, by which time it was reduced to a bar of considerable length, about 1/2 an inch thick, from a   not more than a foot or 9 inches long, and perhaps 2. inches square. To make these, they hammer out a half bloom to the proper thickness, and then cut in into such lengths, which they call, As soon as it had passed through the rollers, it went to the slitting Mill without any other heating, and there cut into rods for nails.

Mr. Watt has contrived a Furnace to burn the Smoke, which he means to apply to the Steam Engine. The draught of Air is conducted backward.

Chapter 12
Cavendish and Grey Family Trees

This study of Cavendish's personality necessarily takes into account his social position. Both Cavendish's paternal and maternal grandparents were dukes and duchesses. The family trees in this chapter begin with the grandparents and their siblings, followed by his parents' generation, then his own, and finally the one after his. By definition he was related to them all, though in most cases the relationship was not close, and he probably met only a small fraction of them, but they were a part of his world, and he would have known about them. Although he chose to associate mainly with persons drawn from another society, that of scientific colleagues, he did not abandon the one he was given, nor could he have. We recognize the Cavendish in the man by his style of living, his property, and his will, which left his vast fortune to Cavendishes close to the center of the clan. The dukedom would eventually pass to descendants of his principal heir Lord George Augustus Henry Cavendish, who appears in the last column of the Cavendish family tree. Unless readers of this book are specialists in this period, the names are unfamiliar to them, but they cannot miss the titles that go with them. Dukes and earls are commonplace, as is often great wealth. The family trees reflect the obligation of the head of the family to reproduce himself, which took the form of large families to insure a male heir. They also show the fragility of life at the time, even for the most privileged. Five children of the second duke of Devonshire died before their parents. Five children of the first duke, his father, not shown, died before their parents. All five sons of the Duke of Kent died before him, extinguishing the Kent line. Marriages normally, though not always, took place between persons of more or less the same social standing. Not shown on the family charts are illegitimate children. The first duke of Devonshire had several, born of unions with an actress and an aspiring

R. McCormmach, *The Personality of Henry Cavendish - A Great Scientist with Extraordinary Peculiarities*, Archimedes 36, DOI 10.1007/978-3-319-02438-7_12,

thespian. The family trees show that many members of the extended Cavendish family did not marry. As a wealthy, single aristocrat, Cavendish was not out of the ordinary. The two family trees in the following pages are nearly complete and, within the limits of their sources, accurate.[1]

[1] The main sources used in developing the family trees are the following. Printed books such as Burke's, Cokayne's, and Debrett's peerages and the *Dictionary of National Biography*. Online genealogical resources such as "The Peerage." Wills from the main probate court in England, the Prerogative Court of Canterbury. Memorials: for the Cavendish family, a record of dates of death and ages of members of the family interred by custom in All Saints Church in Derby; for the Grey family, photographs taken inside the de Grey Mausoleum in Flitton, Bedfordshire. *Burke's Peerage and Baronetage*, 106th ed., ed. C. Mosley, 2 vols. (Crans, Switzerland: Burke's Peerage, 1999). George Edward Cokayne, *The Complete Peerage of England, Scotland, Ireland, Great Britain and the United Kingdom: Extant, Extinct, or Dormant*, vols. 1–3 (Gloucester: A Sutton, 1982). *Debrett's Peerage and Baronetage*, ed. C. Kidd (Richmond, Surrey: Debrett's Peerage and Baronetage, 2008). John Charles Cox and William Henry St. John Hope, *The Chronicle of Collegiate Church or Free Chapel of All Saints, Derby* (London, 1881). "The Peerage: A Genealogical Survey of the Peerage of Britain as Well as the Royal Families of Europe," compiled by D. Lundy http://www.thepeerage.com. English Heritage, "The de Grey Mauseleum," http://www.Bedfordshire.gov.uk/CommunityandLiving/ArchivesAndRecordOffice/CommunityArchives/Flitton/ThedeGreyMausoleum.aspx. The Cavendish and Grey family trees in this book are improvements of those in Jungnickel and McCormmach, *Cavendish* (1996) and (1999).

Cavendish

- Elizabeth Cavendish (b.abt 1670–1741) m.bef 1710 Sir John Wentworth of Broadsworth (1673–1720)
  - Sir Butler Cavendish Wentworth of North Emsall (1710–1741) m. Bridget Milbanke (d.1774)
  - Two more sons
- William Cavendish 2nd Duke of Devonshire (1671–1729) m.1688 Rachel Russell (1674–1725)
  - William Cavendish (1692–1697)
  - Mary Cavendish (1694–1719 age 25)
  - Rachel Cavendish (1697–1780) m.1723 Sir William Morgan of Tredegar (1701–1731)
  - William Cavendish 3rd Duke of Devonshire (1698–1755) m.1718 Catherine Hoskins (1698–1777 age 79)
  - Elizabeth Cavendish (1699–1747) m.1723 Sir Thomas Lowther of Marske (1699–1745)
  - James Cavendish (d.1741 age 41)
  - Charles Cavendish (1704–1783) m.1729 Anne de Grey (1706–1733)
  - Catherine Cavendish (d.1715 age 9)
  - John Cavendish (d.1720 age 12)
  - Anne Cavendish (1710–1780 age 70)
  - Diana Cavendish (1712–1722)
- Henry Cavendish (1673–1700 age 27) m.1696 Rhoda Cartwright (d.1730)
  - Mary Cavendish (b.abt 1700–1778) m. John Fane 7th Earl of Westmorland (1685–1762)
- James Cavendish (1678–1751) m. Ann Yale (d.1734)
  - William Cavendish (b.abt 1710-1751 age 41) m.1731 Barbara Chandler (d.1786)
  - Elizabeth Cavendish (b.abt 1712–1779 age 67) m.1732 Richard Chandler–Cavendish Son of Bishop of Durham (d.1769)

- William Morgan of Tredegar (1725–1763)
- Rachel Morgan (1726–1739)
- Edward Morgan (b.abt 1727–1743)
- Elizabeth Morgan (1729–1787) m.1767 William Jones of Llanarthy

- Caroline Cavendish (1719–1760 age 41) m.1793 William Ponsonby 2nd Earl of Bessborough (1704–1793)
  - Catherine Ponsonby (1742–1789) m.1763 Aubrey Beauclerk, 5th Duke of St Albans (1740–802)
  - Charlotte Ponsonby (b.abt 1750–1822) m.1770 William Fitzwilliam, 4th Earl Fitzwilliam (1748–1833)
  - Frederick Ponsonby, 3rd Earl of Bessborough (1758–1844) m.1780 Henrietta Frances Spencer [dau of John, 1st Earl of Spencer] (1761–1821)
  - Four brothers and four sisters all died young
- William Cavendish 4th Duke of Devonshire (1720–1764) m.1748 Charlotte Elizabeth Boyle (1731–1754)
  - William Cavendish 5th Duke of Devonshire (1748–1811) m.1:1774 Georgiana Spencer (1757–1806) m.2:1809 Elizabeth Christine Hervey (1759–1824)
  - Dorothy Cavendish (1750–1794) m.1766 William Henry Cavendish-Bentinck, 3rd Duke of Portland (1738–1809)
  - Richard Cavendish (d.1781 age 30)
  - George Augustus Henry Cavendish (1754–1834) m.1782 Elizabeth Compton, Countess of Burlington (Dau of 7th Earl of Northampton) (1760–1835)
- George Augustus Cavendish (b.aft 1720–1794)
- Elizabeth Cavendish (1723–1796) m.1743 John Ponsonby (1713–1789)
  - William Brabazon Ponsonby, 1st Baron Ponsonby (1744–1806) m.1769 Louisa Molesworth (1749–1824)
  - Catherine Ponsonby (b.abt 1746)
  - George Ponsonby (1755–1817) m.1781 Mary Butler (b.abt 1755–1826)
  - Frances Ponsonby (1757–1827) m.1774 Cornelius O'Callaghan, 1st Baron Lismore of Shanbally (1741–1797)
  - Mary Charlotte Ponsonby (b.abt 1760–1781) m.1780 Dennis Bowes Daley (1745–1821)
- Rachel Cavendish (1727–1805) m.1748 Horatio Walpole Earl of Orford (1723–1809)
  - Horatio Walpole, Earl of Orford (1752-1822) m.1:1781 Sophia Churchill (1756–1797) m.2: 1806 Catherine Tunstall (b.abt 1758–1807)
  - William Walpole (1756–1764)
  - Mary Walpole (b.abt 1757–1840) m.1777 Thomas Hussey of Galtrim (1749–1825)
  - George Walpole (1758–1835)
  - Robert Walpole (1760–1835)
- Frederick Cavendish (1729–1803)
- John Cavendish (1732–1796 age 64)

- Sir William Lowther (1727–1756)
- Elizabeth Lowther (1728–1779)

- **Henry Cavendish** (1731–1810 age 78)
- Frederick Cavendish (b.1733–1812)

De Grey

- Henry de Grey
  Duke of Kent
  (1671–1740 age 69)
  m.1: 1695
  Jemima Crewe
  (1675–1728)
  m.2: 1729
  Sophia Bentinck
  (dau of Earl of Portland)
  (b.abt 1701–1748 age 47)
  - (m.1) Anthony de Grey
    Earl of Harold
    (1696–1723)
    m. 1718
    Mary Tufton
    (dau of Earl of Thanet)
    (1701–1785)
  - Henry de Grey
    (1697–1717)
  - Amabell de Grey
    (1698–1727)
    m.1718
    John Campbell
    3rd Earl of Breadalbane
    Lord Glenorchy
    (1696–1782)
  - Jemima de Grey
    (1700–1731)
    m.1724
    John, 1st Earl of Ashburnham
    (1687–1737)
  - Lucas de Grey
    (b.abt 1702–1704)
  - Henrietta de Grey
    (1703–1717)
  - Anne de Grey
    (1706–1733)
    m.1729
    Charles Cavendish
    (1704–1783)
  - Jane de Grey
    (b.abt 1708)
  - George de Grey
    (1714–1717)
  - Carolina de Grey
    (1716–1716)
  - Mary de Grey
    (1719–1761)
    m.1743
    David Gregory
    (1696–1767)
  - (m.2) Anne Sophia de Grey
    (1723–1780)
    m.1748
    John Egerton
    Bishop of Durham
    (1721–1787)
  - George de Grey
    Earl of Harold
    (1732–1733)
- Amabell Grey

Henry Campbell
(1720–1726)

Jemima Campbell
Marchioness Grey
Baroness Lucas of Crudwell
(1722–1797)
m.1740
Philip Yorke
Viscount Royston, later 2nd
Earl of Hardwicke
(1720–1790)

- Amabel Yorke
  Baroness Lucas of Crudwell
  Later Countess de Grey of Wrest
  (1751–1833)
  m.1772
  Alexander Hume-Campbell, Lord Polwarth
  (1751–1781)
- Mary Jemima Yorke
  (1757–1830)
  m.1780
  Thomas Robinson
  2nd Baron Grantham
  (1738–1786)

John, 2nd Earl of Ashburnham
(1724–1812)
m.1756
Elizabeth Crawley
(1727–1781)

- George Ashburnham
  Viscount St. Asaph
  (1758–1758)
- Henrietta Theodosia Ashburnham
  (1759–1847)
- George Ashburnham
  3rd Earl of Ashburnham
  (1760–1830)
  m.1:1784
  Sophia Thynne
  (dau of 1st Marquess of Bath)
  (1763–1791)
  m.2: Charlotte Percy
  (dau of Earl of Beverley)
  (1776–1862)
- Jemima Elizabeth Asbburnham
  (1762–1786)
  m.1785
  Sir James Graham, later 3rd Duke of Montrose
  (1755–1836)
- Elizabeth Frances Ashburnham
  (1763–1854)
- Theodosia Maria Ashburnham
  (1765–1822)
  m.1788
  Robert Vyner, of Gautby
  (1765–1810)

Henry Cavendish
(1731–1810)

Frederick Cavendish
(1733–1812)

Jemima Gregory
(b.abt 1743–)

David Gregory
(b.abt 1748–)

Amelia Egerton
(1751–1809)
m.1771
Sir Abraham Hume
(1749–1838)

John William Egerton
7th Earl of Bridgewater
(1753–1823)
m.1783
Charlotte Catherine Anne Haynes
(1763–1849)

Francis Henry Egerton
8th Earl of Bridgewater
(1756–1829)

Chapter 13
Letters Between Brothers: Henry and Frederick Cavendish

Among Cavendish's business papers are eight short letters between the brothers, six by Frederick and two draft replies by Henry.[1] The time is 1806–1810; they are near the end of their lives. Henry shows concern for his brother's health. He sympathizes with his brother's difficulty with tax forms. Through their common banker, Robert Snow, he facilitates an act of charity by his brother. Cavendish's replies to his brother's requests are as close to personal letters as we have. For his part, Frederick asks his brother for financial advice, and he sends him scientific news he has come across. He reassures his brother that he is in good health and spirits, and he tells him about his daily activities to prove it. The bond of the brothers would seem to have two parts: a mutual interest in the other's well-being, natural to brothers; and a financial connection having to do with Frederick's funds.

Frederick writes his letters at home in the village Market Street, near St. Albans in Hertfordshire. He addresses them to "Honble Henry Cavendish at Clapham." He has gout, and his hand is shaky.

Frederick Cavendish to Henry Cavendish
28 October 1806

Dear Brother,
These people who collect the taxes on property, plague me prodigiously. I had at a former time given them an account of my property in the funds; and have at various times had some of these printed-notices sent to me either informing me it was wrong, or wanting further information, &c. and now have had the letter of the 9th Sept. 1806 with the printed notice (both of which I inclose) sent to me. [Here Frederick explains what confuses him.] … But the manner in which they have put it, added to the former notices I have had on this subject, causes me to send these papers to you, thinking that if [you] will apply to Mr Snow, you can explain this business, and advise me how to act. [Y]ou will be kind enough, to return these papers, as the surveyor will call for them. I should have written this letter to you

[1] The letters are in Devonshire Collections, Chatsworth, 86/comp. 1.

R. McCormmach, *The Personality of Henry Cavendish - A Great Scientist with Extraordinary Peculiarities*, Archimedes 36, DOI 10.1007/978-3-319-02438-7_13,

sooner, but that I had a small touch of the gout, in my right hand, which swelled prodigiously, which rendered me incapable of using it; it continued so for about a fortnight, but my health was remarkably good the whole-time, and I was in great spirits. It had no tendency the whole time to any other part. It is at present in some measure in the wrist. I hope you are in good health.

I am,
Your affectionate brother,
Frederick Cavendish

Henry Cavendish to Frederick Cavendish
Draft letter, no date

Dr B
I have given your 2 inclosed papers to Mr Sn & have desired him to send them back to you with the account of your stock. As far as I can judge by the printed paper you made the mistake of setting down the half yearly dividend instead of the stock. It appears from the paper that this relates to the return made in 1804 so that it is very likely that you may have made the same mistake in 1805 & therefore you would do well to see whether this is the case or not & to rectify it if it is. The printed forms sent both by the commissioners of income & assessed taxes are intricate & not clearly expressed so that I think you would do well to get the assessor or some other person to assist you in making the return.

Frederick Cavendish to Henry Cavendish
10 September 1809

Dear Brother,
I thank you very kindly for your inquiry after my health, which has not been better for many years. I have been as usual visiting my friends or riding out most days. It was but yesterday that I took a ride in the chariot to beyond Dunstable and back. I cannot well conceive how any one can give you intelligence that I am very ill. I hope you are very well. We have had continual rains. The roads towards Dunstable are difficult to travel in. In my newspaper some weeks past, there was an article that Dr Herschell had predicted that there would be, more rain fall in the latter end of this summer than has happen'd at the same period for these hundred years past.

Henry Cavendish to Frederick Cavendish
Draft letter, no date

D. B.
I thank you for your letter & am very glad to find you so well. The information I received & which alarmed me was contained in a letter from a person who it should seem knows something of you <a lady who it appears has received assistance from you> but without mentioning the authority for it.

I saw the story you mention of Dr H. in the paper but I do not imagine he can have said any such thing as I trust he has too much sense to make predictions of the weather.

Frederick Cavendish to Henry Cavendish
18 December 1809

As I believe you attend a good deal to the observation of the barometer, I send the account of this morning; when the glass was the lowest I ever observ'd in my life. At 11:45 the mercury was at the lowest degree marked on the scale or 1° above 0 twenty-eight as it is figured on the barometer. The thermometer was at the above observation 37. Cold foggy weather. I hope you are in good health. The rains are continual.

I am dear brother,
Yours affectionately,
Frederick Cavendish

Frederick Cavendish to Henry Cavendish
5 February 1810

Dear Brother,
There is a case at present in which I would ask your assistance. A daughter of Mrs Wakefield is lately married. She is married to an attorney-at-law in London. He is a young man of good connexions, and has been placed by his friends in partnership with a man in that business in a reputable line. But not having much money, in hand, is obliged to live as close as he can. The present call on him is this. He has been fitting up an apartment for him and has paid some money towards it. But the upholsterer's bill tho' partly paid, is still due, and one hundred and fifty, pounds must be raised immediately which will make him happy. If you will enable me to satisfy this man, I shall be much obliged to you.

I am dear brother,
Your affectionate brother, Frederick C.

Frederick Cavendish to Henry Cavendish
9 February 1810

Dear Brother,
The bills of the last year I have not had money enough to pay off; I shall be much obliged to you if you will place in Mr Snow's hands, two hundred pounds to my account.

Frederick Cavendish to Henry Cavendish
12 February 1810

Dear Brother,
I am greatly obliged to you for your letter, and draft for one hundred and fifty pounds, which I am very confident will do a great deal of good.

[Frederick received the £150 for the young man, who wrote to Henry Cavendish: "Mr. Frederick Cavendish has had the great goodness to present me with that draft for 150 pounds and has informed me how much I am beholden to you for it."[2] Twelve days after Frederick's last letter, Henry Cavendish died. Frederick died two years later.]

[2] George Mariotte to Henry Cavendish, 15 February 1810, Devonshire Collections, Chatsworth, box 86/comp. 2.

Chapter 14
Cavendish's Houses: Clapham Common and Bedford Square

Cavendish's houses come up a number of times in this book. More than most men Cavendish spent his time at home, where his instruments and books were and where his work was. His houses were the private places where he applied himself to natural philosophy, and they were the public face of the wealthy aristocrat. The following description of his houses helps us picture his life.

After his father died in 1783, Cavendish came before the world as a wealthy aristocrat. He was 51, and until then he had lived inconspicuously in a mew, a stables converted into living quarters, behind his father's house on Great Marlborough Street in Westminster, London.[1] He evidently used his father's house as his townhouse for a time before leasing it.[2] In 1782, the year before his father's death, he took a terraced house in Hampstead, which temporarily served as his main residence and country house. Here he carried out experiments on heat, took earth magnetic readings, and generally continued his every-day scientific activities. He remained on the Hampstead rate books for about 4 years.[3] While

[1] Thomas Thomson wrote that Cavendish's "apartments were a set of stables, fitted out for his accommodation." Thomson, *History of Chemistry* 1: 159. Converted stables were not uncommon in London at the time. Thomson was only nine when Cavendish moved away from his father's house, and so he would not have seen Cavendish's living quarters but only heard about them. We have an account of what Thomson probably heard about. Behind Charles Cavendish's house was a garden, at the end of which was an apartment with a passageway to the house. The apartment was described in 1725 as "beautiful" and "newly built," adjoining the coach houses and stabling, with its own entrance and street number, its own plumbing, an underground kitchen, and four rooms on the single floor above. A building at the end of the garden is shown on Richard Horwood, *Horwood's Plan of London … 1792–99* (London: London Topographical Society, 1966) (Fig. 3.6).

[2] "Henry Cavendish to M[r] Joshua Brooks. Counterpart Lease of a Messuage or Tenement with the Appurts. No. in Marlborough Street in the Parish of St James Westminster County Middlesex," 1788, Devonshire Collections, L/38/35. London County Council, *Survey of London*, vol. 31: *The Parish of St. James Westminster*, pt. 2, *North of Piccadilly*, general ed. F. H. W. Shepherd (London: Athlone Press, 1963), 256.

[3] Cavendish's first appearance on the rate books was on 3 January 1782, his last on 19 September 1785. "Hampstead Vestry. Poor Rate," Holborn Public Library, London.

R. McCormmach, *The Personality of Henry Cavendish - A Great Scientist with Extraordinary Peculiarities*, Archimedes 36, DOI 10.1007/978-3-319-02438-7_14,

living there, he bought two houses, one in the country and one in the city, which are described below.

Clapham Common

In his scientific calling, Cavendish followed his father, and as an aristocrat who owned houses he again followed his father. When Lord Charles Cavendish married in 1729, he combined money that was due him from his mother's estate with his wife's portion to buy an estate north of London, about half the distance of Cambridge. Three manors came with it, one of which, Putteridge, Charles made his family home, renting out the other two. He also had a substantial residence on Grosvenor Street off Grosvenor Square, a fashionable location in Westminster.[4] Charles's marriage was short, and ended sadly. In 1733, his wife, Anne, died, and he did not remarry. If his wife had lived, we might expect Charles to have continued the familiar living arrangements of a well-to-do gentleman, owning two homes, one in the city and one in the country. Instead, in 1738, he sold Putteridge and the rest of his country estate, and the same year he bought the townhouse on Great Marlborough Street. So far as we know he lived the rest of his long life without keeping up a second home in the country. His activities were in the city, and he may have felt that as a single man he had no need for a second home. His oldest son, Henry, also had two homes, his second coming late in life. Both of them held to patterns of living typical of men of their station and means.

In 1785, Cavendish bought a country house on Clapham Common, which was his main house to the end of his life (Fig. 3.12). Clapham at the time was a straggling village of handsome homes lying in the Clapham parish, about 4 miles distant from Westminster Bridge in London. When Cavendish arrived, the village had a population of nearly 2,500 and was growing. The best view of the village was from Clapham Common, a triangular piece of ground consisting of 202 acres with houses around its perimeter, lying partly in Clapham parish and partly in a neighboring parish (Fig. 3.10). Twenty-five years before Cavendish moved there, the Common was a morass and the roads were impassable. Chiefly through the efforts of the resident and justice of the peace Christopher Baldwin, the Common was drained and planted with a large number of native and exotic trees, giving it the look of a park. As evidence of the improvement, Daniel Lysons, in his *Environs of London* published in 1792, said that a few years earlier Baldwin had sold 14 acres of land near his house for £5,000, or £357 per acre.[5] The buyer, whose name

[4] Charles Cavendish appears on the poor rolls of Westminster Parish of St. Margaret's in 1728, paying £5.5.0 annually, which is what the duke of Kent, his father-in-law, who had a house in the parish, paid. Westminster Public Libraries, Westminster Collection, Accession no. 10, Document no. 343. Charles Cavendish's address in 1729–32 was 48 Grosvenor Street, a three-story brick terraced house, with four windows on each floor, and with touches of elegance: extensive paneling, marble chimney pieces, and a "Great Stair Case" in the entrance hall. British History Online, "Grosvenor Street South," http://www.british-history.ac.uk.

[5] Daniel Lysons, *The Environs of London: Being an Historical Account of the Towns, Villages, and Hamlets, within Twelve Miles of That Capital*, vol. 1: *County of Surrey* (London, 1792), 159–61. In the legal documents, the land Cavendish bought is said to be fifteen acres, not fourteen.

Lysons did not mention, was Cavendish. Property continued to increase in value. In 1810, the year Cavendish died, Robert Thornton sold his land for £500 per acre.[6] Clapham Common contained many country seats for well-to-do merchants, gentry, and members of Parliament. Cavendish and a woman referred to as "Lady" were the only aristocrats.[7]

We first hear of Cavendish's interest in Clapham Common from letters that passed between him and Baldwin beginning in the spring of 1784. Baldwin and Cavendish were both members of the Monday Club, which may be where Cavendish learned about the property on Clapham Common.[8] They met to discuss Baldwin's property, and at the close of a letter following up their meeting, Baldwin wrote, "I wish among your other learned & very curious investigations in our atmosphere, you would tell me when I may safely begin hay-making, since you are interested in the attempt."[9] All business, Cavendish paid no attention to pleasantries and flatteries of this sort.

Baldwin understood that Cavendish wanted to buy three contiguous parcels of land consisting of about 15 acres adjacent to his house for the purpose of building a house on it presumably for his own use. When he was first approached by Cavendish, he said he was not interested, and he suggested other owners who might sell him land. When difficulties arose with another property, Wright's farm, Cavendish's agent Thomas Hanscomb returned to Baldwin.[10]

Baldwin asked Cavendish to tell him what he would pay for the land. Cavendish offered £5,000. Baldwin said that it did not meet what he called the "market price." Two of the three parcels of land were choice. The remaining "*front* land" on the Common, Baldwin said, could not be value by the acre any more than could land in London or Westminster. To justify his valuation, he pointed to the beauty, the health, and the convenience of the parcels. He said that Cavendish should come look at them himself "before it's too late." The additional money Baldwin asked for should be no consideration for "a gentleman of your high rank & well-known great opulence." He calculated the value for the three parcels separately, the total coming to £5,650, which he said was £1,280 below the market value. He asked Cavendish

Historically, Clapham Common was common land for two parishes, Clapham and Battersea. Anon., "Clapham Common," http://en.wikipedia.org/wiki/Clapham_Common.

[6] T. C. Dale, "History of Clapham," in Clapham Antiquarian Society, *Clapham and the Clapham Sect* (Clapham: Clapham Antiquarian Society, 1927), 1–28, on 1.

[7] Map of Clapham Common, with names of all of the residents. "Perambulation of Clapham Common, in 1800. From C. Smith's 'Actual Survey on the Road from London to Brighthelmston,'" in J. H. Michael Burgess, *The Chronicles of Clapham [Clapham Common]. Being a Selection from the Reminiscences of Thomas Parsons. Sometime Member of the Clapham Antiquarian Society* (London: The Ramsden Press, 1929), opposite page 112. Reproduced by permission of the Bodleian Library.

[8] Verner W. Crane, "The Club of Honest Whigs: Friends of Science and Liberty," *William and Mary Quarterly* 23 (1966): 210–33, on 215.

[9] Christopher Baldwin to Henry Cavendish, 15 June 1784, Devonshire Collections, Chatsworth, 86/comp. 1.

[10] Christopher Baldwin to Henry Cavendish, 3 May 1784, ibid.

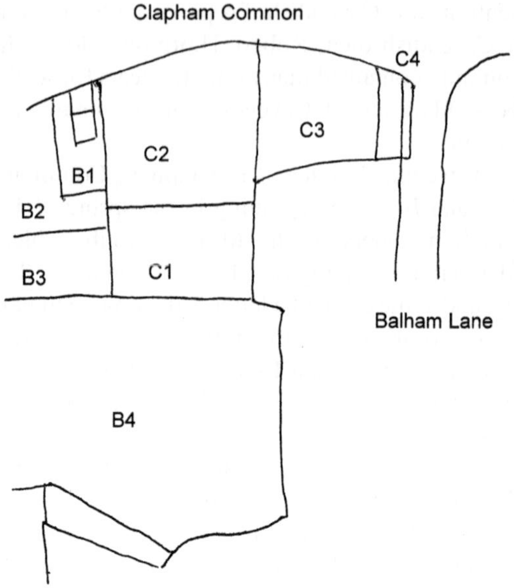

Fig. 14.1 Cavendish's Land on Clapham Common. C1, C2, and C3 are three parcels of land, totaling roughly fifteen acres, which Cavendish bought from Christopher Baldwin in 1784. C4 is a slip of land Cavendish bought from Baldwin later. B1 is Baldwin's house and garden. B2, B3, and B4 are fields owned by Baldwin. "Abstract of the Title of His Grace the Duke of Devonshire to an Estate at Clapham Common in the County of Surrey," 2 November 1784, Devonshire Collections, Chatsworth, 38/78

to come up with "a few hundreds more." He said that he knew Cavendish did not "like long letters," but he wrote then all the same. Cavendish wrote short letters and refused to bargain, and in due course Baldwin accepted his offer.[11] Mortgages on the 15 acres caused delays in closing until the winter of 1784. The purchase was absolute, the parcels belonging to Cavendish and his heirs and assigns forever (Fig. 14.1). Cavendish named Blagden and two other scientific colleagues as trustees to protect the inheritance.[12]

As it turned out Cavendish did not build a house for himself on the 15 acres. He entered into an agreement with the builders Hanscomb, Richard Fothergill, and Thomas Poynder, who would spend a minimum specified amount of money within

[11] Henry Cavendish to Christopher Baldwin, n.d. [After 3 May and 2 June 1784], drafts. Christopher Baldwin to Henry Cavendish, 2 and 7 June, 3 July 1784, ibid.

[12] "Abstract of the Title of His Grace the Duke of Devonshire to the Estate at Clapham Common and in the County of Surrey," 2 November 1784, Devonshire Collections, Chatsworth, 38/78. Blagden and Alexander Dalrymple, hydrographer of the British Admiralty, are named trustees in this document. In a second document with the same date, Alexander Aubert, governor of the London Assurance Company and astronomer, is named a third trustee.

a specified time to erect substantial houses with coach houses and stabling. When the buildings were completed, Cavendish would join with them in granting separate leases for the houses, with covenants prohibiting the building of brick kilns or using any buildings on the property as public houses or shops "for carrying on any noisome or offensive trade or business."[13] The land was intended for up-scale residences, insuring a proper tone. Cavendish arrived at Clapham Common as an eventual land developer and landlord.

It is unclear exactly what Cavendish's plans were for relocating on Clapham Common. As early as May 1784, at the time he began negotiating with Baldwin over the 15 acres, he considered buying an existing house, "Mr. Mount's house," which was probably later "Mrs Mount's house" on property adjacent to the house Cavendish bought the following year.[14] What is certain is that he had his mind fixed on moving to Clapham Common. The location was not smoky like London, an advantage when making astronomical observations, and it was healthier, as Baldwin claimed. His house would be a villa, with spacious grounds, affording quiet and privacy. The pastoral setting was peaceful, with fine trees, pastures, and ponds, again as Baldwin said. The location was advantageous: Baldwin explained that there were good roads, which enabled inhabitants of Clapham Common to travel to London, cross over London Bridge, do business in the city, and return from Westminster, which because of the bridges was no further away from home.[15]

In June 1785, Cavendish bought a house on another side of the Common from his 15 acres. Perhaps the house became available only after he bought the land from Baldwin. Perhaps its readiness appealed to him. By buying an existing house, he did not have to wait, and he avoided the aggravation of building; he could return to his researches with minimum interruption. It is also possible that Cavendish intended from the start to develop the 15 acres rather than to build a house for himself on it, though it is unclear why he would want to.[16]

His house was three-story, double-fronted , and "symmetrically planned and with a central doorway of typical Georgian design." The grounds were considerable.[17]

[13] "Statement of Leases by the Honourable Henry Cavendish of Messuages and Lands at Clapham in Surrey," 1795–1805, Devonshire Collections, Chatsworth, 34/10. "Henry Cavendish Esquire and Messrs Hanscomb, Fothergill and Poynder. Articles of Agreement for a Building Lease," 1791, ibid., L/31/45. "Abstract of the Title of His Grace the Duke of Devonshire to the Estate at Clapham Common in the County of Surrey," ibid., L/38/78. The builders each paid £200 per year rent to Cavendish.

[14] Henry Cavendish to Christopher Baldwin, n.d. [After 3 May 1784]. Mrs. Mount's house is referred to in Henry Cavendish, "Plan of Drains at Clapham & Measures Relating to Bason," Cavendish Scientific Manuscripts, Devonshire Collection, Chatsworth.

[15] Baldwin to Cavendish, 3 May 1784.

[16] In favor of this alternative might be his interest in buying Mrs. Mount's house while he was negotiating with Baldwin about buying the fifteen acres. Also Thomas Hanscomb who dealt with Baldwin as Cavendish's agent would be a builder of houses for Cavendish on the 15 acres.

[17] Clapham Antiquarian Society, "Cavendish House," Occasional Sheet, August 1957. Eric E. F. Smith, *Clapham* (London: London Borough of Lambeth, 1976), 78. Burgess, *Chronicles of Clapham*, 60.

According to the land tax record for 1793, Cavendish owned 10 acres.[18] From a plan of the Common in 1800, we see that with one possible exception, Cavendish's property occupied the largest frontage of the 60-odd residences.

The lease tells us the history of ownership of the house.

> Assignment of lease. 18 June 1785. 1. William Robertson of George Yard. Tower Hill, merchant. 2. Henry Cavendish of Bedford Square Esq. Premises on Clapham, for residue of a term of 29 years granted on the experation of a lease of 22 March 1750 made between William Bridges and Henton Brown. Recitals of subsequent assignment. Consideration £3000.[19]

Henton Brown is thought to be the first owner of the house, perhaps its builder. We know he lived there by 1748, for that year he requested leave of the Vestry to fence a pond he had built on the Common opposite his house, where he kept a pleasure boat. This was Mount Pond, probably at first a gravel pit for road making filled with water surrounding an existing mound. The excavated earth was added to the mound, making it higher and improving the view. It was a fashion at this time at Clapham to build summerhouses on viewing mounds, and Brown built one in the pagoda style on top of the Mount to entertain his guests. Brown, an owner of a bank in London, died in 1775, and his bank failed a few years after.[20] A second person named in the lease, William Bridges, was the freeholder or head lessee, who sublet the property to Brown. A third person was a merchant in Surrey, William Robertson, who probably acquired a lease for the property after Brown. In June 1785, he sold his interest in the house to Cavendish for £3,000. We see from the document that Cavendish did not buy the house freehold; instead he bought a lease for 20 years or so.[21]

Cavendish's house has been called a mansion, but a better description of it from the time is "a tolerable good house, built with red brick."[22] Later owners of the house greatly changed the appearance of the house inside and out, making it difficult to get an idea of the original layout, the number and uses of the rooms, and other details.[23]

---

[18] Clapham "Land Tax Assessment for Land Alone June 1793," Lambeth Archives.

[19] Surrey Deeds (Index), Lambeth Archives, 14.171.

[20] Before Cavendish's arrival on the Common, a scientific experiment had been performed on Mount Pond by Cavendish's colleague in electricity Benjamin Franklin, who was at the time staying with Christopher Baldwin on the Common. Brown's will is in the National Archives, PROB 11/1011/362. Clapham Antiquarian Society, "Cavendish House," *Occasional Sheet* (August 1957). Michael Green, "Mount Pond, Clapham Common: Archaeology and History," The Clapham Society Local History Series 7, http://www.claphamsociety.com/Articles/article7.html.

[21] In our biography *Cavendish* (1999), we said that Cavendish rented his house on Clapham Common. We correct ourselves here: Cavendish bought the house. I am grateful to Colin Thom for clarifying the purchase.

[22] James Edwards, *Companion from London to Brighthelmston* (London, c.1790), 11. Burgess, *Chronicles of Clapham*, 57.

[23] There is a document at Chatsworth I originally thought applied to the house Cavendish bought at Clapham Common, which if so would give us an idea of the number of rooms in the house. But now I doubt that it applies. The inventory is a room-by-room list of bookcases, curtains, stoves, and the like, which were to be valued to the person who bought the estate. The year was 1732. In pencil, Cavendish located each room in the house: "west wing back," etc. Mr. and Mrs. E. Collinson had lived in the house. The fixtures belonged to Mr. Collinson and Mr. Tritton of Clapham. The name

We know better some of the alterations Cavendish made to the house: conversion of the drawing room into a laboratory, of the room next to it into a forge, of an upstairs room into an astronomical observatory, and of the exterior of the house into a scaffold for assorted meteorological instruments. A tree behind the house was used as a platform for making scientific observations,[24] and in the field beyond the house there might be other instruments. Soon after Cavendish arrived, he erected an 80-foot-tall ship's mast, with a horizontal arm, for mounting an aerial telescope.[25] This most conspicuous feature of Cavendish's property would have alerted the neighbors, if they did not already know, that the new resident on the Common was different. He acquired a local reputation as a wizard.

In the middle years of the decade, the 1780s, Cavendish was kept busy moving from one house to another. He explained to Joseph Priestley that a reason he was so long in replying to a letter was that he had been prevented from making any experiments during the summer by "the trouble of removing my house."[26] The move to Clapham, was particularly disruptive of his normally regular life. In June 1785, he postponed the beginning of his journey with Blagden to Wales by 3 weeks because of repairs to his new house on Clapham Common.[27] In September of that year, Blagden wrote to John Michell, who had invited him and Cavendish to visit him in Yorkshire: Cavendish "cannot spare time for another journey this year, as it will give him full employment till winter to bring his new country-house of Clapham into order. He is but just removed thither: & all of his pursuits are interrupted till his books, instruments &c can be brought out of the confusion in which they lie at present."[28] Two months later, Blagden wrote to P. S. Laplace that "Mr Cavendish will not soon have another paper ready, his apparatus having been deranged by moving to another house."[29] Given Cavendish's attachment to scientific activity, his desire to move had to be strong to accept this extended interruption.

Tritton suggests a connection to Cavendish's house: Anna Maria Brown, daughter of Henton Brown, thought to be the first owner of Cavendish's house, married Thomas Tritton (1717–86); she lived on Clapham Common. If the inventory was of fixtures not in Cavendish's house, but in another, Cavendish might have been interested in buying some of them for his house. There is a list of fixtures in his house on Clapham Common, but it is very short, only four items. The puzzling document is "An Inventory of Fixtures Belonging to Mess^{r} Collinson and Tritton of Clapham in Surrey to be Valued to the Purchaser of the Estate May 13th, 1732," Devonshire Collections, Chatsworth, 86/comp. 1. "An Inventory of Fixtures in the House Purchased by Mr Cavendish of Mr Robertson," ibid. "Anna Maria Brown," "The Peerage."

24 Wilson, *Cavendish*, 164.

25 Edwards, *Companion*, 11.

26 Henry Cavendish to Joseph Priestley, 20 December 1784, draft, in Joseph Priestley, *A Scientific Autobiography of Joseph Priestley (1733–1804)*, ed. R. E. Schofield (Cambridge, Mass.: MIT Press, 1966), 239–40, on 239. In 1784, Cavendish would have been moving the contents of his apartment at his father's house to his new house at Hampstead.

27 Charles Blagden to William Lewis, 20 June 1785, draft, Blagden Letter book, Yale.

28 Charles Blagden to John Michell, 13 September 1785, draft; in McCormmach, *Weighing the World*, 395–400, on 399.

29 Charles Blagden to P. S. Laplace, 16 November 1785, draft, Blagden Letters, Royal Society 7:733.

In his letter to Laplace, Blagden said that Cavendish would have "conveniences for carrying on his experiments to still greater perfection" in his new house.[30] That may have been, but Cavendish's most important work was done in his first 25 years, when he lived behind his father's house in town. There is no surprise in this; with rare exceptions, scientists do their best work when they are young. If we think of Cavendish's active career as spanning 50 years, 1760 to his death in 1810, his move to Clapham, falls exactly in the middle. Cavendish filled the last 25 years of his life at Clapham, as he had the first in London, with scientific activity, but with the important exception of his experiment of weighing the world it did not make a notable difference to science.

In the last two decades of Cavendish's life, he shared Clapham Common with evangelical members of the Church of England known as the Clapham Sect, distinguished for their spiritual intensity and for their social activism, especially in moral causes of such as abolition of the slave trade and reform of the penal system. They were troubled by what people did on Sundays, which they insisted should be dedicated to quiet devotion.[31] Cavendish was unaffected; Sunday was a work day like any other. After his death, his house had a brush with the movement. John Thornton, son of Samuel, a member of the Clapham Sect, lived in the house for a few years. Subsequent owners greatly enlarged the house, adding a great reception room, another servants' wing, a terrace along the garden frontage, and an area for hanging paintings. At some point, the original red brick central block, the house as Cavendish knew it, was stuccoed over. In 1880, the house was described by an auctioneer as containing "an elegant drawing room, noble dining room, handsome library, morning room and billiard room, a large conservatory, and seventeen bedrooms," and the park-like grounds were similarly sumptuous. Cavendish would have been hard-pressed to recognize the sensible building he made into a house of science in 1785. In 1905, the estate was sold and the house was torn down, replaced by rows of red brick villas.[32] Cavendish Road, originally Dragmire Lane, memorializes the place where Cavendish's house once stood.

Bedford Square

We do not know why Cavendish did not keep the house on Great Marlborough Street after his father died.[33] In any case, he moved to a house on Bedford Square. Perhaps this house had better arrangements for the library he intended for it. Perhaps he preferred the location of Bedford Square next to the British Museum, where he regularly attended meetings as a trustee. It may have had an intrinsic attraction too

[30] Ibid.

[31] R. de M. Rudolph, "The Clapham Sect," in Clapham Antiquarian Society, *Clapham and the Clapham Sect,* ed. E. Baldwin (Clapham: Clapham Antiquarian Society, 1927), 89–142, on 89.

[32] Smith, *Clapham*, 78.

[33] In his will, Charles Cavendish left his personal estate to Henry; he said nothing about his real estate. He named Henry as his sole executor. In Henry Cavendish, "List of Papers Classed," under "Mine," there is an entry "agreement about house in M.S.," probably "Marlborough Street," where his father's house was. Charles Cavendish's will, signed 1 August 1756, probated 28 May 1783, Devonshire Collections, Chatsworth, L/69/12.

as the first garden square in London to exhibit perfect uniformity and symmetry in its architecture, features which may have appealed to the mathematician in him (Fig. 3.9).

Exactly when he relocated can be clarified here. The rate books for the house give the occupants: 1782–84 Dr. Tye, 1784–86 Hon. John Cavendish, and 1786- Hon. Henry Cavendish.[34] The second occupant, "Hon. John Cavendish" in 1784–86, would have been Henry Cavendish's first cousin the "Right Honourable," though commonly called "Lord," John Cavendish. However this identification is ruled out by the following exchange. In August 1785, Lord John Cavendish wrote to Henry Cavendish, "The last time I came to Marlborough Street, & found your house so compleately shut up that I took it for granted you had quitted it." Henry Cavendish replied, "I am moved to the corner house of Be[dford] Sq[uare] & Gower street on the East side."[35] If Henry had bought the house from his cousin, his explanation would have been unnecessary. The rate books evidently were in error: John Cavendish is not among the occupants of the house. The original 99-year lease for the house was to William Scott and Robert Grews in 1775. In late 1783, the two men leased it to the physician Dr. Michael Teighe for a period of 8 years.[36] By an indenture between Dr. Teighe and Henry Cavendish, registered on 21 May 1784, Cavendish acquired the house, with an absolute purchase for £3,250.[37] With this clarification, we see that he moved to Bedford Square when he quit his house on Great Marlborough Street, and that he was looking for a country house on Clapham Common at this same time. Five months after Cavendish bought the house on Bedford Square, Blagden, who was by then Cavendish's associate, moved to a house on Gower Street, just off Bedford Square, a few houses from Cavendish's.[38] At age 52, while still a Londoner and still a solitary, Cavendish was less narrowly a Londoner, at the point of removing his main home permanently to a country suburb, and less rigorously a solitary, having taken on an associate. In addition, by giving up his father's house for a new house on Bedford Square, he stepped out of his father's shade, though we have no reason to think that this was a motive.

Bedford Square was relatively new when Cavendish moved there. Laid out in 1775–80, it was one of a number of squares built in the West End of London starting in the late seventeenth century. An early form of town planning, the squares imposed a degree of order on an otherwise sprawling metropolis. They came about as joint ventures between owners of large estates and builders, who were granted low-rent,

[34] London County Council, *Survey of London*, vol. 5: *The Parish of St. Hiles-in-the-Fields*, part 2, ed. E. Riley and L. Gromme (London: London County Council, 1914), 162.

[35] Lord John Cavendish to Henry Cavendish, 25 August 1785. Henry Cavendish to Lord John Cavendish, n.d., draft, Devonshire Collections, Chatsworth.

[36] Bedford estate archive, NMR 16/21/3.

[37] Middlesex Deed Register, MDR/1784/2/353.

[38] Charles Blagden to John Blagden Hale, n.d., draft, Blagden Letterbook, Yale. In this letter Blagden told his brother that he was moving his house to Gower St. at the end of next week. He said that he watched Blanchard's balloon on the day he wrote the letter, which dates it, 17 October 1784.

long-term leases. Bedford Square was built on the estate of the duke of Bedford. According to a historian of eighteenth-century London, Bedford Square was "probably the most important of the planned aristocratic building ventures of the century."[39] The houses followed a specified design, lending the square a standard appearance from all approaches. They were three-story with basements and attics, terraced, and built of brick, with wrought-iron balconies to the first-floor windows. The entrance doors were decorated by Coade stone, with rounded fanlights above them. Each side of the square was a block of houses considered as a single unit, the center house of which was distinguished by an ornamented stuccoed feature. Houses built on streets radiating out from the square were covered by the same specifications. Bounded by broad streets, the square was spacious, 520 by 320 ft between facing houses, and in its center was a large garden for use of the residents.[40]

Cavendish's house, No. 11 Bedford Square, is located at the end of Gower Street. Being on a corner of the square, it is not joined to a block. The house carries a bronze tablet, a gift of the duke of Bedford, identifying it as Henry Cavendish's, and giving the years of his birth and death. Today it is used for offices by nearby University of London. Bedford Square is one of the best preserved garden squares in London (Fig. 3.9).

After Cavendish's death, an appraiser wrote of the house, "I have scarce ever met with a more substantial or better built House, and the whole Edifice is finished with the best material." The floors of the two main stories were made of Norway oak, the staircase was made of Portland stone, and the dining and drawing rooms had carved marble chimney pieces.[41] All three stories and the attic for servants had bowed windows in the back looking out over a deep garden leading to the stables and coach house. The house had the quality, elegance, and expense expected of a wealthy Cavendish.

What is unusual is the use Cavendish made of it, a library for his books, which he lent to qualified borrowers. He took the lending seriously, as a duty of service. Normally, a librarian handled requests for books, but when for a time he was without a librarian, he spent a day each week in the library keeping it open. At some point it was his wish that after his death his library would be continued as a lending library.[42] His investment in books was enormous. At his death, his books at Bedford Square were valued at £7,000. To put this in perspective, his heirs sold the Bedford Square house for £3,530.[43]

---

[39] George Rudé, *Hanoverian London 1714–1808* (Berkeley and Los Angeles: University of California Press, 1971), 14.

[40] London County Council, *Survey of London*, vol. 5, pt. 2, p. 150. Anon., "Bloomsbury Squares & Gardens. Bedford Square," http://bloomsburysquares.wordpress.com/bedford-square.

[41] "J. Willcock's Valuation of House & Stables in Bedford Square," 30 December 1813, Devonshire Collections, Chatsworth.

[42] 3 and 4 March 1810, Charles Blagden Diary, Royal Society, 429, 429 (back), 430.

[43] "6 Sept. 1810. Mr Paynes Valuation of Books £7000." "29 April &c. 1814 Account Respecting the Sale of a Leasehold House at the North East Corner of Bedford Square," Devonshire Collections, Chatsworth.

Cavendish altered the house on Bedford Square at least as extensively as he did his house on Clapham Common. When the house was evaluated for sale after his death, it was estimated that because of its long use as "Libraries, and Museums," it would need renovations costing one third of the value of the house to make it "fit for the residence of a family."[44] We can picture the interior as Cavendish left it from a list of the fixtures, furniture, plate, and other contents of the house. Twenty-one rooms, which included a hallway and two closets, were inventoried. Inside the entrance, a semi-octagonal bay opened onto a hall at the end of which was a stone staircase leading to the upper floors. Off the hall to the left was a library room, which appears to have been used as a dining room, and to the right was a bow-window dining room, which appears to have been used as a library room, off of which were two smaller bow-window sitting or dressing rooms. Except for the dining room, all of the rooms on this floor were used for books. The floor above, the principal floor, consisted of two large drawing rooms, front and back, and a small side bow-window sitting room. The drawing room with the bow window was not used for books, but the rest of the floor was. The next floor up, the two pair floor, consisted of two bedrooms to the front, and a bow-window bedroom and dressing room to the back. All four rooms on this floor, which included Cavendish's bedroom, contained books. Only the attic story, which had two bedrooms, a bow-window nursery, and a dressing room for servants, was not used for books. Bookcases were built of handsome uprights, with plinths and cornices, and sliding shelves. There were around 700 sliding shelves all told in the house, the front drawing room on the principal floor holding the largest number, 268.

The house contained various pieces of furniture, evidently of the same quality as the house, many relating to what readers and writers require. The front drawing room on the principal floor had a pair of low steps, a pair of high steps, and a step ladder for reaching high shelves. It also had a glass-topped table, a column-and-claw table, four cushioned banister back chairs, two side desks, two black Wedgwood inkstands, and a table clock. The library room on the ground floor had in addition to shelving ten banister back chairs, a glass-top table with fly leaves, a table desk, and a black inkstand. The two smaller rooms adjoining it, formerly sitting or dressing rooms, contained in addition to shelving a copying machine with double roller and apparatus by Watt & Co., a cupboard for maps, a bracket minute clock by John Skelton, a barometer, and a thermometer. The hall and staircase had a thermometer and an astronomical timepiece by George Graham. The back drawing room on the principal floor, which had no bookshelves, had twelve Japanned elbow chairs, two oval mahogany tables, one of which was a dining table, and silk-covered fire screens. The dining room on the ground floor, the other large room without bookshelves, contained three dining tables and ten banister back chairs. The interior of the house was unified by the use of the color green throughout: mahogany blinds lined with green transparent canvas, curtains of green moreen, green fire screens and chair back screens, and green chair covers. The furniture was mostly mahogany, the main exception being the sliding shelves,

[44] "Willcock's Valuation of House & Stables in Bedford Square."

which were made with less expensive deal, or fir. Cavendish has been called an ascetic, but his property would argue otherwise, as would his wine cabinet, which at the time of his death contained 22 bottles of port, tokay, and white wine, and 10 dozen empty bottles. A contemporary of Cavendish's said that the "sole furniture" of his house on Bedford Square was a library.[45] This was an exaggeration – two large rooms of the house were used for other purposes, as we have seen – but it gave the correct feel of the house. A visitor touring the house when Cavendish lived in it would have concluded that it was a house of knowledge. The house would also have told him that its owner was a wealthy aristocrat who was proud of his lineage. It contained six paintings, one a landscape, the others all portraits of Cavendishes, one of an earl of Devonshire before there was a dukedom.[46]

Cavendish sometimes stayed at his house on Bedford Square, and he kept appointments there, but his needs were less than they were at his country house. He employed seven servants at Clapham Common, and an eighth if an instrument maker is counted. He employed only three at Bedford Square: a porter, a housemaid, and a cook, and a fourth if the librarian is counted. His two houses supported the two main activities of his life, reading and research. Complementing one another, his Bedford Square house treated scientific knowledge as recorded in publications, and his Clapham Common house treated it as knowledge in progress. Cavendish kept his books at Bedford Square and his instruments at Clapham Common, and although the division was not absolute, at the end of his life the value placed on his instruments at Clapham Common was £545 and at Bedford Square nothing.[47]

Other comparisons of Cavendish's possessions in his two houses at the end of his life reinforce our picture of his life. The value of the furniture in each house was essentially the same, £645 at Clapham Common and £633 at Bedford Square, but his plate, China, and linen at Bedford Square were valued at £700, and at Clapham Common £168. He kept his small wardrobe and his carriage and harness at Clapham Common, his pictures and wine at Bedford Square.[48]

---

45 Barrow, *Royal Society*, 148.

46 "Inventory of Sundry Fixtures, Household Furniture, Plate, Linen &c. &c. The Property of the Late Henry Cavendish Esquire at His Late Residence in Bedford Square. Taken the 2nd Day of April 1810," Devonshire Collections, Chatsworth, 114/74. "The Particulars of a Capital Leasehold House, and Offices Situate at the North East Corner of Bedford Square… Sold by Auction, by Mr. Willcock on Friday the Twenty-ninth of April 1814," Devonshire Collections, Chatsworth. There were in addition to the five family portraits in the house ten damaged ones in the lumber room over the stables.

47 "Extracts from Valuation of Furniture &c.," Devonshire Collections, Chatsworth.

48 Ibid.

Chapter 15
Cavendishes, Lowthers, and Holker Hall

Holker Hall is a grand manor on the northwest coast of England, in the county of Cumbria, formerly Lancashire. It is situated among splendid gardens on hilly park-like grounds with woodlands overlooking Morecambe Bay. Built in the sixteenth century, it was altered in the 1780s and again in the next century. Today it belongs to the Cavendish family and is open to the public. Its library contains many books from Henry Cavendish's library (Figs. 15.1 and 15.2).

Late in life, Cavendish had a conversation with a colleague John Barrow about Holker Hall. Barrow thought it belonged to Lord George [Augustus] Cavendish. Cavendish corrected him: "It did belong to him, Sir; but he left it to my father, from whom it descended to me, and will next go to another Lord George [Augustus Henry Cavendish]."[1] Barrow's recollection of the conversation is detailed and plausible, but it raises questions.

It is at odds with published sources, which agree on a succession of ownership of Holker Hall, in which Charles and Henry Cavendish do not figure. According to this version, Holker Hall came into the Cavendish family in 1756, when Lord George Augustus Cavendish acquired it from a Lowther cousin. When Lord George died in 1794, it passed to his brother Lord Frederick. When Lord Frederick died in 1803, it passed to Lord George Augustus Henry Cavendish, who held it until his death in 1834. At the end of this section, we return to Holker Hall and to Henry Cavendish's claim of ownership. In the meantime, we look at the tangled affairs of the Cavendish and Lowther families. They may shed light on the confusion over Holker Hall and how it came about.

The relevant history begins with the last Lowther to live at Holker Hall, Sir Thomas Lowther. He was the son and heir of William Lowther, a large landholder in Lancashire and Yorkshire, who had been raised to a baronetcy at the end of the seventeenth century. Sir Thomas Lowther was a country gentleman and Member of Parliament for Lancashire, described as an independent country Whig. Because of his duties, he spent part of his time in London, the rest mainly at his country house

[1] Barrow, *Royal Society*, 146.

R. McCormmach, *The Personality of Henry Cavendish - A Great Scientist with Extraordinary Peculiarities*, Archimedes 36, DOI 10.1007/978-3-319-02438-7_15,

Fig. 15.1 Holker Hall. Near the village of Cartmel, in Cumbria, on the west coast of England. The house was built in the sixteenth century, and was substantially altered in the eighteenth and nineteenth centuries. It came into possession of the Cavendish family in 1756, and through a complex of legal documents, it passed to Henry Cavendish late in his life. He had an ambitious plan for developing the estate, but he did not live to realize it. His improbable connection with Holker Hall is the subject of Chap. 14 of this book. The sketch of Holker Hall is from a book published in 1880 by James Morris, *The County Seats of the Noblemen and Gentleman of Great Britain and Ireland.* (Wikimedia Commons)

Fig. 15.2 Morecambe Bay. Holker Hall overlooks the Bay, into which five rivers flow. Today there is a good deal of reclaimed land around the Bay, and there was in Cavendish's time; his ambition was to reclaim land for the Holker estate. (Wikimedia Commons)

and family seat Holker Hall, near the village of Cartmel. The rectory and manor of Cartmel also belonged to his estate, as did an abbey and considerable land in Furness, at some distance from Holker. His Yorkshire estate at Marske contained another large tract. Thomas received rents from his lands and returns from crops, timber, and minerals. Despite the extent of his lands – many thousands of acres – he was constantly in debt. He regularly borrowed money from his estate steward, John Fletcher, an uncommon kind of dependency.[2]

In 1723, he married Lady Elizabeth Cavendish, Lord Charles Cavendish's sister. The marriage settlement shows that she brought £6,000 to the marriage, a welcome addition to Thomas's precarious finances. Charles often saw his sister Elizabeth at Holker Hall or at Chatsworth, the duke of Devonshire's country house in Derbyshire. He was named godfather to Elizabeth's second child, a daughter who lived only a short time.[3] In a letter telling of their daughter's death, Thomas wrote that Elizabeth was "in very great concern & trouble."[4] Beginning around this time, in his letters Thomas included regards from his sisters but no longer regularly from his wife, as he had in the past. The spunky Elizabeth, who wished she had been a boy so she could have gone abroad with her youthful brothers Charles and James, was placed in the hands of physicians "to try what effect it will have upon her to make her of better behaviour." She was considered insane by the time she died. Her husband, Thomas, a sportsman who was fond of horse racing, a kind but improvident man, lapsed into heavy drinking and more debt.[5] In 1745 he died without a will. In the month after his death, at his surviving child William's request, Charles Cavendish together with the duke of Devonshire and another relative Lord Lonsdale agreed to serve as guardians during William's minority.[6]

Elizabeth declined the executorship, and William asked Charles Cavendish to be administrator of the estate for his benefit.[7] To carry out his responsibility, Cavendish corresponded with the steward at Marske in Yorkshire and with the steward at Holker in Lancashire, John Fletcher. From them, he requested full information about the estate, which included a variety of properties in addition to buildings and

[2] The first survey of the Lancashire estate in 1775, 30 years after Sir Thomas's death, put Cartmel-Holker at 2,860 acres and Furness at 3,559 acres. J. V. Beckett, "The Lowthers at Holker: Marriage, Inheritance and Debt in the Fortunes of an Eighteenth-Century Landowning Family," *Transactions of the Historic Society of Lancashire and Cheshire*, 127 (1978): 47–64, on 47–51.

[3] Thomas Lowther to James Lowther, 8 August 1728, Cumbria Record Office, Carlisle, D/Lons/W/39.

[4] Thomas Lowther to James Lowther, 26 September 1728, ibid.

[5] His debt was £4,880 at his death. Beckett, "Lowthers at Holker," 51.

[6] Edward Butler to John Fletcher, 16 May 1745, Lancashire Record Office, DDCa 22/3/1.

[7] Charles Cavendish to John Fletcher, 18 July 1745, draft, Devonshire Collections, Chatsworth, L/43/14. Fletcher was steward of the Lancashire estate, Holker and Furness. Cavendish was sworn in as administrator on 30 July 1745. Charles Cavendish to John Fletcher, draft, 30 July 1745, Devonshire Collections, Chatsworth, box 43/14. This bundle contains one notebook of Cavendish's guardian account for William and two notebooks of his administrator's accounts and correspondence for Thomas Lowther's estate. Drafts of his letters to the estate stewards and copies (probably incomplete) of their letters to him are contained in this correspondence, 1745–48. Administrator appointment, 17 August 1745, Devonshire and Collections, box 31/11.

land such as iron pits and a fishery. In his many letters and lengthy notebooks relating to his administration, Cavendish considered a range of issues, including debts, arrears, rents, bonds, interest, dividends, furniture, pictures, books, household expenses, repairs, taxes, corn, hay, pigs, asses, cattle, and horses. Having learned that the "proper method" for an administrator was to publish a sale, he pressed Fletcher for valuations of everything that was to be sold. He overlooked nothing. "As to the dogs you [Fletcher] say that people are more inclined to beg than to buy, but my business is to sell & not to give & therefore I desire you will inquire whether you can get any thing for them." He supposed there would be no point in selling the dogs to the guardians for William to use, for by the time he came of age, "they will most of them being worn out."

Cavendish was both administrator and one of the guardians, which added a level of complexity. He and the other two guardians were ordered to obtain information similar to what the administrator needed: arrears, value of lands, money owed, and servants needed at Holker Hall during William's childhood. To avoid any dispute, Cavendish was concerned that the right things were charged to the guardians' accounts and the right things to the administrators'.[8] On various points, he obtained an opinion from the attorney general. As the executor, he was well-organized, thorough, and insistent on adhering to the methods he set out.

Problems naturally came up, the first one being Fletcher, who caused Cavendish "a great deal of trouble."[9] Fletcher was told to prepare as soon as possible a "perfect state of all the effects whatsoever belonging to Sir Thomas at his death & all of the sums due from him at that time."[10] He was slow to understand and made mistakes in his accounts. Cavendish wrote to him, "I can't suppose you think it [the account Fletcher sent him] such an account as I asked for, nor such as is necessary for me to have in order to know the true state of Sr Thomas's affairs." The next month he wrote again, explaining how to make up his accounts. "I think this method necessary for the regularity of my own accounts in which I must enter a state of all moneys due to the personal Estate of S^r^ Th. Lowther at the time of his death & of all debts then due out of it."[11] Cavendish repeated his instructions over and over. Fletcher was old and ill, and in the spring of 1746, he died. He was succeeded by his capable son-in-law, William Richardson, easing Cavendish's work.[12] Cavendish told the new steward that in dealing with Sir Thomas's creditors, "I have laid it down for a rule to pay every body in proportion as every creditor has an equal right & I suppose is equally desirous to receive his money, & if I depart from that rule in one case there will be no end of solicitations, so that though I am very sorry any

[8] Charles Cavendish to John Fletcher, 27 July 1745, Lancashire Record Office, DDCa 22/5.

[9] Charles Cavendish to William Richardson, 13 March 1746, draft, Devonshire Collections, Chatsworth, box 43/14.

[10] Charles Cavendish to John Fletcher, 20 July [1745], Lancashire Record Office, DDCa 22/5.

[11] Charles Cavendish to John Fletcher, 13 August 1745, ibid.

[12] Charles Cavendish to William Richardson, 13 March 1746, draft, Devonshire Collections, Chatsworth, box 43/14.

person that wants his money should be kept out of it I see no help for it."[13] In the case of creditors who refused to accept only part of the principal, "unless they will agree each of them to take a part of their debt I must offer the whole to some of them & I should chuse to do it to those who make the most difficulty & I desire you will acquaint them with it."[14] In the case of tenants who were in arrears and who would not immediately pay what was due from them, Cavendish directed the steward to distrain their effects. Where this method was not legally allowed, he would recover arrears by legal action; "I must take that method," Cavendish told the steward, and asked him to send him the names of persons calling for that method.[15] A year and a half after he had taken charge of the estate, Cavendish could write to the steward, "I can now be pretty certain that when Sir William comes of Age there will be money enough to pay all the debts, & it will save some trouble."[16] Cavendish understood business well, and he was good at it. In his decisions, he was firm and clear, and he got results.

Cavendish's sister and now widow Elizabeth needed care. He paid sums to "Dr Mead," likely the London physician Richard Mead, the head of his profession, "Dr Wilmot," "Dr Monroe" who received an "allowance," and "Heskins" an apothecary.[17] Elizabeth did not long outlive her husband, Thomas, dying in 1747, while Cavendish was still active as administrator. In the same year, another Lowther, John, died, leaving most of his estate to William on the death of or in jointure with his mother, and Cavendish had to sort out the details of this property as well.[18]

Cavendish kept on friendly terms with his ward. When Sir William – like his father he was baronet – was at the university, Cavendish sent him books he asked for. He introduced William to his world, inviting him to dinner at his house with scientific friends.[19] In 1753 William was appointed lord lieutenant of Westmoreland, and in 1755 he succeeded his relative Sir James Lowther in the Cumberland seat.

Sir James Lowther was born in London and educated at Oxford and Middle Temple. Through inheritance, he became owner of valuable collieries and other properties around Whitehaven in Cumberland, on the northwest coast of England. He expanded his estate, lived frugally, and grew immensely rich, and was believed to be the richest commoner in England. He made important improvements in the extraction and trading of coal, and he encouraged the production of iron in Cumberland. He improved the harbor at Whitehaven, making it a major port for shipping coal. He adopted technical improvements at his collieries. After a visit to

[13] Charles Cavendish to William Richardson, 20 May 1746, draft, ibid.

[14] William Richardson to Charles Cavendish, 2 May 1746, copy, ibid.

[15] Charles Cavendish to William Richardson, 21 June 1746, Lancashire Record Office, DDCa 22/7.

[16] Charles Cavendish to William Richardson, 27 December 1746, ibid.

[17] On his sister's behalf he also paid "Mr. Duffield," who received regular payments up to £180 each time, and "Mrs. Potter." Various dates in "Guardians Account" and in an untitled notebook containing six pages of accounts, 1745–48, Devonshire Collections, Chatsworth, box 43/14.

[18] On 9 January 1747, the steward, Danby, for the Yorkshire estate informed Charles that John Lowther had died. "Sr W. Lowther's Estate," Devonshire Collections, Chatsworth, box 43/14.

[19] Thomas Birch Diary (5 June 1753).

Whitehaven, Richardson said that he "did not imagine to have found so many new contrivances."[20] Lowther was the first to install a Newcomen steam engine in Cumberland. He dug the first undersea coal mine in England, Saltom Pit. Thomas Lowther wrote to James that a shipment of coal from Saltom had arrived, and everyone was saying that these were the "finest coals that ever came into this country." Excavation of the mine released a great quantity of explosive firedamp, which we know to be mainly methane. William Brownrigg, a physician in Whitehaven who took a medical interest in the firedamp that miners breathed, was "earnestly solicited" by Lowther to study the problem.[21] Brownrigg developed ways of transferring and collecting firedamp and provided Lowther with bladders filled with it to submit to the Royal Society.[22] In 1736, Lowther was elected to the Royal Society, with Charles Cavendish's support.[23] Among Henry Cavendish's scientific manuscripts is a paper on firedamp written by Brownrigg for Lowther.[24] Progressive and scientifically minded – a friend mentioned Lowther's "old Acquaintance Sr Isaac Newton"[25] – Lowther was the kind of industrialist Henry Cavendish was drawn to.

Thomas had been close to James. They corresponded regularly, and Thomas paid visits to Whitehaven.[26] James died in January 1755, and having no children of his own, he left his collieries and extensive lands in Cumberland to Thomas's son William. James was not related to the Cavendishes, but William of course was, and his inheritance was viewed as an enrichment of the family. Lord Hartington, soon to be fourth duke of Devonshire, was congratulated, "I must wish y^{r} Lordship Joy of the very great Acquisition made its by your near Relation Sr W. Lowther, which I am credibly informed, is 4,000 £ a year in Land, Coal Mines bringing in 11,000 £ a year, & not less than 400,000 £ in Money. Sr James Lowther has 100,000 £ & an Estate in Middlesex."[27]

In the spring of the following year, 1756, William Lowther contracted scarlet fever. Katherine, wife of the recently deceased third duke of Devonshire, wrote to the fourth duke William that "every body is in great pain for Sr Wm Lowther." He had been ill for a week or 10 days, attended by "Shaw & Heberden." The day she wrote, William had had "a very bad night," and his doctors had called in "Willmot," who ordered more blisters. She wrote a postscript to the letter, saying that Lord

[20] Thomas Lowther to James Lowther, 6 June 1734, Cumbria Record Office, Carlisle, D/Lons/W.

[21] Joshua Dixon, *The Literary Life of William Brownrigg …* (London, 1801), 5.

[22] This was in 1733. Young, "Cavendish," 436. Anon., "Sir James Lowther, 4th Baronet." Anon., "William Brownrigg," http://en.wikipedia.org/wiki/William_Brownrigg.

[23] Cavendish signed Lowther's certificate. 20 May 1736, Journal Book, Royal Society 15: 331.

[24] "Some Observations upon the Several Damps in the Coal Mines near Whitehaven by Dr Willm Brownrig Phisitian of that Town Communicated by Him to Sr James Lowther Bar^{t}," Cavendish Scientific Manuscripts, Devonshire Collections, Chatsworth.

[25] Henry Newman to James Lowther, 26 August 1732, Cumbria Record Office, Carlisle, D/Lons/L1/1/53.

[26] There are many letters from Thomas to James Lowther in the Carlisle archive. Charles Cavendish also visited Whitehaven. Thomas Lowther to James Lowther, 11 July 1734, ibid., D/Lons/W/37.

[27] H. Fox to William, Lord Hartington, 4 January 1755, Devonshire Collections, Chatsworth, No. 330.30.

Charles was just there to tell her that Sir William had died.[28] On the same day, the duke received a consoling letter saying that persons who knew William thought he had "left the Chief part of His fortune to Your Brothers."[29] The "Chief part of His fortune" referred to Hoker Hall, which we return to below.

At this point, a second Sir James Lowther enters our account. He too was remembered in the will of his relative Sir James Lowther of Whitehaven. When William Lowther died, he was 28 and unmarried, and because he had no children, the Cumberland estates, which he had recently inherited, reverted by Sir James Lowther's will to young Sir James Lowther, then age 19.[30] Commenting on this inheritance, the Reverend Theophilus Lindsey wrote to the earl of Huntington of the "immense accession to young Sir James Lowther's own fortunes by the death of Sir William, and the distribution of the unentailed fortunes of the latter among the Cavendishes, Lords John, George and Frederick, his relations."[31] The fortune of young James Lowther caused Horace Walpole to fear that England was becoming the "property of six or seven people."[32]

In his will, William named his former guardian Charles Cavendish as his executor.[33] He left his money, stock, goods, chattels, and personal estate not otherwise specified to Cavendish in trust to pay for his funeral expenses and his legacies and to pay off his debts. What remained of the personal estate after these payments he left to Cavendish as his executor. Because he lived in London, Cavendish depended on the steward at Holker, Richardson, to provide him with the information he needed from William's estate at both Holker and Whitehaven. His letters to Richardson tell us about his actions and problems. Other than for the pictures, which were to remain in Holker Hall, none of the furnishings in any of William's houses was specifically given in his will, so "the whole" belonged to Cavendish. That was the easy part. He needed to know what particulars belonged to William's personal estate and what their values were and which of them young Sir James wanted to buy. Because much of William's estate was in Cumberland, he depended on John Spedding, steward to the late James Lowther and after him to the late William at Whitehaven. To keep the money coming in, Cavendish allowed Spedding to continue to use what he needed from the personal estate to carry on the coal trade. He told Richardson to go to Whitehaven and talk to Spedding to learn what particulars at the collieries belonged

[28] K. Devonshire to the fourth Duke of Devonshire, 15 April 1756, ibid., No. 344.8. I assume the letter writer is Katherine, wife of the recently deceased 3d duke of Devonshire.

[29] Ducannon to William, 4th duke of Devonshire, 15 April 1756, ibid., No. 294.46.

[30] Beckett, "Lowthers at Holker," 52. Also Sir William left all of the buildings at Cockermouth, near Whitehaven in Cumberland, to Charles Cavendish to hold in trust for young James Lowther.

[31] Theophilus Lindsey to Francis, 10th earl of Huntington, 25 May 1756, Great Britain, Historical Manuscripts Commission, *Report on the Manuscripts of the Late Reginald Rawdon Hastings, Esq., of the Manor House, Ashby de la Zouch*, vol. 3, ed. F. Bickley (London: His Majesty's Stationary Office, 1934), 117.

[32] Horace Walpole to George Montagu, 20 April 1756, in Horace Walpole, *Correspondence*, 48 vols., ed. W. S. Lewis (New Haven: Yale University Press, 1941) 9: 183–87, on 185.

[33] Will of William Lowther, dated 7 April 1755, probated 22 April 1756, Devonshire Collections, Chatsworth, L/31/47.

to William's personal estate. He sent him off with a list of particulars that he thought belonged.[34] Cavendish set about with evident total confidence to settle the affairs of this complex estate.

There was a difference of opinion on who owned the steam engines at the pits, and on the value of the ships and of the leasehold collieries and estates. Cavendish confided to Richardson his concern about having to depend on Spedding for valuations. He wanted Richardson to tell him how much trust he could place on the accounts he received from Spedding. He understood that Spedding would be partial to the owner of that estate, who was then young James. He was "intitled to a full discovery [of all Sir Williams personal estate] by Law as well as from the principles of justice."[35] In all disputes of interest, he told Richardson, it was his "desire to act with perfect openness & candour," and he did "not in the least desire to get anything which I am not justly intitled to."[36] He suspected that measurements of the quantities of some stores "may not have done me strict justice," but he did not know what to do about it other than to insist that Spedding give him strong assurances of the "truth" of the inventory before signing an agreement with him. Richardson thought that some of the prices Cavendish demanded were too high. Cavendish told him that he had no objection to lowering them if he saw fit, explaining, "I do not desire to have a farthing more than I have a right to."[37] We recognize Charles Cavendish from the words he uses: "principles of justice," "strict justice," "openness," candor," and "truth." We meet them again in his son Henry's business affairs.

From letters to his steward, we see the settlement of the estate from Cavendish's point of view. We have another view of it from Catherine Lowther through her letters to her son, young James. She told him, "Lord C – is determined to give you all the trouble in his power; you must therefore make the best of it."[38] She told him that they had "great calls for money."[39] She was "very pressing to have the affairs at Whitehaven settled" with him, but Cavendish would not settle until he knew what the personal estate consisted of and what parts of it her son wanted to buy.[40] Cavendish was not without sympathy, but he would not bend his principles. He accommodated her immediate needs by advancing her any money she asked for from William's legacy to her of £6,000, in discharge of all demands of the estate.[41]

---

[34] Charles Cavendish to William Richardson, 27 April, 13, 27 May 1756, Lancashire Record Office, DDCa 22/7. Cavendish's list: arrears of rent; bonds, notes, etc.; furniture, plate, etc.; coal debts; coals raised; wagons, carts, etc.; horses; tools; corn, hay, etc.; timber in yard; timber felled; material for buildings not used; ships; engines; leasehold estates & collieries.

[35] Charles Cavendish to William Richardson, 26 June 1756, Lancashire Record Office, DDCa 22/7.

[36] Charles Cavendish to William Richardson, 29 June 1756, ibid.

[37] Charles Cavendish to William Richardson, 27 July 1756, ibid.

[38] Catherine Lowther to James Lowther, 11 July 1756, Cumbria Record Office, Carlisle, D/Lons/L1/61.

[39] Catherine Lowther to James Lowther, 8 July 1756, ibid.

[40] Charles Cavendish to William Richardson, 8 May 1756, Lancashire Record Office, DDCa 22/7.

[41] Cavendish to Richardson, 27 April 1756.

We come to a major disagreement. Cavendish claimed £30,000 in New South Sea Annuities, which were put in trust to finance the transfer of William's estate to young James. The reason he gave was that the transfer could not take place in the specified time since James was not of age. In July 1756, Cavendish and James agreed that the latter would bring a bill in the Court of Chancery against Cavendish to "have the right relative to the 30,000" and also relative to the leasehold estates and the steam engines. Cavendish and James agreed on two other points: Richardson and Spedding between them would decide the values of the collieries and the furniture in the house at Whitehaven; the legacies would be paid and the personal estate and the stock would be given to James when he came of age, while in the meantime he would receive dividends.[42] Upon reading the agreement, Catherine wrote to her son, "I think most of it very unreasonable," in keeping with "His Lord^{s} conduct."[43] She was certain to be pleased with the outcome of the suit. The case was debated, counsel on both sides was heard, and the judge declared that the £30,000 and the properties belonged to young James Lowther.[44]

Repeatedly in his letters to Richardson, Cavendish used the expression "what belongs to me," or its equivalent. His letters read as though he was furthering his own interests, and that is how I originally read them,[45] but now I have a better understanding of his way of speaking: he meant by it, what belonged to him in trust for uses specified in the will, with anything left over belonging to him as specified in the will. He administered a very large estate, and he went about it with his customary conscientiousness. There is another consideration too. William was generous – he tripled Spedding's pay when he acquired James Lowther's estate[46] – and his will was generous. The specific legacies totaled £79,000.[47] At the time he made out his will, this total, large as it was, would have seemed realistic, owing to his recent inheritance. The elder James Lowther's annual income was well above £20,000 a year, a good portion of which would have gone to William,[48] and his income from his father's estates, of the order of £4,000,[49] would have paid part and perhaps most of his living expenses.[50] He could not have foreseen that he would benefit from

[42] "Heads of What Is Agreed on between Ld Charles Cavendish & Sr James Lowther," [before 19 July 1756], Cumbria Record Office, Carlisle, D/Lons/L1/62.

[43] Catherine Lowther to James Lowther, 19 July 1756, ibid., D/Lons/L1/61.

[44] "Sr W. & Sr J. Lowthers' Wills & Papers Relating to Law Suit between L.C.C. & Sr J. Lowther," Devonshire Collections, Chatsworth, L/31/17. Cavendish appealed the decision concerning the £30,000.

[45] Jungnickel and McCormmach, *Cavendish* (1999), 93–94.

[46] Becker, "Lowthers at Holker," 52.

[47] Plus several small annuities.

[48] Becker, "Lowthers at Holker," 64. Not all of James's income would have gone to William. For example, he left his South Sea annuities to young James, who would have received the dividends. Sir James Lowther's will, 1754, Devonshire Collections, Chatsworth, L/31/17.

[49] Becker, "Lowthers at Holker," 52.

[50] Because of his very short life as a very wealthy man, not much can be made of the record we have of it. His income from 5 July 1755 to 25 May 1756 (the month after his death) was £11,640. His expenses were £8,251, which included large payments to Girolamo Belloni, the head of a family

James Lowther's wealth for so short a time. To realize the intent of William's will, Cavendish would have wanted to claim everything he could as personal estate and turn it into money. In his letters to Richardson, he spoke of his appreciation of his former ward, "a benefactor whose great fortune enabled him to do what the generosity of his temper prompted him to."[51]

When Catherine Lowther informed her son about William's death, she gave him advice about the great wealth coming to him. The "acquisition of fortune, cannot be any recompense for the want of so worthy a friend [William] & will only make you more the subject of envy than you have already been, & can in no shape conduce to y^r^ happiness, either in this world or another, unless you use it, as he did, in doing good, otherwise will only draw upon you, misery in both."[52] Six months later, she reminded him that "it is a debt due to that Great Being, who has made you accountable for so large a portion of this worlds good; wh^ch^ if properly managed, will not only make you happy here, but eternally so." The world at first would look on him favorably "as a person endow'd by providence wth the power of relieving the distress'd, & making happy his fellow creatures," a power denied to a poor man, who can offer only prayer and hope.[53] James disregarded the advice, using his money for a different kind of power. He did some good for Whitehaven, for example, by setting up a manufactory for copper and stockings, but he grew to be one of the "profligate wicked wretches" and "villains" his mother warned him against. He became known throughout the region as the "bad earl," distinguished equally by his unenviable character as by his immense wealth. James Boswell called him a "brutal fellow." Horace Walpole said he was "equally un-amiable in public and private." The Reverend Alexander Carlyle, a leader of the Church of Scotland, said that he was "more detested than any man alive." Through lavish expenditure, he kept mistresses and controlled nine members of Parliament known as "Sir James's Ninepins," who were required to vote as he ordered.[54] Otherwise, he was miserly. To show his contempt for common people, he traveled in a rundown carriage pulled by ungroomed horses.[55] In his attitude toward money, James could hardly be more different than William, his benefactor. Horace Walpole wrote to Montague 5 days after Williams death: "Sir William Lowther has made a charming will, and been as generous at his death, as he was in his short life; he has left thirteen legacies of £5,000 each to friends." He made the comparison: "but what do you think of young Sir James Lowther, who, not of age, becomes master of one or two and forty thousand

bank in Rome. "Sr William Lowther Bar^t^ His Account with Robt Snow & Willm Denne 1755," 5 July 1755 to 25 May 1756, Devonshire Collections, Chatsworth, box 43/14.

[51] Cavendish to Richardson, 8 May 1756. Cavendish directed his steward to continue William's generosity by distributing £50 to persons in the neighborhood who were most in need, as William would have done were he alive.

[52] Catherine Lowther to James Lowther, April 1756, Cumbria Record Office, Carlisle, D/Lons/L1/61.

[53] Catherine Lowther to James Lowther, 28 October 1756, ibid.

[54] "Lowther, James, Earl of Lonsdale (1736–1802)," *DNB*, 1ST ed., 12: 217–20, on 219.

[55] Donaldson, *Brewer's Rogues,* 409.

pounds a year."[56] We do not know what Cavendsh thought. Through his execution of William's will, he helped make possible this outcome, but he had no responsibility for it. That rested with Sir James's character and the forces that shaped it.

To this point, we have not looked at what William placed at the head of his will and gave most attention to, Holker Hall. William left this house along with other manors, buildings, and lands to William Cavendish third duke of Devonshire and his eldest son "to the several uses upon the trusts." Upon the trusts, Holker Hall was to go first to his own male offspring, of which he had none. In that event, it was to go to his aunt Catherine Lowther for her "use" over the course of her life; and upon her death, the estate was to pass to Lord George Augustus Cavendish for his use during his life; after his death, it was to pass to his younger brother Lord Frederick Cavendish for his use during his life; and after his death, it was to pass to the youngest brother Lord John Cavendish for his use during his life.[57] The three brothers were the younger sons of the third duke of Devonshire, and they were Charles Cavendish's nephews and Henry Cavendish's first cousins. None of the three brothers married.

Not long after William died, Cavendish heard from friends of Catherine Lowther "that she has thoughts of making over the estate to Lord George for a proper consideration."[58] This evidently was soon done. Lord George Augustus Cavendish became the first male Cavendish to live at Holker Hall, making it his home for nearly 40 years, until his death in 1794. In his final will he spoke of "the person or persons who shall upon my decease succeed and become entitled to the said House [Holker Hall] and Estate at Holker."[59] The wording might suggest that there was uncertainty about his successor, but as directed by William Lowther's will, Holker Hall went next to Lord Frederick Cavendish, who held it until his death in 1803.

Nowhere in William's will is Charles Cavendish said to be entitled to Holker Hall, nor is he in George Augustus Cavendish's and Frederick Cavendish's wills. If what Henry Cavendish told John Barrow is correct, that Holker Hall was left to his father and his father left it to him, it is unlikely that his father acquired it from Lord George Augustus Cavendish, as Henry also told Barrow; by Sir Williams's will, Frederick Cavendish was next in line. When Frederick died, his younger brother John was already dead, and the beneficiaries named in William's will came to an end. If there was uncertainty, it may have come at this juncture.

56 Horace Walpole to George Montagu, 20 April 1756, in Walpole, *Correspondence* 9: 183–87, on 184–85.

57 William Lowther's will, 7 April 1755, probated 22 April 1756, Devonshire Collections, Chatsworth, L/36/47. He died on 15 April 1756.

58 Charles Cavendish to William Richardson, 28 December 1756, Lancashire Record Office, DDCa 22/7.

59 George Augustus Cavendish's will, signed 9 March 1792, probated 12 July 1794, Public Record Office, National Archives, Prob 12/1247. He died on 2 May 1794. He used the same expression for his estates in the county of Huntington: "at the time of my decease into the person or persons who shall upon my death succeed or become entitled to those estates."

So far as we can judge from his will, Frederick did not think there was any uncertainty. He treated Holker Hall as his property. With the exception of special legacies, he left "the Capital messuage or mansion house of Holker Hall with the park lands and hereditamenti" in the parish of Cartmel, Lancashire, together with his other properties to his nephew Lord George Augustus Henry Cavendish and his heirs and assigns.[60] This Lord George was also Henry Cavendish's principal heir and the Lord George that Henry told Barrow he was going to leave Holker Hall to. The land tax returns for Lower Holker, which includes Holker Hall, list Lord Frederick Cavendish through 1803, the year he died. In 1804, the name changes to Lord George [Augustus Henry] Cavendish.[61] Henry Cavendish's name does not appear. If he was entitled to Holker Hall, he did not occupy it and he did not pay land taxes on it. By the time Frederick died, Charles had been dead for 10 years, and Henry had 7 years to live. Barrow was elected to the Royal Society in 1805, and their conversation was unlikely to have taken place before then, 5 years before Henry died. Other than in contemplation, Henry Cavendish had no occasion to enjoy the splendor of the mansion overlooking Morecambe Bay.[62]

There are three possible reasons why Henry Cavendish's ties to Holker Hall remain elusive. One is that we have missed something, either a document that has not yet been found or a right that a lawyer would understand. Another is that Barrow's recollection is wrong, though it seems unlikely that he would remember Cavendish having said that he owned the manor if he did not say it. Third, Cavendish was confused about the ownership. He was a very accurate man, and we do not consider this possibility lightly. But let us see. To begin with, he certainly knew about his father's involvement with the Lowthers. When Charles Cavendish was appointed administrator of Thomas Lowther's estate in 1745, when he was Sir William's guardian in 1745–48, and when he became executor of Sir William just state in 1756, Henry was 14–17, and 25. He was away at school for part of the time, but when he was 25 he was home all of the time. He would have known that his father made journeys in connection with this work. Later, he himself was involved. Charles Cavendish and after him Henry were trustees of Cartmel Rectory, part of the Lowther estate. The bishop of Chester leased Cartmel Rectory to Henry Cavendish in trust for the persons entitled to it under Sir William Lowther's will, who were the persons entitled to Holker Hall: George Augustus Cavendish and Frederick Cavendish, followed by George Augustus Henry Cavendish.[63] After his father's death, Henry made an inventory of the contents of a cabinet he kept in his

[60] Frederick Cavendish's will, signed 24 January 1797, probated 29 October 1803, Prerogative Court of Canterbury, PROB 112/1399/369.

[61] The 1803 land tax return was dated 7 July. The 1804 land tax return was dated 28 June. George Augustus Henry Cavendish's name is listed from 1804 through the year of Henry Cavendish's death, 1810, and beyond. Lancaster County Archives, QDL/LN/23.

[62] From his conversation with Barrow, it seems that Cavendish knew the manor and its setting. Possibly his father brought him there on one or more of his many visits. In 1786, on a journey with Blagden, he passed into Cumbria, but there is no mention of Holker Hall. Blagden to Banks, 4 September 1786.

[63] The documents are in Devonshire Collections, Chatsworth, L/36/62.

own bedchamber; the cabinet contained William Lowther's and James Lowther's wills and papers relating to the lawsuit between Charles Cavendish and James Lowther.[64] Henry made a list of his father's papers; the papers contained letters about William Lowther's executorship.[65] Henry made a list of keys; they included keys to Sir William Lowther's chest of drawers and trunk.[66] Henry lived among the relics of his father's dealings with the Lowther family. Henry had all the paperwork at hand, but he may never have looked at it. It is written in legal language and is extensive, and the transfer of Lowther property was, as we have seen, complicated. In light of the history of the Lowther estate, if Henry made a mistake about it, he is forgiven. He may have remembered incorrectly or misinterpreted something his father once told him about Holker Hall.

[64] Henry Cavendish, "Walnut Cabinet in Bed Chamber," Devonshire Collections, Chatsworth.

[65] Henry Cavendish, "List of Papers Classed," Devonshire Collections, Chatsworth.

[66] Henry Cavendish, "Keys at London," Devonshire Collections, Chatsworth.

Chapter 16
Autism, Background

Much research has been directed to the way autistic persons think, how they understand other people's minds and behavior, and how they experience sounds, sights, taste, smell, and touch.[1] Beginning in the 1990s, the number of publications in English on autism has risen sharply. These include testimonials by persons diagnosed with the disorder, one of whom is Temple Grandin, an animal scientist who was first diagnosed with autism, later with a less severe variant, Asperger's syndrome. She writes: "Not everyone understands that autism and Asperger's Syndrome are neurological conditions that a child is born with, nor is it generally understood that autism contains a continuum of traits from normal to abnormal."[2] The distinction between autism and Asperger's syndrome and the meaning of "continuum of traits" are taken up below. Because the definitions, diagnostics, and terminology of autism are constantly in flux, we begin with a brief history.

In 1943, Leo Kanner, a child psychiatrist trained in Berlin and at the time employed at Johns Hopkins University, published a case study of 11 children with similar behaviors. The first behavior was the children's aloofness from society, for which he coined the phrase "autistic aloneness." The word "autism" comes from the Greek *autos*, meaning "self," being absorbed in oneself, and it had been used for schizophrenia. The word with its new meaning was well-chosen, for persons with autism have "profound difficulty in appreciating another person's different perspective, as if one's own perspective were the true, correct view."[3] The second behavior was a desire for sameness in life as opposed to variety.[4] Specific behaviors included abnormal or no speech, tantrums, fending off, sensitivity to touch, and rocking

[1] Roth, *Autism Spectrum*, 27.

[2] Grandin quoted in Norm Ledgin, *Asperger's and Self-Esteem: Insight and Hope through Famous Role Models* (Arlington, Texas: Future Horizons, 2002), x.

[3] Baron-Cohen, *Autism and Asperger Syndrome*, 16.

[4] Roth, *Autism Spectrum*, 7–8.

R. McCormmach, *The Personality of Henry Cavendish - A Great Scientist with Extraordinary Peculiarities*, Archimedes 36, DOI 10.1007/978-3-319-02438-7_16,

movements.[5] Kanner, who worked within the frame of medical psychiatry, named the strange behaviors a syndrome, a disorder with a set of symptoms. His observations were very good, and are still up-to-date. We differ from him mainly in recognizing a greater variation in behaviors and their severity.[6]

Around the same time, in 1944, the Viennese pediatrician Hans Asperger described a similar pattern of behavior in children. Although he was unaware of Kanner's work until much later, he invented a similar name for the behavior, "autistic psychopathy." Traits that Asperger observed in the children included language and speech difficulty, lack of empathy, preoccupation, clumsiness, and sensitivity to sounds, smells, textures, and touch. Because the parents of the children had similar traits, he concluded that the cause of autism was primarily genetic rather than psychological or environmental. He saw autism as lying on a continuum of characteristics that merge with the normal. The disorder, he believed, was permanent, stable, and free of the disintegration of personality found in schizophrenics. He also observed that some of the children had notable talents.[7]

Kanner's work was recognized quickly. Asperger's was not, at least outside Austria, partly because he wrote in German. In 1981, Lorna Wing published an article calling attention to the similarity of the work of the two men. She also called attention to a difference between the children they studied. Because Asperger's children, but not Kanner's, were high-functioning, she proposed another name for the disorder, "Asperger's syndrome." Her work too was only slowly taken up. In 1991 Uta Frith edited a book with the new name as its title, *Asperger Syndrome*. In 1993–94 the World Health Organization and the American Psychiatric Association included Asperger syndrome as a diagnostic classification of a medical condition.[8] That same year the term "high-functioning autism" was first used. Today, both terms are used interchangeably. If a person has an IQ in the normal range, usually he is given a diagnosis of Asperger syndrome, a part of the autism continuum.[9]

For many years, clinicians and scientists have looked for the underlying causes of autism, which are believed to be complicated and physical, an abnormality of the brain.[10] Autism can come about through accident during pregnancy or birth or in infancy, but inheritance is a commoner path. A study found that 46 % of first-degree relatives of autistic persons have autistic-like traits. The traits in relatives usually appear in a subclinical degree, and are more a personality description than a disorder or syndrome. Hormones are a factor, since many more males than females are

[5] Grandin, *Thinking in Pictures*, 45.

[6] Roth, *Autism Spectrum*, 8.

[7] Ibid., 9. Attwood, *Asperger's Syndrome*, 12–13.

[8] Baron-Cohen, *Autism and Asperger Syndrome*, vii. Roth, *Autism Spectrum*, 25. Attwood, *Asperger's Syndrome*, 36.

[9] Attwood, *Asperger's Syndrome*, 44–45.

[10] In technical terms, there is a dysfunction of the "social brain" located in the frontal and temporal cortex, and probably also in the amygdala, the basil ganglia, and the cerebellum. There is probably also a dysfunction of the right hemispherical cortex and an abnormality of the dopamine system. Attwood, *Asperger's Syndrome*, 327–28.

autistic. There is a range of environmental factors too. Biomedical research into the causes of autism, especially into genetic factors, is extensive today.[11]

With autism seen as a medical problem, its terminology is correspondingly medical. The medical model has been criticized by those who see autism as a difference, not a "disorder." They accept that autistic persons think and perceive differently than typical people, but deny that the differences constitute a medical condition. Autistic persons may have poor social skills, but they often have superior nonsocial skills, which are valued. Autistic adults often find a niche in society where they do well, where their autistic limitations no longer seriously interfere with their lives. Critics find support in statements by persons with Asperger's syndrome or high-functioning autism that they would not choose to become normal.[12] Grandin is one of those persons, but she also uses the medical word "disorder," and this book follows her example. The advantage of using the psychopathologist Simon Baron-Cohen's alternative "condition" in the case of Henry Cavendish is discussed in Chap. 8 of this book.

When people are diagnosed with Asperger's syndrome, they are informed of their opportunities as well as their difficulties. The clinical psychologist Tony Attwood lists 15 qualities that recommend an Asperger person for employment. Of these, 12 apply to Cavendish and the other 3 are not relevant; we note that the relevant qualities apply equally to an individual who works mainly by himself, as Cavendish did, as to a person who works in a group. The 12 qualities are: persistence, perfectionism, ability to identify errors, technical ability, a sense of social justice and integrity, accuracy, attention to detail, logic, conscientiousness, knowledgeableness, originality at problem solving, and honesty.[13] Attwood lists 17 qualities that do not recommend an autistic person for employment; they do not apply to Cavendish because he was self-employed, and even if he were not self-employed they would not necessarily interfere. If Cavendish were living today, he would be readily employable in a research institute

The foregoing discussion is based on *DSM*-IV. In a forthcoming new edition of the manual, the definition of autism will be narrowed, with consequences for diagnostics. A motive for the change is vagueness in the criteria for the disorder, which has led to an inflation of the numbers of persons diagnosed with it; in 2012, in the most comprehensive US study to that date, the Centers for Disease Control and Prevention estimated that 1 child out of 88 is autistic.[14] A proposed change in definition will consolidate the sub-types of autism under a single umbrella classification, "autism spectrum disorder." "Spectrum" means that the disorder is a matter of degree, and that there is no distinct boundary between normal and autistic people. "Asperger syndrome" will not appear as a category in the revised manual.

---

[11] Ledgin, *Asperger's and Self-Esteem*, 30. Attwood, *Asperger's Syndrome*, 21, 328. Baron-Cohen, *Autism and Asperger Syndrome*, viii, 85–86, 92, 95.

[12] Roth, *Autistic Spectrum*, 277–8.

[13] Attwood, *Asperger's Syndrome*, 295.

[14] Eugene, Oregon, *Register-Guard*, 30 March 2012.

The proposed new criteria will eliminate most high-functioning persons currently diagnosed with a form of autism.[15]

As this book goes to press, the new revision *DSM*-V has come out. As anticipated, Asperger's syndrome and pervasive developmental disorder not otherwise specified are subsumed under "autism spectrum disorder." Diagnostic criteria are grouped under two categories, instead of three as before: (1) Persistent deficits in social communication and interaction. (2) Restricted, repetitive patterns of behavior and interests. The symptoms must appear at an early age, and they must be clinically significant and not better explained by other causes. There are three levels of severity. The one requiring the least support applies to persons who have some difficulty initiating social interaction and responding to social overtures and may have little interest in social interactions; and to persons who have difficulty because of inflexible behavior or poor organizing and planning. A diagnosis of the disorder is not equivalent to a need for treatment, that being up to clinical judgment.[16] I leave unchanged what is written in the book. The sources predate DSM-V, and they are not rendered obsolete because of it; definitions and diagnostic criteria may have changed but the neurodevelopmental reality has not.

---

[15] *New York Times*, 20 January 2012. Roth, *Autism Spectrum*, 275.

[16] American Psychiatric Association, *Desk Reference to the Diagnostic Criteria from DSM-5* (Washington DC, London: American Psychiatric Publishing, 2013), 3–6, 27–31.

Chapter 17
Physical Theory and Theory of Autism

In the histories Cavendish usually appears foremost as an experimentalist. He was that certainly, but he was no less a theorist. Occasionally, the fact has been brought forward. James Clerk Maxwell fully recognized this side of Cavendish.[1] Larmor, a physicist and editor of Cavendish's mathematical and dynamical manuscripts, said that if Cavendish "had no other claim to renown he would be entitled to rank high among the theoretical physicists of his period."[2] The historian of science James Crowther recognized the "penetration of Cavendish's theoretical insight," and he observed that Cavendish's "experiments were always guided by a theoretical idea, and intended to collect data bearing on it."[3] The author of this book has written about Cavendish's mastery of the art of theory.[4]

Cavendish's cultivation of physical theory illustrates an understanding of the work of the natural philosopher at the time. It was to advance knowledge of the physical world by increasing the body of physical facts and equally by establishing their connectedness. Cavendish's contemporaries stated this common understanding in various ways. Natural philosophy cannot advance one step without experiment, but unless experiment is guided by philosophy the result is an endless collection of facts, which is not philosophy: "the disposing of one fact, that is, the putting it into its proper place in science or the general order of one's knowledge, is doing more for natural philosophy, than a thousand experiments made without that order of connection or relation which is to inform the understanding."[5] An investigator who increases the public store of experimental facts "is an experimentalist; but he alone, who, by examining the nature and absorbing the relations of facts, arrives at general truths, is a philosopher. A moderate share of industry may suffice for the former: patient attention, deep reflection, and acute penetration are necessary in the

[1] In his edition, Cavendish, *Electrical Researches*.

[2] Larmor, in Cavendish, *Scientific Papers* 2: 399.

[3] Crowther, *Scientists in the Industrial Revolution*, 302, 316.

[4] McCormmach, *Speculative Truth*.

[5] James Hutton, *A Dissertation upon the Philosophy of Light, Heat, and Fire* (Edinburgh, 1794), xi.

R. McCormmach, *The Personality of Henry Cavendish - A Great Scientist with Extraordinary Peculiarities*, Archimedes 36, DOI 10.1007/978-3-319-02438-7_17,

latter. It is therefore no wonder, that amongst many experimentalists there should be few philosophers."[6] "Men of fact" are content to register one fact after another, whereas "men of theory" regard "general results as the great and dignified objects of Science," valuing facts only as they illustrated general theorems.[7] Experimentalists view nature as "disunited"; systematic writers view it as "united."[8] Natural historians, investigators of the living parts of nature, thought about their work similarly to natural philosophers. A naturalist who only named and collected specimens was not a "philosopher"; a philosopher was one who considered general ideas. Things living and nonliving were not studied individually but as parts of a "system of Nature."[9] Words appearing in these statements – order of connection, general truths, theory, united, and system – relate to a way of thinking about nature. Cavendish carried out experiments to acquire the necessary empirical details, and he established laws, or general rules, or theories to connect the details.

The "philosopher" in these quotations could be the person with a less severe form of autism in the systematizing and empathizing theory of autism, as presented by Baron-Cohen, and described in Chap. 8 above. From physical hypotheses about the nature of electricity and heat, Cavendish developed mathematical theories that connected the phenomena.[10] Nothing can be inferred about Cavendish's normality or abnormality from his interest in physical theory and his success in crafting it, but if we knew that Cavendish was autistic, we would not be surprised at his interest and success. He would naturally have been drawn to physical theory, a systematizing tool of natural philosophy.

[6] William Enfield, *Institutes of Natural Philosophy, Theoretical and Experimental* … (London, 1785), vi–vii.

[7] William Nicholson, *A Dictionary of Chemistry* …, 2 vols. in 1 (London, 1795), v.

[8] George Adams, *Natural and Experimental Philosophy*, 5 vols. (London, 1794) 1: 126, 129.

[9] William Smelie, *The Philosophy of Natural History*, 2 vols. (Edinburgh, 1790, 1791) 1: 523, 525.

[10] His chemical researches were guided by theory too, though the theory was not mathematical. Phlogiston, the central concept of the theory, brought much of the phenomena of chemistry into a system.

Bibliography

DNB: Dictionary of National Biography
Philosophical Transactions: Philosophical Transactions of the Royal Society of London

Adams, George. 1794. *Natural and experimental philosophy*. 5 vols. London.
American Psychiatric Association. 2013. *Desk reference to the diagnostic criteria from DSM-5*. Washington, DC/London: American Psychiatric Publishing.
Anderson, R.G. See P. J. Hartog.
Anon. 1804. *Eccentric biography, or, memoirs of remarkable characters*…. 2nd rev. ed. London edition; first American edition, Boston.
Anon. 1812. Memoirs of the Late Frederick Cavendish, Esquire. *Gentleman's Magazine* 82: 289–291.
Anon. *1750–1795 in fashion*. http://www.en.wikipedia.org/wiki/1750.
Anon. *1795–1820 in fashion*. http://www.en.wikipedia.org/wiki/1795.
Anon. *A leader among British eccentrics*. http://www.suite101.com/article/a-leader-among-british-eccentrics.
Anon. *Biographical sketches of eccentric characters*. N.p. 1832.
Anon. *Bloomsbury squares & gardens. Bedford Square*. http://bloomsburysquares.wordpress.com/bedford-square.
Anon. *Chaise*. http://en.wikipedia.org/wiki/Chaise.
Anon. *Charles Waterton*. http://en.wikipedia.org/wiki/Charles_Waterton.
Anon. *Clapham Common*. http://en.wikipedia.org/wiki/Clapham_Common.
Anon. *Cyfarthfa ironworks*. http://en.wikipedia.org/wiki/Cyfarthfa_Ironworks.
Anon. *Eccentricity (Behavior)*. http://en.wikipedia.org/wiki/Eccentricity_(behavior).
Anon. *Edward Wortley Montagu*. http://en.wikipedia.org/wiki/Edward_Wortley_Montagu.
Anon. *Extroversion and introversion*. http://en.wikipedia.org/wiki/extroversion_and_introversion.
Anon. *Forge, furnace, what's the difference?* http://www.engr.psu.edu/mtah/essays/forge_furnace.htm.
Anon. *Garth Hill*. http://en.wikipedia.org/wiki/Garth Hill.
Anon. *John Wilkinson (industrialist)*. http://en.wikipedia.org/wiki/John_Wilkinson.
Anon. *Potting and stamping*. http://en.wikipedia.org/wiki/Potting_and_stamping.
Anon. *Samuel Homfray*. http://en.wikipedia.org/w/index.php?title=Samuel_Homfray&oldid=513293991.
Anon. *Schizoid personality disorder*. http://www.mayoclinic.com/health/schizoid- personality-disorder.
Anon. *Sir James Lowther, 4th Baronet*. http://en.wikipedia.org/wiki/Sir_James_Lowther,_4th-Baronet.
Anon. *Social anxiety disorder*. http://www.socialanxietyinstitute.org/dsm.html.
Anon. *Social anxiety disorder*. http://www.wikipedia.org/wiki/Social_anxiety_disorder.
Anon. *Social Phobia*. http://www.behavenet.com/social-phobia.
Anon. *Thomas Manning*. http://en.wikipedia.org/wiki/Thomas_Manning_(sinologist).

R. McCormmach, *The Personality of Henry Cavendish - A Great Scientist with Extraordinary Peculiarities*, Archimedes 36, DOI 10.1007/978-3-319-02438-7,

Anon. *Tricorne*. http://www.en.wikipedia.org/wiki/Tricorne.
Anon. *Wig*. http://www.en.wikipedia.org/wiki/Wig.
Anon. *William Brownrigg*. http://en.wikipedia.org/wiki/William_Brownrigg.
Anon. *William Cavendish-Scott-Bentinck, fifth Duke of Portland*. http://www.en.wikipedia.org/wiki/William_Cavendish-Scott-Bentinck_5th_Duke_of_Portland.
Asendorpf, Jens. 1990. The expression of shyness and embarrassment. In *Shyness and Embarrassment*, ed. W. Ray Crozier, 87–118. Cambridge/New York: Cambridge University Press.
Ashton, T.S. 1996. *The industrial revolution 1760–1830*. Oxford/New York: Oxford University Press.
Asperger Management. http://www.aspergermanagement.com/personal-appearance.
Attwood, Tony. 2007. *The complete guide to Asperger's syndrome*. London/Philadelphia: Jessica Kingsley Publishers.
Aymes-Stokes, Sophie, and Laurent Mellet. 2012. Introduction. In *In and out: Eccentricity in Britain*, ed. S. Aymes-Stokes and L. Mellet, 1–32. Newcastle upon Tyne: Cambridge Scholars Publishing.
Babbage, Charles. 1830. *Reflections on the decline of science in England, and on some of its causes*. London.
Baigent, Elizabeth. Manning, Thomas (1772–1840). *DNB* 36: 509–510.
Baron, Sean. See Temple Grandin.
Baron-Cohen, Simon. 2003. *The essential difference: Male and female brains and the truth about autism*. New York: Basic Books.
Baron-Cohen, Simon. 2008. *Autism and Asperger syndrome*. Oxford: Oxford University Press.
Baron-Cohen, Simon. 2012. Autism and the technical mind. *Scientific American* 307: 72–75.
Barrow, John. 1849. *Sketches of the Royal Society and Royal Society Club*. London.
Batten Jr., Charles L. 1978. *Pleasurable instruction: Form and convention in eighteenth-century travel literature*. Berkeley/Los Angeles/London: University of California Press.
Bayne-Powell, Rosamond. 1951. *Travelers in eighteenth-century England*. London: John Murray.
Beckett, J.V. 1978. The Lowthers at Holker: Marriage, inheritance and debt in the fortunes of an eighteenth-century Landowning Family. *Transactions of the Historic Society of Lancashire and Cheshire* 127: 47–64.
BehaveNet. *Social Phobia*. http://www.behavenet.com/social-phobia.
Bickley, Francis. 1911. *The Cavendish family*. London: Constable.
Biot, J.-B. 1813. Cavendish, Henri. In *Biographie Universelle*, vol. 7, 272–273. Paris.
Birch, Alan. *The economic history of the British iron and steel industry, 1784–1879*. London: Frank Cass, 1967; Abingdon/Oxon: Routledge, 2006.
Blagden, Charles. See George Augustus Henry Cavendish.
Boswell, James. 1821. *The life of Samuel Johnson, LL.D.…* Rev. ed. 5 vols. London.
Bradley, James. A letter… Concerning an apparent motion observed in some of the fixed stars. *Philosophical Transactions*: 1–43. In Cavendish, *Scientific Papers* 2: 127–144.
Bret, P. See E. Grison.
Briggs, Stephen R. See Jonathan M. Cheek.
British History Online. *Grosvenor Street South*. http://www.british-history.ac.uk.
Brougham, Henry. 1845. *Lives of men of letters and science who flourished in the time of George III*, vol. 1. London/Philadelphia.
Brown, Anna Maria. The peerage: A genealogical survey of the peerage of Britain as well as the royal families of Europe. http://www.thepeerage.com/p2986.htm.
Bucholtz, R.O. Seymour, Charles, Sixth Duke of Somerset (1662–1748). *DNB* 49: 857–860.
Burgess, J.H.M. 1929. *The chronicles of Clapham [Clapham Common]. Being a selection from the reminiscences of Thomas Parsons. Sometime member of the Clapham Antiquarian Society*. London: The Ramsden Press.
Buss, Arnold H. 1984. A conception of shyness. In *Avoiding communication: Shyness, reticence and communication apprehension*, ed. J.A. Daley and J.C. McCrosky, 39–49. Beverley Hills: Sage.
Buss, Arnold H. 1990. The theory of shyness. In *Shyness: Perspectives on research and treatment*, ed. W.R. Crozier, 39–46. Cambridge: Cambridge University Press.

Cannon, John. 1984. *Aristocratic century: The peerage of eighteenth-century England*. Cambridge/London/New York: Cambridge University Press.
Caroe, Gwendy. 1985. *The Royal Institution: An informal history*. London: John Murray.
Caroll, Victoria. 2008. *Science and eccentricity: Collecting, writing and performing science for early nineteenth-century audiences*. London: Pickering and Chatto.
Cattell, Raymond B. 1971. *Abilities: Their structure, growth, and action*. Boston: Houghton Mifflin.
Cavendish, Lord George Augustus (1727–94), of Holker, Lancs. In *The history of parliament: The house of commons 1790–1820*, ed. R. Thorne. London: Boydell and Brewer, 1986. http://www.historyofparliamentonline.org/volume/1790-1820/cavendish-george-augustus.
Cavendish, Henry. 1751. Luctus. In *Academiae Cantabrigiensis Luctus in Obitum Frederici celsissimi Walliae Principis*. Cambridge: Cambridge University.
Cavendish, Henry. 1766. Three papers, containing experiments on factitious air. *Philosophical Transactions* 56: 141–184. In Cavendish, *Scientific Papers* 2: 77–101.
Cavendish, Henry. 1771. An attempt to explain some of the principal phaenomena of electricity, by means of an elastic fluid. *Philosophical Transactions* 61: 1–50. In *The electrical researches of the Honourable Henry Cavendish*, ed. J.C. Maxwell, 1–43. Cambridge, 1879. Reprinted in *The scientific papers of the Honourable Henry Cavendish, F.R.S.* Vol. 1: *The electrical researches*, ed. J.C. Maxwell. Revised by J. Larmor. Cambridge: Cambridge University Press, 1921.
Cavendish, Henry. 1776. An account of the meteorological instruments used at the Royal Society House. *Philosophical Transactions* 66: 375–401. In Cavendish, *Scientific Papers* 2: 112–126.
Cavendish, Henry. 1783. An account of a new eudiometer. *Philosophical Transactions* 73: 106–135. In Cavendish, *Scientific Papers* 2: 127–144.
Cavendish, Henry. 1788. On the conversion of a mixture of dephlogisticated and phlogisticated air into nitrous acid. *Philosophical Transactions* 78: 261–76. In Cavendish, *Scientific Papers* 2: 224–232.
Cavendish, Henry. 1790. On the height of the luminous arch which was seen on Feb. 23, 1784. *Philosophical Transactions* 80: 101–105. In Cavendish, *Scientific Papers* 2: 233–235.
Cavendish, Henry. 1798. Experiments to determine the density of the Earth. *Philosophical Transactions* 88: 469–526. In Cavendish, *Scientific Papers* 2: 249–286.
Cavendish, Henry. 1921. *The scientific papers of the Honourable Henry Cavendish*. 2 vols, ed. E. Thorpe. Cambridge: Cambridge University Press.
Cavendish, William. 1982. *The Devonshire Diary: William Cavendish, Fourth Duke of Devonshire, Memoranda on State of Affairs, 1759–1762*, ed. P.D. Brown and K.W. Schweizer. Camden Fourth Series, vol. 27. London: Royal Historical Society.
Cavendish, George Augustus Henry, and Charles Blagden. 1810, March. *Gentleman's Magazine*, 292. Family obituary of Henry Cavendish.
Chappell, Edgar L. 1995. *Historic Melingriffith: An account of Pentyrch Iron Works and Melingriffith Tinplate Works*. 2d ed. Merton: Merton Priory Press.
Cheek, Jonathan M., and Stephen R. Briggs. 1990. Shyness as a personality trait. In *Shyness and Embarrassment*, ed. W. Ray Crozier, 315–337. Cambridge/New York: Cambridge University Press.
Child, Ernest. 1940. *The tools of the chemist*. New York: Reinhold.
Clapham Antiquarian Society. 1957, August. Cavendish House. *Occasional Sheet*.
Clarke, Edward Daniel. 1793. *A tour through the South of England, Wales, and part of Ireland, made during the summer of 1791*. London.
Clarke, T. E. 1848. *Guide to Merthyr-Tydfil*. London.
Cloninger, C. Robert. 1994. Temperament and personality. *Current Biology* 4: 266–273.
Cokayne, George Edward. 1982. *The complete peerage of England, Scotland, Ireland, Great Britain and the United Kingdom: Extant, extinct, or dormant*, vol. 1–3. Gloucester: A Sutton.
Coley, N.G. Fordyce, George (1736–1802). *DNB* 20: 358–360.
Colonial Williamsburg Foundation. A colonial gentleman's clothing: A glossary of terms. http://www.history.org/history/clothing/men/mglossary.cfm.
Cook, A.H. 1987. Experiments on gravitation. In *Three hundred years of gravitation*, ed. S.W. Hawking and W. Israel, 51–79. Cambridge: Cambridge University Press.
Cooper, John M. 2012a. Ancient philosophies as ways of life. *Oxford Philosophy* (summer): 16.

Cooper, John M. 2012b. *Pursuits of wisdom: Six ways of life in ancient philosophy from Socrates to Plotinus*. Princeton/Oxford: Princeton University Press.

Cox, John Charles, and William Henry St. John Hope. 1881. *The Chronicle of Collegiate Church or Free Chapel of All Saints, Derby*. London.

Crowther, James Gerald. 1962. *Scientists in the industrial revolution: Joseph Black, James Watt, Joseph Priestley, Henry Cavendish*. London: Crescent Press.

Crozier, W. Ray. ed. 1990a. *Shyness and embarrassment: Perspectives from social psychology*. Cambridge: Cambridge University Press. The short title *Shyness and Embarrassment* is used for specific contributions to this volume.

Crozier, W. Ray. 1990b. Social psychological perspective on shyness, embarrassment and shame. In *Shyness and Embarrassment*, ed. W. Ray Crozier, 19–50. Cambridge/New York: Cambridge University Press.

Dale, T.C. 1927. History of Clapham. In *Clapham and the Clapham Sect*, ed. Clapham Antiquarian Society, 1–28. Clapham: Clapham Antiquarian Society.

Davis, Lennard J. 2008. *Obsession, A history*. Chicago/London: Chicago University Press.

Davy, Humphry. 1812. *Elements of Chemical Philosophy*. London.

Davy, John. 1836. *Memoirs of the Life of Sir Humphry Davy, Bart*. 2 vols. London.

Davy, Humphry. 1839. *The collected works of Sir Humphry Davy, Bart*, ed. J. Davy, vol. 7. London.

Deluc, J. A. 1809. *An Elementary Treatise on Geology: Determining Fundamental Parts in That Science, and Containing an Examination of Some Modern Geological Systems, and Particularly of the Huttonian Theory of the Earth*. Trans H. De la Fite. London.

Dictionary of National Biography. 60 vols, ed. H.C.G. Matthew and B. Harrison. Oxford/New York/Auckland: Oxford University Press, 2004. 1st ed. Edited by L. Stephen and S. Lee. New York: Macmillan, 1908–1909.

Donaldson, William. 2002. *Brewer's Rogues, Villains, and Eccentrics: An A-Z of Roguish Britons through the ages*. London: Cassell.

Elms, Alan C. 1994. *Uncovering lives: The uneasy alliance of biography and psychology*. Oxford: Oxford University Press.

Elms, Alan C. The psychologist as biographer. http://www.ulmus.net/aceworks/psycholbiog.html.

Enfield, William. 1785. *Institutes of natural philosophy, theoretical and experimental....* London.

Farmelo, Graham. 2009. *The strangest man: The hidden life of Paul Dirac, mystic of the atom*. New York: Basic Books.

Feist, Gregory J. 2006. *The psychology of science and the origins of the scientific mind*. New Haven/London: Yale University Press.

Fitzgerald, Michael. 2004. *Autism and creativity: Is there a link between autism in men and exceptional ability?* Hove/New York: Brunner-Routledge.

Fleisher, Marc. 2003. *Making sense of the unfeasible: My life journey with Asperger syndrome*. London/New York: Jessica Kingsley.

Foote, Yolanda. *Waterton, Charles (1782–1865)*. *DNB* 57: 573–575.

Franklin, Benjamin. 1996. *The papers of Benjamin Franklin*, ed. L.W. Labaree, vol. 10. New Haven: Yale University Press.

Friends of Melingriffith Water Pump. *Melingriffith tin works*. http://friendsofmelingriffith-waterpump.org.

Frith, Uta. 2003. *Autism: Explaining the Enigma*, 2 ed. Malden/Oxford/Victoria: Blackwell.

Frith, Uta. See Rab Houston.

Geikie, Archibald. 1917. *Annals of the Royal Society Club: The record of a London dining-club in the eighteenth and nineteenth centuries*. London: Macmillan.

Georgian Index. *Carriages*. http://www.georgianindex.net/horse and carriage.html.

Gill, Miranda. *Rethinking eccentricity*. http://www.cam.ac.uk/research/features/rethinking-eccentricity.

Gloucestershire County Council. *Gloucestershire archives: Online Catalog*. http://www.gloucestershire.gov.uk.

Golinski, Jan. 2002. Conversations on chemistry: Talk about phlogiston in the Coffee House Society, 1780–1787. In *Discussing chemistry and steam: The minutes of a Coffee House Philosophical Society, 1780–1787*, ed. T.H. Levere and G.L.E. Turner, 191–205. Oxford/New York: Oxford University Press.

Goode, Erica. Cases; A disorder far beyond eccentricity. http://www.nytimes.com/2001/10/09/health/cases-a-disorder-far-beyond eccentricity.html.

Gordon, Alexander, rev. Anita McConnell. Elwes, John (1714–1789). *DNB* 18: 370–371.

Grandin, Temple. 1995. *Thinking in pictures and other reports from my life*. New York: Doubleday.

Grandin, Temple. 2011. *The way I see it*. 2d rev. ed. Arlington: Future Horizons.

Grandin, Temple, and Sean Barron. 2005. *The unwritten rules of social relationships,* ed. V. Zysk. Arlington: Future Horizons.

Great Britain, Historical Manuscripts Commission. 1913. *Report on manuscripts in various collections*, vol. 8. *The manuscripts of the Hon. Frederick Lindley Wood; M. L. S. Clements, Esq.; S. Philip Unwin, Esq*, 188–189. London: His Majesty's Stationary Office.

Great Britain, Historical Manuscripts Commission. 1934. *Report on the manuscripts of the late Reginald Rawdon Hastings, Esq., of the Manor House, Ashby de la Zouch,* ed. F. Bickley, vol. 3. London: His Majesty's Stationary Office.

Green, Candida Lycett. *Sapperton canal tunnel, Gloucestershire*. http://www.candidalycettgreen.co.uk.

Green, Michael. *Mount Pond, Clapham Common: Archaeology and history*. The Clapham Society Local History Series 7. http://www.claphamsociety.com/Articles/article7.html.

Grundy, Isobel. Montague, Edward (1713–1776). *DNB* 38: 714–715.

Hamilton, S.B. 1958. Building and civil engineering construction. In *History of technology*, vol. 4. *The industrial revolution, c1750 to c1850*, ed. C. Singer, E.J. Holmyard, A.R. Hall, and T.I. William, 442–519. New York/London: Oxford University Press.

Handsworth. 1780s pumping engine for the Coalbrook Dale Company. http://www.search.digital-handsworth.org.uk/engine.

Hare, Edward. 1991. The history of 'Nervous Disorders' from 1600 to 1840, and a comparison with modern views. *British Journal of Psychiatry* 159: 37–45.

Harris, J.R. Wilkinson, John (1728–1808). *DNB* 58: 1010–1013.

Hartog, P.J., rev. R. G. Anderson. Wilson, George (1818–1859). *DNB* 59: 536–539.

Harvey, R.A. 1980. The private library of Henry Cavendish. *The Library* 2: 281–284.

Hayman, Richard. 2003. *The Shropshire Wrought-Iron Industry c 1600–1900: A study of technological change*. PhD thesis, University of Birmingham.

Hayman, Richard. n.d. *Working Iron in Merthyr Tydfil*. Merthyr Tydfil: Merthyr Tydfil Heritage Trust.

Heberden, William. 1765. Some account of a salt found on the Pic of Teneriffe. *Philosophical Transactions* 55: 57–60.

Heilbron, John. 1979. *Electricity in the 17th and 18th centuries: A study of early modern physics*. Berkeley: University of California Press.

History World. *History of transport and travel*. http://www.historyworld.net/wrldhis/PlainTextHistories.

Hobsbawm, Eric. 1999. *Industry and empire: The birth of the industrial revolution*. New York: The New Press.

Hope, St. John, William Henry. See John Charles Cox.

Horwood, Richard. 1966. *Horwood's Plan of London … 1792–99*. London: London Topographical Society.

Houston, Rab, and Uta Frith. 2000. *Autism in history: The case of Hugh Blair of Borgue*. Oxford/Malden: Blackwell.

Huddleston, David. Hamilton, James, Eighth Earl of Abercorn (1712–1789). *DNB* 24: 854.

Hunter, John. 1788. Some observations on the heat of wells and springs in the Island of Jamaica, and on the temperature of the Earth below the surface in different climates. *Philosophical Transactions* 78: 53–65.

Hutton, Charles. 1795–96. *A mathematical and philosophical dictionary….* 2 vols. London.

Hyson, Marion C. See Carroll E. Izard.

Ince, Laurence. 1993. *The South Wales iron industry 1750–1885*. Merton: Merton Priory Press.

Izard, Caroll E., and Marion C. Hyson. 1986. Shyness as a discrete emotion. In *Shyness: Perspectives on research and treatment*, ed. W.H. Jones, J.M. Cheek, and S.R. Briggs, 147–160. New York/London: Plenum Press.

James, Ioan. 2006. *Asperger's syndrome and high achievement: Some very remarkable people*. London: Jessica Kingsley.

James, Jamie. See David Weeks.

Jungnickel, Christa, and Russell McCormmach. 1996. *Cavendish*. Philadelphia: American Philosophical Society.

Jungnickel, Christa, and Russell McCormmach. 1999. *Cavendish, the experimental life*. Lewisburg: Bucknell University Press.

Kagan, Jeremy, J. Steven Reznick, and Nancy Snidman. 1988. Biological bases of childhood shyness. *Science* 240: 167–171.

Keay, John. 1982. *Eccentric travellers*. London: John Murray.

Kemp, Anthony E. 1996. *Musical temperament: Psychology and personality of musicians*. Oxford: Oxford University Press.

Kidd. C. ed. 2008. *Debrett's Peerage and Baronetage*. Richmond, Surrey: Debrett's Peerage and Baronetage.

Kirby, R.S. 1820. *Kirby's wonderful and eccentric museum*; *or, magazine of remarkable characters....* vol. 2. London: R.S. Kirby.

Kirwan, Richard. See Louis-Bernard Guyton de Morveau.

Korte, Barbara. 2000. *English Travel Writing from Pilgrimages to Postcolonial Explorations*. Trans. K. Matthias. New York: St. Martin's Press.

Lamont-Brown, Raymond. 1984. *A book of British eccentrics*. Newton Abbot: David & Charles.

Langford, Paul. 2000. *Englishness identified: Manners and character 1650–1850*. Oxford/New York: Oxford University Press.

Larmor, Joseph. 1921. Preface to volume 1. In *The scientific papers of the Honourable Henry Cavendish, F.R.S.* Vol. 1. *The electrical researches,* ed. J.C. Maxwell and J. Lamor, v–x. Cambridge: Cambridge University Press.

Ledgin, Norm. 2000. *Diagnosing Jefferson: Evidence of a condition that guided his beliefs, behavior, and personal associations*. Arlington: Future Horizons.

Ledgin, Norm. 2001. *Asperger's and self-esteem: Insight and hope through famous role models*. Arlington: Future Horizons.

Lépine, Pierre, and Jacques Nicoll. 1964. *Sir Henry Cavendish: l'homme qui a pesé la terre*. Paris: Seghers.

Levere, T.H. 2002. Introduction. In *Discussing chemistry and steam: The minutes of a Coffee House Philosophical Society, 1780–1787*, ed. T.H. Levere and G.L.E. Turner, 1–15. Oxford: Oxford University Press.

Lewis, William. 1963. *Commercium philosophico-technicum*; *or, the philosophical commerce of arts*: *Designed as an attempt to improve arts, trades and manufactures*. London.

Lidbetter, Hugo. 2009. Henry Cavendish and Asperger's syndrome: A new understanding of the scientist. *Personality and Individual Differences* 46: 784–793.

Lloyd, John. 1906. *The early history of the Old South Wales ironworks 1760–1840*. London: Bedford Press.

London County Council. 1914. *Survey of London*, vol. 5. *The Parish of St. Giles-in-the-Fields,* ed. E. Riley and L. Gromme. Pt. 2. London: London County Council.

London County Council. 1963. *Survey of London*, vol. 31. *The Parish of St. James Westminster*. Pt. 2. *North of Piccadilly,* ed. F.H.W. Shepherd. London: Athlone Press.

Lubbock, Constance. 1933. *The Herschel Chronicle: The life story of William Herschel and his sister Caroline Herschel*. Cambridge: Cambridge University Press.

Lucas, Philip, and Anne Sheeran. 2006. Asperger's syndrome and the eccentricity and genius of Jeremy Bentham. *Journal of Bentham Studies* 8: 1–37.

Ludlam, William. 1769. *Astronomical observations made in St. John's College, Cambridge, in the years 1767 and 1768*: *With an account of several astronomical instruments*. Cambridge.

Lysons, Daniel. 1792. *The environs of London: Being an historical account of the towns, villages, and hamlets, within twelve miles of that capital*, vol. 1. *County of Surrey*. London.
Manuel, Frank E. 1968. *A portrait of Isaac Newton*. New York: Da Capo Press.
Marshall, John R. 1995. *Social Phobia: From shyness to stage fright*. New York: Basic Books.
Mayo Clinic. *Schizoid personality disorder*. http://www.mayoclinic.com/health/schizoid-personality-disorder.
McConnell, Anita. See Alexander Gordon.
McCormmach, Russell. 2004. *Speculative truth: Henry Cavendish, natural philosophy, and the rise of modern theoretical science*. Oxford: University Press.
McCormmach, Russell. 2012. *Weighing the world: The Reverend John Michell of Thornhill*. Dordrecht/Heidelberg/London: Springer.
McCormmach, Russell. See Christa Jungnickel.
Melingriffith Water Pump. Melingriffith Tin works. http://friendsofmelingriffithwaterpump.org.
Mellet, Laurent. See Sophie Aymes-Stokes.
Mills, Stephen. 1998. Fromebridge Mill, Frampton-on-Severn, Gloucestershire. *Gloucestershire Society for Industrial Archaeology*, 6–22.
Milne, Lors, and Margery Milne. 1962. *The senses of animals and men*. New York: Atheneum.
Milne, Margery. See Lors Milne.
Morgan, Rod. Howard, John (1726–1790). *DNB* 28: 390–394.
Morson, Geoffrey V. Porson, Richard (1759–1808). *DNB* 44: 919–924.
Morveau, Louis-Bernard Guyton de, and Richard Kirwan. 1994. *A scientific correspondence during the chemical revolution: Louis-Bernard Guyton de Morveau and Richard Kirwan 1782–1802*, ed. E. Grison, M. Sadoun-Goupil, and P. Bret. Berkeley: Office for History of Science and Technology, University of California at Berkeley.
Mosley, C., ed.1999. *Burke 's Peerage and Baronetage*. 106th ed. 2 vols. Crans: Burke's Peerage.
Musson, A.E., and E. Robinson. 1969. *Science and technology in the industrial revolution*. Toronto: University of Toronto Press.
Newton, Isaac. 1952. *Opticks; or a treatise of the reflexions, refractions, inflexions and colours of light*. 4th ed., 1730. Rep.. New York: Dover Publications.
Nicholson, William. 1795. *A dictionary of chemistry*.... 2 vols. in 1. London.
Owen, John A. *Merthyr Tydfil – Iron Metropolis 1790–1860*. http://himedo.net/The HopkinThomas-Project/TimeLine/Wales/MerthyrTydfil/OwenMerthyrTydfil.
Pearson, John. 1983. *The Serpent and the Stag: The Saga of England's powerful and glamorous Cavendish family from the age of Henry the eighth to the present*. New York: Holt, Rinehart and Winston.
Pennant, Thomas. 1778–93. *A Tour in Wales MDCCLXX*. 2 vols. London.
Penn State College of Engineering. *Forge, furnace, what's the difference?* http://www.engr.psu.edu/mtah/essays/forge_furnace.htm.
Pervin, Lawrence A. 1993. *Personality, theory and research*. New York/Chichester/Brisbane: Wiley.
Phipps, Constantine John. 1774. *A Voyage towards the North Pole, Undertaken by His Majesty's Command, 1773*. London.
Pickover, Clifford A. 1999. *Strange brains and genius: The secret lives of eccentric scientists and madmen*. New York: Quill.
Playfair, John. 1822. *The works of John Playfair*, ed. J.G. Playfair. 4 vols. Edinburgh.
Porter, Roy. 1977. *The making of geology: Earth science in Britain, 1660–1815*. Cambridge: Cambridge University Press.
Priestley, Joseph. 1966. *A scientific autobiography of Joseph Priestley (1733–1804)*, ed. R.E. Schofield. Cambridge, MA: MIT Press.
Reznick, Steven. See Jeremy Kagan.
Riden, Philip. 1992. Early ironworks in the Lower Taff Valley. *Morganwg* 36: 69–93. http://welsh-journals.llgc.org.uk/browse/viewpage/llgc-id:1169834.
Robison, John. 1797. Physics. In *Encyclopaedia Britannica*. 3d ed, vol. 14, 637–659. Edinburgh.
Robinson, E. See A. E. Musson.

Ross, Jeralyn. 1993. Social Phobia: The consumers perspective. *Journal of Clinical Psychiatry* 54, Supplement, 5–9, on 5–7.
Roth, Ilona. 2010. *The autism spectrum in the 21st century: Exploring psychology, biology and practice*. London/Philadelphia: Jessica Kingsley.
Rowley, William. 1804. *A treatise on madness and suicide, with the modes of determining with precision mental affections*.... London.
Rudé, George. 1971. *Hanoverian London 1714–1808*. Berkeley/Los Angeles: University of California Press.
Rudolph, R. de M. 1927. The Clapham Sect. In *Clapham and the Clapham Sect*, ed. Clapham Antiquarian Society and E. Baldwin, 89–142. Clapham: Clapham Antiquarian Society.
Sacks, Oliver. 1995. *An anthropologist on Mars*. New York: Vintage.
Sacks, Oliver. 2001. Henry Cavendish: An early case of Asperger's syndrome? *Neurology* 57: 1347.
Sacks, Oliver. 2008. *Musicophilia: Tales of music and the brain*. Rev. ed. New York: Vintage Books.
Sacks, Oliver. *Uncle Tungsten: Memories of a chemical boyhood*. Toronto: Random House, 2001; New York: Vintage Books, 2002.
Sadoun-Goupil, M. See E. Grison.
Schubert, H.R. 1958. Extraction and production of metals: Iron and steel. In *History of technology*. Vol. 4. *The industrial revolution, c1750 to c1850*, eds. C. Singer, E.J. Holmyard, A.R. Hall, and T.I. Williams, 99–117. New York/London: Oxford University Press.
Sheeran, Anne. See Philip Lucas.
Sheffield Trades Historical Society. A welcome to Wortley Ironworks. http://www.topforge.co.uk/1955%20Guide.htm.
Short, James. 1753. An account of a book, intitled, P. D. Pauli Frisii Mediolanensis, &c. Disquisitio mathematica .... *Philosophical Transactions* 48: 5–17.
Sitwell, Edith. *English eccentrics*. London: Faber and Faber, 1933; Harmondsworth: Penguin Books, 1976.
Sitwell, Edith. 1965. *Taken care of: The autobiography of Edith Sitwell*. New York: Atheneum.
Skempton, A. W., ed. 2002. *Biographical dictionary of civil engineers in Great Britain and Ireland*, vol. 1. *1500–1830*. London: Thomas Telford.
Smeaton, John. 1754. Description of a new pyrometer, with a table of experiments made therewith. *Philosophical Transactions* 48: 598–613.
Smiles, Samuel. 1874. *Lives of the engineers. Harbours – Bridges. Smeaton and Rennie*. Rev. ed. London.
Smith, Eric E.F. 1972. *Clapham*. London: London Borough of Lambeth.
Smith, Crosbie. 1976. 'Mechanical Philosophy' and the emergence of physics in Britain: 1800–1850. *Annals of Science* 33: 3–29.
Snidman, Nancy. See Jeremy Kagan.
Spacks, Patricia Meyer. 2003. *Privacy: Concealing the eighteenth-century self*. Chicago/London: University of Chicago Press.
Stafford, Barbara Maria. 1984. *A voyage into substance: Art, science, nature, and the illustrated travel account, 1760–1840*. Cambridge/London: MIT Press.
Steiner, Lewis H. 1855. *Henry Cavendish and the discovery of the chemical composition of water*. New York.
Stewart, Dugald. 1854–60. *The collected works of Dugald Stewart, Esq., F.R.SS*, ed. W. Hamilton. 11 vols. Edinburgh.
Stitch 'N Save. *Colonial Wigs*. http://www.stitchnsave.com/wigs/Colonial_Wigs.asp.
Sullivan, Richard Joseph. 1780. *Observations made during a tour through parts of England, Scotland, and Wales. In a series of letters*. London.
Tammet, Daniel. 2007. *Born on a blue day: Inside the extraordinary mind of an Autistic Savant, a Memoir*. New York/London/Toronto: Free Press.
The Peerage: A genealogical survey of the peerage of Britain as well as the royal families of Europe. Compiled by D. Lundy. http://www.thepeerage.com.

Thomas, Bruce. 1992. Merthyr Tydfil and early ironworks in South Wales. In *The company town: Architecture and society in the early industrial age*, ed. J.S. Garner, 17–42. Oxford/New York/Toronto: Oxford University Press.
Thomas, Peter D. G. Cavendish, Sir Henry, Second Baronet (1732–1804). *DNB* 10: 627–628.
Thomson, Thomas. 1812. *History of the Royal Society from its institution to the end of the eighteenth century*. London.
Thomson, Thomas. 1830–31. *The history of chemistry*. 2 vols. London.
Thorne, James. 1876. *Environs of London*. London.
Thorne, Roland. Hastings, Francis Rawdon, First Marquess of Hastings and Second Earl of Moira (1754–1826). *DNB* 25: 744–749.
Thorpe, Edward. Introduction. In Cavendish, *Scientific Papers* 2: 1–74.
Timbs, John. 1866. *English eccentrics and eccentricities*. 2 vols. London.
Timpson, John. 1991. *Great English eccentrics*. Norwich: Jarrold Publishing.
Topham, Jonathan R. Egerton, Francis Henry, Eighth Earl of Bridgewater (1756–1829). *DNB* 17: 994–995.
Victorian Web. *Francis Egerton, 8th Earl of Bridgewater*. http://www.victorianweb.org/science/science_texts/bridgewater/b1.htm.
Walpole, Horace. 1941. *Correspondence*, ed. W.S. Lewis. 48 vols. New Haven: Yale University Press.
Watson, William. 1750. A letter… Declaring that he as well as many others have not been able to make odours pass thro' glass by means of electricity…. *Philosophical Transactions* 46: 348–356.
Watson, William. 1751. An account of Professor Winkler's experiments relating to odours passing through electrified globes and tubes.…. *Philosophical Transactions* 47: 231–241.
Weeks, David, and Jamie James. 1995. *Eccentrics: A study of sanity and strangeness*. New York: Villard.
Welch, H. T. Barrett, John (1753/4–1821). *DNB* 4: 45–46.
William Cavendish-Scott-Bentinck, Fifth Duke of Portland. http://www.en.wikipedia.org/wiki/William_Cavendish-Scott- Bentinck,_5th_Duke_of_Portland.
Williams, Trevor I. Wollaston, William Hyde (1766–1828). *DNB* 59: 991–993. http://www.aspergermanagement.com/personal-appearance.
Wilson, G.H. 1807. *The eccentric mirror, reflecting a faithful and interesting delineation of male and female characters….* vol. 3. London.
Wilson, George. 1860. *The five senses*; *or*, *gateways to knowledge*. Philadelphia.
Wilson, George. 1862. *Religio Chemici. Essays*, ed. J.A. Wilson. London.
Wilson, George. *The life of the Hon*ble *Henry Cavendish*. London: Cavendish Society, 1851; Rep. New York: Arno Press, 1975.
Wilson, Jessie Aitkin. 1860. *Memoir of George Wilson*. Edinburgh/London/Cambridge.
Wolff, Sula. 1995. *Loners: The life path of unusual children*. London: Routledge.
Wortley Top Forge. History of iron making at Wortley. http://www.topforge.co.uk/History%20Notes.htm.
Young, Arthur. *Tours in England and Wales, selected from the annals of agriculture*. http://www.visionofbritain.org.uk/text/contents_page.jsp?t_id=Young.
Young, Thomas. Life of Cavendish. *Encyclopaedia Britannica* Supplement (1816–24). Rep. in Cavendish, *Scientific Papers* 1: 435–447.
Young, Thomas. 1802. *A syllabus of a course of lectures on natural and experimental philosophy*. London.
Zimbardo, Philip G. 1977. *Shyness: What it is, what to do about it*. Reading: Addison Wesley.

Index

R. McCormmach, *The Personality of Henry Cavendish - A Great Scientist with Extraordinary Peculiarities*, Archimedes 36, DOI 10.1007/978-3-319-02438-7,

MIX
Papier aus verantwortungsvollen Quellen
Paper from responsible sources
FSC® C105338

If you have any concerns about our products,
you can contact us on
ProductSafety@springernature.com

In case Publisher is established outside the EU,
the EU authorized representative is:
Springer Nature Customer Service Center GmbH
Europaplatz 3, 69115 Heidelberg, Germany

Printed by Libri Plureos GmbH
in Hamburg, Germany